THIRD EDITION

STUDENT'S SOLUTIONS MANUAL

Helen Burrier
Kirkwood Community College

Elementary Algebra for College Students

Allen R. Angel
Monroe Community College

PRENTICE HALL
Englewood Cliffs, New Jersey 07632

Production Editor: *Valerie Zaborski*
Acquisitions Editor: *Priscilla McGeehon*
Supplements Acquisitions Editor: *Susan Black*
Prepress Buyer: *Paula Massenaro*
Manufacturing Buyer: *Lori Bulwin*

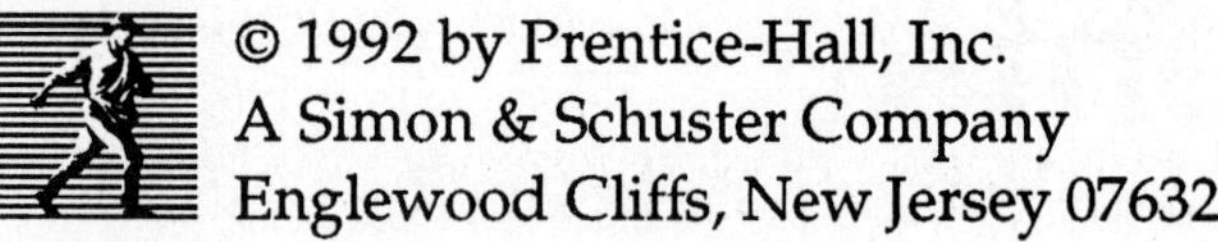

A Simon & Schuster Company
Englewood Cliffs, New Jersey 07632

Printed in the United States of America

10 9 8 7 6 5 4

ISBN 0-13-259607-5

Prentice-Hall International (UK) Limited, *London*
Prentice-Hall of Australia Pty. Limited, *Sydney*
Prentice-Hall Canada Inc., *Toronto*
Prentice-Hall Hispanoamericana, S.A., *Mexico*
Prentice-Hall of India Private Limited, *New Delhi*
Prentice-Hall of Japan, Inc., *Tokyo*
Simon & Schuster Asia Pte. Ltd., *Singapore*
Editora Prentice-Hall do Brasil, Ltda., *Rio de Janeiro*

Contents

CHAPTER ONE

Exercise Set 1.1

Study skills: Answers will vary campus to campus.

Exercise Set 1.2

1. $\frac{5 \div 5}{15 \div 5} = \frac{1}{3}$

3. $\frac{9 \div 3}{12 \div 3} = \frac{3}{4}$

5. $\frac{18 \div 18}{36 \div 18} = \frac{1}{2}$

7. $\frac{15 \div 5}{35 \div 5} = \frac{3}{7}$

9. $\frac{40 \div 8}{64 \div 8} = \frac{5}{8}$

11. $\frac{8}{15}$ is in lowest terms.

13. $\frac{96 \div 24}{72 \div 24} = \frac{4}{3}$ or $1\frac{1}{3}$

15. $\frac{1}{2} \cdot \frac{3}{5} = \frac{3}{10}$

17. $\frac{5}{\underset{4}{\cancel{8}}} \cdot \frac{\overset{1}{\cancel{2}}}{7} = \frac{5}{28}$

19. $\frac{\overset{1}{\cancel{3}}}{\underset{4}{\cancel{8}}} \cdot \frac{\overset{1}{\cancel{2}}}{\underset{3}{\cancel{9}}} = \frac{1}{12}$

21. $\frac{1}{3} \div \frac{1}{5} = \frac{1}{3} \cdot \frac{5}{1} = \frac{5}{3}$

    or $1\frac{2}{3}$

23. $\frac{5}{12} \div \frac{4}{3} = \frac{5}{\underset{4}{\cancel{12}}} \cdot \frac{\overset{1}{\cancel{3}}}{4} = \frac{5}{16}$

25. $\frac{10}{3} \div \frac{5}{9} = \frac{\overset{2}{\cancel{10}}}{\underset{1}{\cancel{3}}} \cdot \frac{\overset{3}{\cancel{9}}}{\underset{1}{\cancel{5}}} = 6$

27. $\frac{\overset{1}{\cancel{4}}}{\underset{3}{\cancel{9}}} \cdot \frac{\overset{5}{\cancel{15}}}{\underset{4}{\cancel{16}}} = \frac{5}{12}$

29. $\frac{4}{15} \div \frac{12}{13} = \frac{\overset{1}{\cancel{4}}}{15} \cdot \frac{13}{\underset{3}{\cancel{12}}} = \frac{13}{45}$

31. $\frac{\overset{1}{\cancel{12}}}{7} \cdot \frac{19}{\underset{2}{\cancel{24}}} = \frac{19}{14}$ or $1\frac{5}{14}$

33. $1\frac{4}{5} \cdot \frac{20}{3} = \frac{\overset{3}{\cancel{9}}}{\underset{1}{\cancel{5}}} \cdot \frac{\overset{4}{\cancel{20}}}{\underset{1}{\cancel{3}}} = 12$

35. $(\frac{3}{5})(1\frac{2}{3}) = \frac{\overset{1}{\cancel{3}}}{\underset{1}{\cancel{5}}} \cdot \frac{\overset{1}{\cancel{5}}}{\underset{1}{\cancel{3}}} = 1$

37. $3\frac{2}{3} \div 1\frac{5}{6} = \frac{11}{3} \div \frac{11}{6}$

$$= \frac{\overset{1}{\cancel{11}}}{\underset{1}{\cancel{3}}} \cdot \frac{\overset{2}{\cancel{6}}}{\underset{1}{\cancel{11}}} = 2$$

39. $\frac{2}{5} + \frac{1}{5} = \frac{2 + 1}{5} = \frac{3}{5}$

41. $\frac{5}{12} - \frac{2}{12} = \frac{\overset{1}{\cancel{3}}}{\underset{4}{\cancel{12}}} = \frac{1}{4}$

43. $\frac{9}{13} + \frac{4}{13} = \frac{9 + 4}{13} = \frac{13}{13} = 1$

45. $\frac{21}{29} - \frac{18}{29} = \frac{21 - 18}{29} = \frac{3}{29}$

47. $\frac{2}{5} + \frac{5}{6} = \frac{12 + 25}{30} = \frac{37}{30}$

or $1\ \frac{7}{30}$

49. $\frac{4}{12} - \frac{2}{15} = \frac{20 - 8}{60} = \frac{\overset{1}{\cancel{12}}}{\underset{5}{\cancel{60}}} = \frac{1}{5}$

51. $\frac{2}{10} + \frac{1}{15} = \frac{1}{5} + \frac{1}{15} = \frac{3 + 1}{15}$

$= \frac{4}{15}$

53. $\frac{8}{9} - \frac{4}{6} = \frac{32 - 24}{36} = \frac{\overset{2}{\cancel{8}}}{\underset{9}{\cancel{36}}} = \frac{2}{9}$

55. $\frac{5}{6} + \frac{9}{24} = \frac{20 + 9}{24} = \frac{29}{24}$

or $1\ \frac{5}{24}$

57. $\frac{11}{12} - \frac{3}{4} = \frac{11 - 9}{12} = \frac{\overset{1}{\cancel{2}}}{\underset{6}{\cancel{12}}} = \frac{1}{6}$

59. $3\ \frac{1}{4} + \frac{2}{3} = \frac{13}{4} + \frac{2}{3} = \frac{39 + 8}{12}$

$= \frac{47}{12}$ or $3\ \frac{11}{12}$

61. $2\ \frac{1}{2} + 1\ \frac{1}{3} = \frac{5}{2} + \frac{4}{3} = \frac{15 + 8}{6}$

$= \frac{23}{6}$ or $3\ \frac{5}{6}$

63. $4\ \frac{2}{3} - 1\ \frac{1}{5} = \frac{14}{3} - \frac{6}{5}$

$= \frac{70 - 18}{15} = \frac{52}{15}$ or $3\ \frac{7}{15}$

65. $1\ \frac{4}{5} - \frac{3}{4} + 3 = \frac{9}{5} - \frac{3}{4} + 3$

$= \frac{36 - 15 + 60}{20} = \frac{81}{20}$

or $4\ \frac{1}{20}$

67. $5 \cdot 2\ \frac{3}{4} = \frac{5}{1} \cdot \frac{11}{4} = \frac{55}{4}$

or $13\ \frac{3}{4}$ yards

69. $16\ \frac{3}{4} - 3\ \frac{1}{16} = \frac{67}{4} - \frac{49}{16}$

$= \frac{268 - 49}{16}$

$= \frac{219}{16}$ or $13\ \frac{11}{16}$ inches

71. $3\ \frac{3}{8} + 5\ \frac{1}{16} = \frac{27}{8} + \frac{81}{16}$

$= \frac{54 + 81}{16} = \frac{135}{16}$

or $8\ \frac{7}{16}$ feet

73. $20\frac{3}{4} - 8\frac{7}{8} = \frac{83}{4} - \frac{71}{8}$

$= \frac{166 - 71}{8} = \frac{95}{8}$

or $11\frac{7}{8}$ feet

75. $4\frac{1}{2} \cdot \frac{3}{4} = \frac{9}{2} \cdot \frac{3}{4} = \frac{27}{8}$

or $3\frac{3}{8}$ tsp.

77. $4\frac{5}{8} \div 3 = \frac{37}{8} \cdot \frac{1}{3} = \frac{37}{24}$

or $1\frac{13}{24}$ acres

79. a) $\frac{5}{\cancel{6}_{2}} \cdot \frac{\cancel{3}^{1}}{8} = \frac{5}{16}$

b) $\frac{5}{6} \div \frac{3}{8} = \frac{5}{\cancel{6}_{3}} \cdot \frac{\cancel{8}^{4}}{3} = \frac{20}{9}$

or $2\frac{2}{9}$

c) $\frac{5}{6} + \frac{3}{8} = \frac{20 + 9}{24} = \frac{29}{24}$

or $1\frac{5}{24}$

d) $\frac{5}{6} - \frac{3}{8} = \frac{20 - 9}{24} = \frac{11}{24}$

81. To multiply fractions, multiply the numerators together and multiply the denominators together.

83. To add or subtract fractions, write the fractions with a common denominator. Then add or subtract the numerators while keeping the same denominator.

85. To convert a fraction whose numerator is greater than its denominator, divide the numerator by the denominator. The quotient is the whole number, and the remainder is the numerator. The denominator of the mixed number is the same as the denominator of the fraction in the mixed number.

Just for fun

One serving: Rice 1/3 cup, Water 1/3 cup, Salt 1/8 tsp., Butter 1/2 tsp. Three servings: Rice 1 cup, Water 1 cup, Salt 3/8 tsp., and Butter 1 1/2 tsp.

Exercise Set 1.3

1. (...-3,-2,-1,0,1,2,3,...)

3. (1,2,3,...)

5. (...,-3,-2,-1)

7. True

9. True

11. False

13. False

15. True

17. False

19. True

21. False

23. True

25. True

27. False

29. True

31. True

33. True

35. False

37. False

39. a) 7, 9
b) 0, 7, 9
c) -6, 0, 7, 9
d) $-6, -2\frac{1}{4}, \frac{-9}{5}$
0, 0.35, 7, 9, 12.4
e) $\sqrt{3}, \sqrt{7}$
f) All are real numbers.

41. a) 5 b) 5 c) -300
d) 5, -300 e) -300,
$-9\frac{1}{2}$, -1.67, $\frac{5}{12}, \frac{1}{2}$,
$4\frac{1}{2}$, 5
f) $-\sqrt{2}, \sqrt{2}$
g) All are real numbers.

43. $-\frac{2}{3}, \frac{1}{5}, \frac{3}{2}$,
are among an infinite number of possible solutions.

45. $-\sqrt{2}, \sqrt{3}, \sqrt{5}$,
are among an infinite number of possible solutions.

47. -3, 0, 1,
are among an infinite number of possible solutions.

49. -4, -3, -1,
are among an infinite number of possible solutions.

51. 0, 3.4, $\frac{15}{17}$
are among an infinite number of possible solutions.

53. $-\frac{2}{3}$, 0, 3,
are among an infinite number of possible solutions.

Cumulative Review Exercises

55. $6\frac{2}{3} = \frac{20}{3}$

57. $\frac{3}{5} + \frac{5}{8} = \frac{24 + 25}{40} = \frac{49}{40}$
or $1\frac{9}{40}$

Exercise Set 1.4

1. 2 < 3 since 2 is farther left than 3.

3. -3 < 0 since -3 is farther left than 0.

5. $\frac{1}{2} > -\frac{2}{3}$ since $\frac{1}{2}$ is farther

right than $-\frac{2}{3}$.

7. $0.2 < 0.4$ since 0.2 is farther left than 0.4.

9. $\frac{2}{5} > -1$ since $\frac{2}{5}$ is farther right than -1.

11. $4 > -4$ since 4 is farther right than -4.

13. $-2.1 < -2$ since -2.1 is farther left than -2.

15. $\frac{5}{9} > -\frac{5}{9}$ since $\frac{5}{9}$ is farther right than $-\frac{5}{9}$.

17. $-\frac{3}{2} < \frac{3}{2}$ since $-\frac{3}{2}$ is farther left that $\frac{3}{2}$.

19. $0.49 > 0.43$ since 0.49 is farther right than 0.43.

21. $5 > -7$ since 5 is farther right than -7.

23. $-0.006 > -0.007$ since -0.006 is farther right than -0.007.

25. $\frac{3}{5} < 1$ since $\frac{3}{5}$ is farther left than 1.

27. $-\frac{2}{3} > -3$ since $-\frac{2}{3}$ is farther right than -3.

29. $8 > |-7|$ since $|-7| = 7$ and 8 is farther right than 7.

31. $|0| < \frac{2}{3}$ since $|0| = 0$ and 0 is farther left than $\frac{2}{3}$.

33. $|-3| < |-4|$ since

$|-3| = 3$, $|-4| = 4$ and 3 is farther left than 4.

35. $4 < \left|-\frac{9}{2}\right|$ since $\left|-\frac{9}{2}\right|$

$= \frac{9}{2} = 4\frac{1}{2}$ and 4 is

farther left than $4\frac{1}{2}$.

37. $\left|-\frac{6}{2}\right| > \left|-\frac{2}{6}\right|$ since

$\left|-\frac{6}{2}\right| = \frac{6}{2} = 3$,

$\left|-\frac{2}{6}\right| = \frac{2}{6} = \frac{1}{3}$ and 3 is

farther right than $\frac{1}{3}$.

39. 4, -4

41. 2, -2

43. The absolute value of a number is the distance between the number and 0 on the number line.

Cumulative Review Exercises

45. {0,1,2,3,4...}

47. a) 5
b) 5,0
c) 5,-2,0

d) $5, -2, 0, \frac{1}{3}, -\frac{5}{9}, 2.3$

e) $\sqrt{3}$

f) $5, -2, 0, \frac{1}{3}, \sqrt{3}, -\frac{5}{9}, 2.3$

Exercise Set 1.5

1. The opposite of 18 is -18.

3. The opposite of -32 is 32.

5. The opposite of 0 is 0.

7. The opposite of $\frac{5}{3}$ is $-\frac{5}{3}$.

9. The opposite of $\frac{3}{5}$ is $-\frac{3}{5}$.

11. The opposite of 0.63 is -0.63.

13. The opposite of $2\frac{1}{2}$ is $-2\frac{1}{2}$.

15. The opposite of -3.1 is 3.1.

17. $4 + 3 = 7$

19. $4 + (-3) = 1$

21. $-4 + (-2) = -6$

23. $6 + (-6) = 0$

25. $-4 + 4 = 0$

27. $-8 + (-2) = -10$

29. $-3 + 3 = 0$

31. $-3 + (-7) = -10$

33. $0 + 0 = 0$

35. $-6 + 0 = -6$

37. $22 + (-19) = 3$

39. $-45 + 36 = -9$

41. $18 + (-9) = 9$

43. $-14 + (-13) = -27$

45. $-35 + (-9) = -44$

47. $4 + (-30) = -26$

49. $-35 + 40 = 5$

51. $180 + (-200) = -20$

53. $-105 + 74 = -31$

55. $184 + (-93) = 91$

57. $-452 + 312 = -140$

59. True

61. False

63. False

65. \$193 + (-\$112) = \$81

67. 940 + (-486) = 454 m

69. 2400 feet + 200 feet = 2600 feet

71. To add two numbers with like signs, add their absolute values. The sum has the same sign as the numbers being added.

Cumulative Review Exercises

73. $\left(\frac{3}{5}\right)\left(1\frac{2}{3}\right) = \frac{\overset{1}{\cancel{3}}}{\underset{1}{\cancel{5}}} \cdot \frac{\overset{1}{\cancel{5}}}{\underset{1}{\cancel{3}}} = 1$

75. $|-3| > 2$

Exercise Set 1.6

1. $6 - 3 = 6 + (-3) = 3$

3. $4 - 5 = 4 + (-5) = -1$

5. $3 - 3 = 3 + (-3) = 0$

7. $(-7) - (-4) = -7 + 4 = -3$

9. $-3 - 3 = -3 + (-3) = -6$

11. $3 - (-3) = 3 + 3 = 6$

13. $0 - 6 = 0 + (-6) = -6$

15. $0 - (-6) = 0 + 6 = 6$

17. $-3 - 5 = -3 + (-5) = -8$

19. $-5 + 7 = 2$

21. $5 - 3 = 5 + (-3) = 2$

23. $6 - (-3) = 6 + 3 = 9$

25. $8 - 8 = 8 + (-8) = 0$

27. $-8 - 10 = -8 + (-10) = -18$

29. $-5 - (-3) = -5 + 3 = -2$

31. $(-4) - (-4) = -4 + 4 = 0$

33. $6 - 6 = 6 + (-6) = 0$

35. $8 - 8 = 8 + (-8) = 0$

37. $4 - 5 = 4 + (-5) = -1$

39. $-2 - 3 = -2 + (-3) = -5$

41. $-25 - 16 = -25 + (-16)$
$= -41$

43. $37 - 40 = 37 + (-40) = -3$

45. $-100 - 80 = -100 + (-80)$
$= -180$

47. $-20 - 90 = -20 + (-90)$
$= -110$

49. $70 - (-70) = 70 + 70 = 140$

51. $87 - 87 = 87 + (-87) = 0$

53. $-45 - 37 = -45 + (-37)$
$= -82$

55. $9 - 4 = 9 + (-4) = 5$

57. $-15 - 3 = -15 + (-3) = -18$

59. $-8 - 8 = -8 + (-8) = -16$

61. $18 - 8 = 18 + (-8) = 10$

63. $-5 - (-3) = -5 + 3 = -2$

65. $9 - (-4) = 9 + 4 = 13$

67. $18 - 18 = 18 + (-18) = 0$

69. $8 - 12 = 8 + (-12) = -4$

71. $-4 - (-15) = -4 + 15 = 11$

73. $45 - (-36) = 45 + 36 = 81$

75. $6 + 5 - (+4) = 11 - (+4)$
$= 11 + (-4)$
$= 7$

77. $-3 + (-4) + 5 = -7 + 5$
$= -2$

79. $-13 - (+5) + 3$
$= -13 + (-5) + 3$
$= -18 + 3$
$= -15$

81. $-9 - (-3) + 4$
$= -9 + 3 + 4$
$= -6 + 4$
$= -2$

83. $5 - (+3) + (-2)$
$= 5 + (-3) + (-2)$

$= 2 + (-2)$
$= 0$

85. $25 + (+12) - (-6)$
$= 37 + 6$
$= 43$

87. $-4 + (-7) + 5 = -11 + 5$
$= -6$

89. $-4 + 7 - 12 = 3 + (-12)$
$= -9$

91. $45 - 3 - 7$
$= 45 + (-3) + (-7)$
$= 42 + (-7)$
$= 35$

93. $-9 - 4 - 8$
$= -9 + (-4) + (-8)$
$= -13 + (-8)$
$= -21$

95. $-4 - 13 + 5$
$= -4 + (-13) + 5$
$= -17 + 5$
$= -12$

97. $-9 - 3 - (-4) + 5$
$= -9 + (-3) + 4 + 5$
$= -12 + 4 + 5$
$= -8 + 5$
$= -3$

99. $32 + 5 - 7 - 12$
$= 37 + (-7) + (-12)$
$= 30 + (-12)$
$= 18$

101. $-7 - 4 - 3 + 5$
$= -7 + (-4) + (-3) + 5$
$= -11 + (-3) + 5$
$= -14 + 5$
$= -9$

103. 2000 ft + 1500 ft = 3500 ft

105. 29,028 ft + 36,198 ft
= 65,226 ft

107. a) yes

b) $a + (-b) \stackrel{?}{=} a - b$

$(-3) + (-5) \stackrel{?}{=} -3 - 5$
$-8 = -8$
yes

109. a) Add the opposite of 6, −6, to −9.
b) $-9 - (6) = -9 - 6$
$= -15$

111. The set of real numbers consists of the set of rational and irrational numbers.

113. $|-6| < |-7|$

Just for fun

1. $1 - 2 + 3 - 4 + 5 - 6 + 7 - 8 + 9 - 10 = (1-2) + (3-4) + (5-6) + (7-8) + (9-10)$
$= (-1) + (-1) + (-1) + (-1) + (-1) = -5$

2. $1 - 2 + 3 - 4 + 5 - 6 + \ldots + 99 - 100 = (1-2) + (3-4) + (5-6) + \ldots + (99-100) = (-1) + (-1) + (-1) + \ldots + (-1)$
$= -50$

3. $-1 + 2 - 3 + 4 - 5 + 6 - \ldots - 99 + 100 = (-1+2) + (-3+4) + (-5+6) + \ldots + (-99+100) = (1) + (1) + (1) + (1) + \ldots + (1) = 50$

Exercise Set 1.7

1. $(-4)(-3) = 12$

3. $3(-3) = -9$

5. $(-4)(8) = -32$

7. $9(-1) = -9$

9. $-4(-3) = 12$

11. $-9(-4) = 36$

13. $-6 \cdot 5 = -30$

15. $5(-12) = -60$

17. $-4(0) = 0$

19. $(-4)(-4) = 16$

21. $8 \cdot 3 = 24$

23. $5(-3) = -15$

25. $8(12) = 96$

27. $-9(-9) = 81$

29. $-2(5) = -10$

31. $(-6)(2)(-3) = -12(-3) = 36$

33. $0(3)(8) = 0(8) = 0$

35. $(-1)(-1)(-1) = 1(-1) = -1$

37. $-5(-3)(8)(-1) = 15(8)(-1)$
$= 120(-1) = -120$

39. $4(3)(1)(-1) = 12(1)(-1)$
$= 12(-1) = -12$

41. $(-4)(3)(-7)(1)$
$= -12(-7)(1)$
$= 84(1) = 84$

43. $\left[\frac{-1}{2}\right]\left[\frac{3}{5}\right] = \frac{-3}{10}$

45. $\left[\frac{-8}{9}\right]\left[\frac{-7}{12}\right] = \frac{-\overset{2}{\cancel{8}}}{9}\left[\frac{-7}{\underset{3}{\cancel{12}}}\right] = \frac{14}{27}$

47. $\left[\frac{6}{-3}\right]\left[\frac{4}{-2}\right] = \left[\frac{\overset{3}{\cancel{6}}}{3}\right]\left[\frac{4}{\underset{1}{\cancel{2}}}\right]$

$= 1\left[\frac{4}{1}\right] = 4$

49. $\left[\frac{5}{-7}\right]\left[\frac{6}{8}\right] = \left[\frac{5}{-7}\right]\left[\frac{\overset{3}{\cancel{6}}}{\underset{4}{\cancel{8}}}\right] = \frac{-15}{28}$

51. $\frac{6}{2} = 3$

53. $-16 \div (-4) = 4$

55. $\frac{-36}{-9} = 4$

57. $\frac{-16}{4} = -4$

59. $\frac{18}{-1} = -18$

61. $-15 + (-3) = 5$

63. $\frac{-6}{-1} = 6$

65. $\frac{-25}{-5} = 5$

67. $\frac{1}{-1} = -1$

69. $\frac{-48}{12} = -4$

71. $\frac{-18}{-2} = 9$

73. $\frac{0}{-1} = 0$

75. $-30 \div (-30) = 1$

77. $0 \div 3 = 0$

79. $20 \div (-5) = -4$

81. $-30 \div (-10) = 3$

83. $-60 \div 5 = -12$

85. $80 \div (-20) = -4$

87. $-90 \div (-2) = 45$

89. $\frac{5}{12} \div \left[\frac{-5}{9}\right] = \frac{\overset{1}{\cancel{5}}}{\underset{4}{\cancel{12}}}\left[\frac{\overset{-3}{\cancel{-9}}}{\underset{1}{\cancel{5}}}\right] = \frac{-3}{4}$

91. $\frac{3}{-10} \div (-8) = \frac{3}{-10} \div \left[\frac{-8}{1}\right]$

$= \frac{3}{-10} \cdot \left[\frac{1}{-8}\right] = \frac{3}{80}$

93. $\frac{-15}{21} \div \left[\frac{-15}{21}\right] = \frac{\overset{1}{\cancel{-15}}}{\underset{1}{\cancel{21}}}\left[\frac{\overset{1}{\cancel{21}}}{\underset{1}{\cancel{-15}}}\right] = 1$

95. $(-12) \div \left[\frac{5}{12}\right] = \frac{-12}{1}\left[\frac{12}{5}\right]$

$= \frac{-144}{5}$

97. $6 \div \left[\frac{-5}{6}\right] = \frac{6}{1} \cdot \frac{6}{-5} = \frac{-36}{5}$

99. $0 \div 6 = 0$

101. $\frac{0}{0}$ indeterminate

103. $\frac{0}{1} = 0$

105. $8 \div 0$ undefined

107. $\frac{0}{-6} = 0$

109. False

111. True

113. True

115. False

117. False

119. True

121. The sign of the product or quotient of two numbers with like signs is positive. The sign of the product or quotient of two numbers with unlike signs is negative.

123. The product will be a negative number. There are 17 pairs of numbers ($34 \div 2 = 17$). The product of each pair of numbers is a negative number. The product of 17 negative numbers (an odd number of negative numbers) is a negative number.

Cumulative Review Exercises

125. $-20 - (-18) = -20 + 18 = -2$

127. $5 - (-2) + 3 - 7$
$= 5 + 2 + 3 - 7 = 3$

Just for fun

1. $\frac{1-2+3-4+5 - \ldots + 99-100}{1-2+3-4+5 - \ldots + 99-100} = 1$

2. $\frac{-1+2-3+4-5 + \ldots - 99-100}{1-2+3-4+5 - \ldots + 99-100}$

$= \frac{(-1)(1-2+3-4+5-\ldots+99-100)}{(1-2+3-4+5-\ldots+99-100)}$

$= (-1)(1)$
$= -1$

Exercise Set 1.8

1. $3^2 = 9$

3. $2^3 = 8$

5. $3^3 = 27$

7. $6^3 = 216$

9. $(-2)^3 = (-2)(-2)(-2) = -8$

11. $(-1)^3 = (-1)(-1)(-1) = -1$

13. $(3)^3 = 27$

15. $-6^2 = -1(6)(6) = -36$

17. $(-6)^2 = (-6)(-6) = 36$

19. $2^4 = 2 \cdot 2 \cdot 2 \cdot 2 = 16$

21. $5^1 = 5$

23. $(-2)^4 = (-2)(-2)(-2)(-2)$
$= 16$

25. $-2^4 = -1(2)(2)(2)(2) = -16$

27. $(-4)^3 = (-4)(-4)(-4) = -64$

29. $5^2 \cdot 3^2 = 5 \cdot 5 \cdot 3 \cdot 3$
$= 25(9)$
$= 225$

31. $5(4^2) = 5(4)(4) = 80$

33. $2^1 \cdot 4^2 = 2 \cdot 4 \cdot 4 = 32$

35. $3(-5^2) = 3(-1)(5)(5) = -75$

37. $3^2 \cdot 2^4 = 3 \cdot 3 \cdot 2 \cdot 2 \cdot 2 \cdot 2 = 144$

39. $x \cdot y \cdot z \cdot z = xyz^2$

41. $xxxxz = x^4z$

43. $aabbab = a^3b^3$

45. $x \cdot x \cdot y \cdot z \cdot z = x^2yz^2$

47. $x \cdot x \cdot x \cdot y \cdot y = x^3y^2$

49. $xyyyy = xy^4$

51. $5 \cdot 5yyz = 5^2y^2z$

53. $x^2y = xxy$

55. $xy^3 = xyyy$

57. $xy^2z^3 = xyyzzz$

59. $3^2yz = 3 \cdot 3yz$

61. $2^3x^3y = 2 \cdot 2 \cdot 2 \cdot xxxy$

63. $(-2)^2y^3z = (-2)(-2)yyyz$

65. a) $x^2 = (3)^2 = 9$

b) $-x^2 = -(3)^2 = -(3)(3)$
$= -9$

67. a) $x^2 = (4)^2 = 16$

b) $-x^2 = -(4)^2 = -(4)(4)$
$= -16$

69. a) $x^2 = (-2)^2 = 4$

b) $-x^2 = -(-2)^2 = -(-2)(-2)$
$= -4$

71. a) $x^2 = (7)^2 = 49$

b) $-x^2 = -(7)^2 = -(7)(7)$
$= -49$

73. a) $x^2 = (-1)^2 = (-1)(-1)$
$= 1$

b) $-x^2 = -(-1)^2 = -(-1)(-1)$
$= -1$

75. a) $x^2 = [\frac{-1}{2}]^2 = \frac{1}{4}$

b) $-x^2 = -[\frac{-1}{2}]^2$

$= -[\frac{-1}{2}] \cdot [\frac{-1}{2}] = \frac{-1}{4}$

77. False

79. False

81. True

83. True

85. True

87. Any nonzero number will be positive when squared.

89. $(-1)^{100}$ will be positive, because an even number of negative numbers are being multiplied.

91. $-4 - 3 + 9 - 7 = -5$

93. $\frac{0}{4} = 0$

Just for fun

1. $2^2 \cdot 2^3 = 2^5$

2. $3^2 \cdot 3^3 = 3^5$

3. $x^m \cdot x^n = x^{m+n}$

4. $\frac{2^3}{2^2} = \frac{2 \cdot 2 \cdot 2}{2 \cdot 2} = 2$

5. $\frac{3^4}{3^2} = \frac{3 \cdot 3 \cdot 3 \cdot 3}{3 \cdot 3} = 3^2$

6. $\frac{x^m}{x^n} = x^{m-n}$

7. $[2^3]^2 = 2^3 \cdot 2^3 = 2^6$

8. $[3^3]^2 = 3^3 \cdot 3 = 3^6$

9. $[x^m]^n = x^{m \cdot n}$

10. $(2x)^2 = (2x)(2x) = 2^2 \cdot x^2$

11. $(3x)^2 = (3x)(3x) = 3^2 \cdot x^2$

12. $(ax)^2 = (ax)(ax) = a^2 \cdot x^2$

Exercise Set 1.9

1. $3 + 4 \cdot 5 = 3 + 20 = 23$

3. $2 - 2 + 5 = 0 + 5 = 5$

5. $1 + 3 \cdot 2^2$
$= 1 + 3 \cdot 4$
$= 1 + 12$
$= 13$

7. $-3^2 + 5 = -9 + 5 = -4$

9. $(4 - 3)(5 - 1)^2$
$= (4 - 3)(4)^2$
$= (4 - 3) \cdot 16$
$= 1(16)$
$= 16$

11. $(3 \cdot 7) + (4 \cdot 2)$
$= 21 + 8 = 29$

13. $[1 - 4(5)] + 6$
$= (1 - 20) + 6$
$= -19 + 6$
$= -13$

15. $4^2 - (3 \cdot 4) - 6$
$= 16 - (3 \cdot 4) - 6$
$= 16 - 12 - 6$
$= 4 - 6$
$= -2$

17. $-3[-4 + (6 - 8)]$
$= -3[-4 + (-2)]$
$= -3(-6)$
$= 18$

19. $(6 + 3)^3 + (4^2 + 8)$
$= (2)^3 + (16 + 8)$
$= 8 + 2$
$= 10$

21. $-4^2 + (8 + 2 \cdot 5) + 3$
$= -16 + 4(5) + 3$

$= -16 + 20 + 3$
$= 4 + 3$
$= 7$

23. $3 + [4^2 - 10]^2 - 3$

$= 3 + (16 - 10)^2 - 3$

$= 3 + (6)^2 - 3$
$= 3 + 36 - 3$
$= 39 - 3$
$= 36$

25. $-[12 - (-4 - 5)]^2$

$= -[12 - (-9)]^2$

$= -(12 + 9)^2$

$= -(21)^2$
$= -441$

27. $[3^2 - 1] \div (3 + 1)^2$

$= (9 - 1) \div (4)^2$
$= 8 \div 16$

$= \frac{1}{2}$

29. $2[(36 \div 9) + 1]$
$= 2[(4) + 1]$
$= 2(5)$
$= 10$

31. $2[3(8 - 2^2) - 6]$
$= 2[3(8 - 4) - 6]$
$= 2[3(4) - 6]$
$= 2(6)$
$= 12$

33. $10 - [8 - (3 + 4)]^2$

$= 10 - [8 - (7)]^2$

$= 10 - (8 - 7)^2$

$= 10 - (1)^2$
$= 10 - 1$
$= 9$

35. $[4 + ((5 - 2)^2 \div 3)^2]^2$

$= [4 + ((3)^2 \div 3)^2]^2$

$= [4 + (9 \div 3)^2]^2$

$= [4 + (3)^2]^2$

$= [4 + 9]^2$
$= [13]^2$
$= 169$

37. $[-2(4 - 6)^2]^2$
$- 4[-3(6 \div 3)^2]$

$= [-2(-2)^2]^2 - 4[-3(2)^2]$

$= (-2(4)]^2 - 4[-3(4)]$

$= [-8]^2 - 4[-12]$

$= 64 + 48$
$= 112$

39. $(14 \div 7 \cdot 7 \div 7 - 7)^2$

$= (2 \cdot 7 \div 7 - 7)^2$

$= (14 \div 7 - 7)^2$

$= (2 - 7)^2$

$= (-5)^2$
$= 25$

41. $(8.4 + 3.1)^2 - (3.64 - 1.2)$

$= (11.5)^2 - 2.44$
$= 132.25 - 2.44$
$= 129.81$

43. $(4.3)^2 + 2(5.3) - 3.05$
$= 18.49 + 10.6 - 3.05$
$= 26.04$

45. $[\frac{2}{7} + \frac{3}{8}] - \frac{3}{112}$

$= \frac{16 + 21}{56} - \frac{3}{112}$

$= \frac{37}{56} - \frac{3}{112}$

$= \frac{74 - 3}{112}$

$= \frac{71}{112}$

47. $\frac{3}{4} - 4 \cdot [\frac{5}{40}] = \frac{3}{4} - \frac{\overset{1}{\cancel{20}}}{\underset{2}{\cancel{40}}}$

$= \frac{3}{4} - \frac{1}{2}$

$= \frac{3 - 2}{4}$

$= \frac{1}{4}$

49. $2(3 + \frac{2}{5}) \div (\frac{3}{5})^2$

$= 2(\frac{15 + 2}{5}) \div \frac{9}{25}$

$= \frac{2}{1}(\frac{17}{5}) \div \frac{9}{25}$

$= \frac{34}{\underset{1}{\cancel{5}}} \cdot \frac{\overset{5}{\cancel{25}}}{9}$

$= \frac{170}{9}$ or $18\frac{8}{9}$

51. $[(6 \cdot 3) - 4 - 2]$
$= 18 - 4 - 2$
$= 14 - 2$
$= 12$

53. $(20 \div 5 + 12 - 8) \cdot 9$
$= (4 + 12 - 8) \cdot 9$
$= (16 - 8) \cdot 9$
$= 72$

55. $[\frac{4}{5} + \frac{3}{7}] \cdot \frac{2}{3} = [\frac{28 + 15}{35}] \cdot \frac{2}{3}$

$= [\frac{43}{35}] \cdot \frac{2}{3}$

$= \frac{86}{105}$

57. $x + 4 = -2 + 4$
$= 2$

59. $3x - 2 = 3(4) - 2$
$= 12 - 2$
$= 10$

61. $x^2 - 6 = (-3)^2 - 6$
$= 9 - 6$
$= 3$

63. $-3x^2 - 4 = -3(1)^2 - 4$
$= -3(1) - 4$
$= -3 - 4$
$= -7$

65. $-4x^2 - 2x + 5$

$= -4(-3)^2 - 2(-3) + 5$
$= -4(9) - 2(-3) + 5$
$= -36 + 6 + 5$
$= -30 + 5$
$= -25$

67. $3(x - 2)^2 = 3[(7) - 2]^2$

$= 3(5)^2$
$= 3(25)$
$= 75$

69. $2(x - 3)(x + 4)$
$= 2(1 - 3)(1 + 4)$
$= 2(-2)(5)$
$= -4(5)$
$= -20$

71. $-6x + 3y = -6(2) + 3(4)$
$= -12 + 12$
$= 0$

73. $x^2 - y^2 = (-2)^2 - (-3)^2$
$= 4 - 9$
$= -5$

75. $4(x + y)^2 + 4x - 3y$

$= 4[(2) + (-3)]^2 + 4(2) - 3(-3)$

$= 4(-1)^2 + 4(2) - 3(-3)$
$= 4(1) + 4(2) - 3(-3)$
$= 4 + 8 + 9$
$= 12 + 9$
$= 21$

77. $3(a + b)^2 + 4(a + b) - 6$

$= 3[(4) + (-1)]^2 + 4[(4) + (-1)] - 6$

$= 3(3)^2 + 4(3) - 6$
$= 3(9) + 4(3) - 6$
$= 27 + 12 - 6$
$= 39 - 6$
$= 33$

79. $x^2y - 6xy + 3x$

$= (2)^2(3) - 6(2)(3) + 3(2)$
$= 4(3) - 6(2)(3) + 3(2)$
$= 12 - 12(3) + 3(2)$
$= -24 + 6$
$= -18$

81. $6x^2 + 3xy - y^2$

$= 6(2)^2$

$+ 3(2)(-3) - (-3)^2$
$= 6(4) + (-18) - (9)$
$= 24 - 18 - 9$
$= -3$

83. $5(2x - 3)^2 - 4(6 - y)^2$

$= 5[2(-2) - 3)]^2$

$- 4[6 - (-1)]^2$

$= 5[-4 - 3]^2 - 4[6 + 1]^2$

$= 5(-7)^2 - 4(7)^2$
$= 5(49) - 4(49)$
$= 245 - 196$
$= 49$

85. 1. Evaluate within parentheses or brackets, innermost parentheses first.
2. Evaluate exponents.
3. Multiply and/or divide in order from left to right.
4. Add and/or subtract in order from left to right.

87. a) 1. Substitute 5 in for each x.
2. Evaluate exponent.
3. Evaluate multiplications from left to right.
4. Evaluate addition and subtractions from left to right.

b) $-4x^2 + 3x - 6$

$= -4(5)^2 + 3(5) - 6$
$= -4(25) + 3(5) - 6$
$= -100 + 15 - 6$
$= -91$

Cumulative Review Exercises

89. a) $x^2 = (-5)^2 = 25$

b) $-x^2 = -(-5)^2 = -25$

91. $-2^4 = -16$

Just for fun

1. $4[[3(x-2)]^2 + 4]$

$= 4[[3(4-2)]^2 + 4]$

$= 4[(3(2))^2 + 4]$

$= 4[(6)^2 + 4]$
$= 4(36 + 4)$
$= 4(40)$
$= 160$

2. $[(3 - 6)^2 + 4]^2 + 3\cdot 4 - 12 \div 3$

$= [(-3)^2 + 4]^2 + 3\cdot 4 - 12 \div 3$

$= (9 + 4)^2 + 3\cdot 4 - 12 \div 3$

$= (13)^2 + 3\cdot 4 - 12 \div 3$
$= 169 + 12 - 4$
$= 181 - 4$
$= 177$

3. $-2[[3x^2 + 4]^2 - [3x^2 - 2]^2]$

$= -2[[3(-2)^2 + 4]^2 - [3(-2)^2 - 2]^2]$

$= -2[[3(4) + 4]^2 - [3(4) - 2]^2]$

$= -2[(12 + 4)^2 - (12 - 2)^2]$

$= -2[(16)^2 - (10)^2]$
$= -2(256 - 100)$
$= -2(156)$
$= -312$

Exercise Set 1.10

1. Distributive property

3. Commutative property of multiplication

5. Distributive property

7. Associative property of multiplication

9. Distributive property

11. $3 + 4 = 4 + 3$

13. $-6\cdot(4\cdot 2) = (-6\cdot 4)\cdot 2$

15. $(6)(y) = (y)(6)$

17. $1(x + y) = x + y$

19. $4x + 3y = 3y + 4x$

21. $5x + 5y = 5(x + y)$

23. $(x + 2)3 = 3(x + 2)$

25. $(3x + 4) + 6 = 3x + (4 + 6)$

27. $3(x + y) = (x + y)3$

29. $4(x + y + 3) = 4x + 4y + 12$

31. Commutative property of addition

33. Distributive property

35. Commutative property of addition

37. Distributive property

39. Yes, order results in the same final outcome.

41. No, order results in different outcomes.

43. $2\frac{3}{5} + \frac{2}{3} = \frac{13}{5} + \frac{2}{3}$

$= \frac{39 + 10}{15} = \frac{49}{15}$ or $3\frac{4}{15}$

45. $12 - 24 \div 8 + 4\cdot 3^2$
$= 12 - 3 + 4\cdot 9$
$= 12 - 3 + 36$
$= 45$

Review Exercises

1. $\frac{3}{5}\cdot\frac{5}{6} = \frac{\overset{1}{\cancel{3}}}{\underset{1}{\cancel{5}}}\cdot\frac{\overset{1}{\cancel{5}}}{\underset{2}{\cancel{6}}} = \frac{1}{2}$

3. $\frac{5}{12} \div \frac{3}{5} = \frac{5}{12} \cdot \frac{5}{3} = \frac{25}{36}$

5. $\frac{3}{8} - \frac{1}{9} = \frac{27}{72} - \frac{8}{72} = \frac{19}{72}$

7. {1,2,3,...}

9. {...,-3,-2,-1,0,1,2,3,...}

11. {all numbers that can be represented on the real number line}

13. a) {1} b) {1}
c) {-9,-8}
d) {-9,-8,1}
e) $\{-9,-8,-2.3,-\frac{3}{17},1,1\frac{1}{2}\}$
f) $\{-9,-8,-2.3,-\frac{3}{17},1,1\frac{1}{2}, -\sqrt{2},\sqrt{2}\}$

15. $-2 < 1$

17. $-2.6 > -3.6$

19. $4.6 > 4.06$

21. $5 > |-3|$

23. $|-2.5| = |\frac{5}{2}|$

25. $-4 + (-5) = -9$

27. $4 + (-9) = -5$

29. $-10 + 4 = -6$

31. $-9 - (-4) = -9 + 4$
$= -5$

33. $0 - 2 = 0 + (-2)$
$= -2$

35. $2 - 12 = 2 + (-12)$
$= -10$

37. $2 - 7 = 2 + (-7)$
$= -5$

39. $-7 - 5 = -7 + (-5)$
$= -12$

41. $-5 + 7 - 6 = -5 + 7 + (-6)$
$= 2 + (-6)$
$= -4$

43. $-2 + (-3) - 2$
$= -2 + (-3) + (-2)$
$= -5 + (-2)$
$= -7$

45. $7 - (4) - (-3)$
$= 7 + (-4) + 3$
$= 3 + 3$
$= 6$

47. $4 - (-2) + 3$
$= 4 + 2 + 3$
$= 6 + 3$
$= 9$

49. $(-9)(-3) = 27$

51. $-2(3) = -6$

53. $\left[\frac{10}{11}\right]\left[\frac{3}{-5}\right] = \left[\frac{\overset{-2}{\cancel{10}}}{11}\right]\left[\frac{3}{\underset{1}{\cancel{-5}}}\right]$

$= \frac{-6}{11}$

55. $0 \cdot \frac{4}{9} = \frac{0}{1} \cdot \frac{4}{9} = 0$

57. $(-1)(-3)(4) = 3(4)$
$= 12$

59. $(-3)(-4)(-5) = 12(-5)$
$= -60$

61. $(-4)(-6)(-2)(-3)$
$= 24(-2)(-3)$
$= -48(-3)$
$= 144$

63. $6 \div (-2) = \frac{\overset{3}{\cancel{6}}}{1}\left[\frac{-1}{\underset{1}{\cancel{2}}}\right] = -3$

65. $-36 \div (-2) = \frac{\overset{-18}{\cancel{-36}}}{1} \cdot \frac{-1}{\underset{1}{\cancel{2}}} = 18$

67. $0 \div (-4) = \frac{0}{1} \cdot \frac{1}{4} = 0$

69. $-40 \div (-8) = \frac{\overset{-5}{\cancel{-40}}}{1} \cdot \frac{-1}{\underset{1}{\cancel{8}}} = 5$

71. $\frac{15}{32} \div (-5) = \frac{\overset{3}{\cancel{15}}}{32} \cdot \frac{-1}{\underset{1}{\cancel{5}}} = \frac{-3}{32}$

73. $\frac{28}{-3} \div \frac{9}{-2} = \frac{-28}{3} \cdot \frac{-2}{9} = \frac{56}{27}$

75. $\left[\frac{-5}{12}\right] \div \left[\frac{-5}{12}\right] = \frac{\overset{-1}{\cancel{-5}}}{\underset{1}{\cancel{12}}} \cdot \left[\frac{\overset{-1}{\cancel{-12}}}{\underset{1}{\cancel{5}}}\right]$

$= 1$

77. $0 \div (-6) = \frac{0}{1} \cdot \frac{-1}{6} = 0$

79. $-4 \div 0$ Undefined

81. $0 \div 1 = \frac{0}{1} \cdot \frac{1}{1} = 0$

83. $2(4 - 8) = 2(-4)$
$= -8$

85. $(-4 + 3) - (2 - 6)$
$= -1 - (-4)$
$= -1 + 4$
$= 3$

87. $(-4 - 2)(-3) = (-6)(-3)$
$= 18$

89. $9[3 + (-4)] + 5 = 9(-1) + 5$
$= -9 + 5$
$= -4$

91. $(-3 \cdot 4) \div (-2 \cdot 6)$
$= -12 \div (-12)$
$= \frac{\overset{-1}{\cancel{-12}}}{1} \cdot \frac{-1}{\underset{1}{\cancel{12}}}$
$= 1$

93. $[-2(3) + 6] - 4$
$= (-6 + 6) - 4$
$= 0 + (-4)$
$= -4$

95. $6^2 = 6 \cdot 6$
$= 36$

97. $1^5 = 1 \cdot 1 \cdot 1 \cdot 1 \cdot 1$
$= 1$

99. $2^4 = 2 \cdot 2 \cdot 2 \cdot 2$
$= 4 \cdot 2 \cdot 2$
$= 8 \cdot 2$
$= 16$

101. $(-1)^9 = -1$
Negative to an odd power is negative.

103. $\left[\frac{2}{7}\right]^2 = \left[\frac{2}{7}\right] \cdot \left[\frac{2}{7}\right] = \frac{4}{49}$

105. $\left[\frac{2}{5}\right]^3 = \left[\frac{2}{5}\right] \cdot \left[\frac{2}{5}\right] \cdot \left[\frac{2}{5}\right]$

$= \frac{4}{25} \cdot \frac{2}{5} = \frac{8}{125}$

107. $xyy = xy^2$

109. $yyzz = y^2z^2$

111. $5 \cdot 7 \cdot 7xxy = 5 \cdot 7^2x^2y$

113. $x^2y = xxy$

115. $y^3z = yyyz$

117. $-x^2 = (-3)(3)$
$= -9$

119. $-x^3 = -(3)(3)(3)$
$= -3(3)(3)$
$= -9(3)$
$= -27$

121. $3 + 5 \cdot 4 = 3 + 20$
$= 23$

123. $3 \cdot 5 + 4 \cdot 2 = 15 + 8$
$= 23$

125. $6 + 4 \cdot 5 = 6 + 20$
$= 26$

127. $6 - 3^2 \cdot 5 = 6 - 9 \cdot 5$
$= 6 - 45$
$= 6 + (-45)$
$= -39$

129. $[6 - (3 \cdot 5)] + 5$
$= (6 - 15) + 5$
$= [6 + (-15)] + 5$
$= -9 + 5$
$= -4$

131. $[-3^2 + 4^2] + [3^2 \div 3]$
$= (-9 + 16) + (9 \div 3)$

$= (7) + \left[\frac{\overset{3}{\cancel{9}}}{1} \cdot \frac{1}{\underset{1}{\cancel{3}}}\right]$

$= 7 + 3 = 10$

133. $(4 \div 2)^4 + 4^2 \div 2^2$

$= \left[\frac{\overset{2}{\cancel{4}}}{1} \cdot \frac{1}{\underset{1}{\cancel{2}}}\right]^4 + 16 \div 4$

$= 2^4 + \left[\frac{\overset{4}{\cancel{16}}}{1} \cdot \frac{1}{\underset{1}{\cancel{4}}}\right]$

$= 16 + 4$
$= 20$

135. $4^3 \div 4^2 - 5(2 - 7) \div 5$
$= 64 \div 16 - 5\ (-5) \div 5$

$= \frac{\overset{4}{\cancel{64}}}{1} \cdot \frac{1}{\underset{1}{\cancel{16}}} + 25 \div 5$

$= 4 + 5$
$= 9$

137. $4x - 6$ when $x = 5$
$4(5) - 6 = 20 - 6$
$= 20 + (-6)$
$= 14$

139. $6 - 4x$ when $x = -5$
$6 - 4(-5) = 6 + 20$
$= 26$

141. $5y^2 + 3y - 2$ when $y = -1$

$5(-1)^2 + 3(-1) - 2$
$= 5(1) + 3(-1) - 2$
$= 5 + (-3) + (-2)$
$= 0$

143. $-x^2 + 2x - 3$ when $x = -2$

$-(-2)^2 + 2(-2) - 3$
$= -4 + (-4) + (-3)$
$= -8 + (-3)$
$= -11$

145. $3xy - 5x$ when $x = 3$ and $y = 4$
$3(3)(4) - 5(3)$
$= 9(4) - 15$
$= 36 + (-15)$
$= 21$

147. $6 \cdot x = x \cdot 6$ Commutative property of multiplication.

149. $(x + 4)3 = 3(x + 4)$
Commutative property of multiplication.

151. $(x + 7) + 4 = x + (7 + 4)$
Associative property of addition.

Practice Test

1. a) {42} b) {0,42}
   c) {-7,-6,-1,0,42}
   d) $\{-7,-3\frac{1}{2},-6,-1,0,\frac{5}{9},6.52,42\}$
   e) $\{\sqrt{5},\}$
   f) $\{-7,-6,-3\frac{1}{2},-1,0,\frac{5}{9},\sqrt{5},6.52,42\}$

2. $-6 < -3$

3. $|-3| > |-2|$

4. $-4 + (-8) = -12$

5. $-6 - 5 = -6 + (-5) = -11$

6. $4 - (-12) = 4 + 12 = 16$

7. $5 - 12 - 7$
   $= 5 + (-12) + (-7)$
   $= -14$

8. $(-4 + 6) - 3(-2)$
   $= 2 + 6 = 8$

9. $(-4)(-3)(2)(-1) = 12(2)(-1)$
   $= 24(-1)$
   $= -24$

10. $[\frac{-2}{9}] \div [\frac{-7}{8}] = \frac{-2}{9} \cdot \frac{-8}{7} = \frac{16}{63}$

11. $[-12 \cdot \frac{1}{2}] \div 3 = -6 \div 3 = -2$

12. $3 \cdot 5^2 - 4 \cdot 6^2 = 3.25 - 4 \cdot 36$
    $= 75 - 144$
    $= -69$

13. $[4 - 6^2] \div [4(2 + 3) - 4]$
    $= (4 - 36) \div [4(5) - 4]$
    $= -32 \div (20 - 4)$
    $= -32 \div (16)$
    $= -2$

14. $-6(-2 - 3) \div 5 \cdot 2$
    $= -6(-5) \div 5 \cdot 2$
    $= 30 \div 5(2)$
    $= 6(2)$
    $= 12$

15. $(-3)^4 = (-3)(-3)(-3)(-3)$
    $= 9(-3)(-3)$
    $= -27(-3)$
    $= 81$

16. $[\frac{3}{5}]^3 = \frac{3}{5} \cdot \frac{3}{5} \cdot \frac{3}{5}$
    $= \frac{9}{25} \cdot \frac{3}{5}$
    $= \frac{27}{125}$

17. $2 \cdot 2 \cdot 5 \cdot 5yyzzz = 2^2 \cdot 5^2 y^2 z^3$

18. $2^2 \cdot 3^3 \cdot x^4 \cdot y^2 = 2 \cdot 2 \cdot 3 \cdot 3 \cdot 3xxxxyy$

19. $2x^2 - 6 = 2(-4)^2 - 6$
    $= 2(16) - 6$
    $= 32 - 6$
    $= 26$

20. $6x - 3y^2 + 4$
    $= 6(3) - 3(-2)^2 + 4$
    $= 18 - 12 + 4$
    $= 6 + 4$
    $= 10$

21. $-x^2 - 6x + 3$
    $= -(-2)^2 - 6(-2) + 3$
    $= -4 + 12 + 3$
    $= 11$

22. $x + 3 = 3 + x$
    Commutative property of addition.

23. $4(x + 9) = 4x + 36$
    Distributive property.

24. $(2 + x) + 4 = 2 + (x + 4)$
Associative Property of addition.

25. $5(x + y) = (x + y)5$
Commutative property of multiplication.

CHAPTER TWO

Exercise Set 2.1

1. $3x + 5x = 8x$

3. $2x - 3x$
$= 2x + (-3x)$
$= -1x$
$= -x$

5. $12 + x - 3$
$= x + 12 + (-3)$
$= x + 9$

7. $-4x + 7x = 3x$

9. $x + 3x - 7$
$= 1x + 3x + (-7)$
$= 4x + (-7)$
$= 4x - 7$

11. $6 - 3 + 2x$
$= 6 + (-3) + 2x$
$= 3 + 2x$
$= 2x + 3$

13. $-7 + 5x + 12$
$= 5x + (-7) + 12$
$= 5x + 5$

15. $5x + 2y + 3 + y$
$= 5x + 2y + 1y + 3$
$= 5x + 3y + 3$

17. $4x - 2x + 3 - 7$
$= 4x + (-2x) + 3 + (-7)$
$= 2x + 3 + (-7)$
$= 2x + (-4)$
$= 2x - 4$

19. $4 + x + 3$
$= x + 4 + 3$
$= x + 7$

21. $-3x + 2 - 5x$
$= -3x + 2 + (-5x)$
$= -3x + (-5x) + 2$
$= -8x + 2$

23. $5 + 2x - 4x + 6$
$= 5 + 2x + (-4x) + 6$
$= 5 + (-2x) + 6$
$= (-2x) + 5 + 6$
$= -2x + 11$

25. $x - 2 - 4 + 2x$
$= 1x + (-2) + (-4) + 2x$
$= 1x + (-6) + 2x$
$= 1x + 2x + (-6)$
$= 3x + (-6)$
$= 3x - 6$

27. $2 - 3x - 2x + 1$
$= 2 + (-3x) + (-2x) + 1$
$= 2 + (-5x) + 1$
$= (-5x) + 2 + 1$
$= -5x + 3$

29. $2y + 4y + 6 = 6y + 6$

31. $x - 6 + 3x - 4$
$= 1x + (-6) + 3x + (-4)$
$= 1x + 3x + (-6) + (-4)$
$= 4x + (-6) + (-4)$
$= 4x - 10$

33. $4 - x + 4x - 8$
$= 4 + (-1x) + 4x + (-8)$
$= 4 + 3x + (-8)$
$= 3x + 4 + (-8)$
$= 3x + (-4)$
$= 3x - 4$

35. $x + \frac{3}{4} - \frac{1}{3}$

$= x + \frac{9 - 4}{12} = x + \frac{5}{12}$

37. $68.2x - 19.7x + 8.3$
$= 48.5x + 8.3$

39. $x + \frac{1}{2} y - \frac{3}{8} y$

$= x + \frac{4}{8} y - \frac{3}{8} y$

$= x + \frac{1}{8} y$

41. $-4x - 3.1 - 5.2 = -4x - 8.3$

43. $1 + x + 6 - 3x$
$= x - 3x + 1 + 6$
$= -2x + 7$

45. $3x - 7 - 9 + 4x$
$= 3x + (-7) + (-9) + 4x$
$= 3x + (-16) + 4x$
$= 3x + 4x + (-16)$
$= 7x + (-16)$
$= 7x - 16$

47. $4x + 6 + 3x - 7$
$= 4x + 6 + 3x + (-7)$
$= 4x + 3x + 6 + (-7)$
$= 7x + (6) + (-7)$
$= 7x + (-1)$
$= 7x - 1$

49. $-4 + x - 6 + 2$
$= -4 + x + (-6) + 2$
$= -4 + x + (-4)$
$= x + (-4) + (-4)$
$= x + (-8)$
$= x - 8$

51. $-19.36 + 40.02x + 12.25 - 18.3x$
$= 40.02x - 18.3x - 19.36 + 12.25$
$= 21.72x - 7.11$

53. $\frac{3}{5} x - 3 - \frac{7}{4} x - 2$

$= \frac{3}{5} x - \frac{7}{4} x - 3 - 2$

$= \frac{12}{20} x - \frac{35}{20} x - 5$

$= - \frac{23}{20} x - 5$

55. $2(x + 4) = 2 \cdot x + 2 \cdot 4$
$= 2x + 8$

57. $4(x + 5) = 4 \cdot x + 4(5)$
$= 4x + 20$

59. $-2(x - 4) = -2 \cdot x + (-2)(-4)$
$= -2x + 8$

61. $\frac{-1}{2}(2x - 4) = \frac{-1}{2}(2x + (-4))$

$= \frac{-1}{2}(2x) + [\frac{-1}{2}](-4)$

$= \frac{-2x}{2} + \frac{4}{2}$

$= -x + 2$

63. $1(-4 + x) = 1(-4) + 1(x)$
$= -4 + x$

65. $\frac{1}{4}(x - 12) = \frac{1}{4}(x + (-12))$

$= \frac{1}{4} x + \frac{1}{4}(-12)$

$= \frac{1}{4} x + \frac{-12}{4}$

$= \frac{1}{4} x + (-3)$

$= \frac{1}{4} x - 3$

67. $-0.6(3x - 5)$
$= -0.6(3x + (-5)$
$= -0.6(3x) + (-0.6)(-5)$
$= -1.8x + 3$

69. $\frac{1}{2}(-2x + 6) = \frac{1}{2}(-2x) + \frac{1}{2}(6)$

$= -x + 3$

71. $0.4(2x - 0.5)$
$= 0.4(2x) + 0.4(-0.5)$
$= 0.8x - 0.2$

73. $-(-x + y) = -1(-x + y)$
$= -1(-x) + (-1)(y)$
$= x + (-y)$
$= x - y$

75. $-(-2x - 6y + 8)$
$= -1[-2x + (-6y) + 8]$

$= -1(2x) + (-1)(-6y) + (-1)(8)$
$= -2x + 6y + (-8)$
$= -2x + 6y - 8$

77. $3(4 - 2x + y)$
$= 3[4 + (-2x) + y]$
$= 3(4) + 3(-2x) + 3(y)$
$= 12 + (-6x) + 3y$
$= 12 - 6x + 3y$

79. $2(\frac{1}{2} x - 4y + \frac{1}{4})$

$= 2(\frac{1}{2} x) + 2(-4y) + 2(\frac{1}{4})$

$= x - 8y + \frac{1}{2}$

81. $(x + 3y - 9)$
$= 1[x + 3y + (-9)]$
$= 1(x) + 1(3y) + 1(-9)$
$= x + 3y + (-9)$
$= x + 3y - 9$

83. $-(-x + 4 + 2y)$
$= -1(-x + 4 + 2y)$
$= -1(-x) + (-1)4 + (-1)4 + (-1)2y$
$= x + (-4) + (-2y)$
$= x - 4 - 2y$

85. $4(x - 2) - x$
$= 4[x + (-2)] + (-x)$
$= 4(x) + 4(-2) + (-x)$
$= 4x + (-8) + (-x)$
$= 4x + (-x) + (-8)$
$= 3x + (-8)$
$= 3x - 8$

87. $-2(3 - x) + 1$
$= -2[3 + (-x)] + 1$
$= (-2)(3) + (-2)(-x) + 1$
$= -6 + 2x + 1$
$= 2x + (-6) + 1$
$= 2x + (-5)$
$= 2x - 5$

89. $6x + 2(4x + 9)$
$= 6x + 2(4x) + 2(9)$
$= 6x + 8x + 18$
$= 14x + 18$

91. $2(x - y) + 2x + 3$
$= 2[x + (-y)] + 2x + 3$
$= 2x + 2(-y) + 2x + 3$
$= 2x + (-2y) + 2x + 3$
$= 4x + (-2y) + 3$
$= 4x - 2y + 3$

93. $(x + y) - 2x + 3$
$= x + y + (-2x) + 3$
$= x + (-2x) + y + 3$
$= -x + y + 3$

95. $8x - (x - 3)$
$= 8x + (-1)(x + (-3))$
$= 8x + (-1)(x) + (-1)(-3)$
$= 8x + (-1x) + 3$
$= 7x + 3$

97. $2(x - 3) - (x + 3)$
$= 2[x+(-3)] + (-1)(x+3)$
$= 2x + 2(-3) + (-1)(x) + (-1)3$
$= 2x + (-6) + (-1x) + (-3)$
$= 2x + (-1x) + (-6) + (-3)$
$= x + (-6) + (-3)$
$= x + (-9)$
$= x - 9$

99. $4(x - 3) + 2(x - 2) + 4$
$= 4[x+(-3)] + 2[x+(-2)] + 4$
$= 4x + 4(-3) + 2x + 2(-2) + 4$
$= 4x + (-12) + 2x + (-4) + 4$
$= 4x + 2x + (-12) + (-4) + 4$
$= 6x + (-12) + (-4) + 4$
$= 6x + (-16) + 4$
$= 6x - 12$

101. $2(x - 4) - 3x + 6$
$= 2[x+(-4)] + (-3x) + 6$
$= 2x + 2(-4) + (-3x) + 6$
$= 2x + (-8) + (-3) + 6$
$= 2x + (-3x) + (-8) + 6$
$= -x + (-8) + 6$
$= -x + (-2)$
$= -x - 2$

103. $-3(x - 4) + 2x - 6$

$= -3[x+(-4)] + 2x + (-6)$
$= -3(x) + (-3)(-4) + 2x + (-6)$
$= -3x + 12 + 2x + (-6)$
$= -3x + 2x + 12 + (-6)$
$= -x + 6$

105. $4(x - 3) + 4x - 7$
$= 4[x+(-3)] + 4x + (-7)$
$= 4(x) + 4(-3) + 4x + (-7)$
$= 4x + (-12) + 4x + (-7)$
$= 4x + 4x + (-12) + (-7)$
$= 8x + (-12) + (-7)$
$= 8x + (-19)$
$= 8x - 19$

107. $0.4 + (x + 5) - 0.6 + 2$
$= 0.4 + x + 5 - 0.6 + 2$
$= x + 0.4 + 5 + (-0.6) + 2$

109. $9 - (-3x + 4) - 5$
$= 9 + (-1)(-3x + 4) + (-5)$
$= 9 + (-1)(-3x) + (-1)(4) + (-5)$
$= 9 + 3x + (-4) + (-5)$
$= 9 + 3x + (-9)$
$= 3x + 9 + (-9)$
$= 3x + 0$
$= 3x$

111. $4(x + 2) - 3(x - 4) - 5$
$= 4(x + 2) + (-3)(x + (-4)) - 5$
$= 4x + 4(2) + (-3)x + (-3)(-4) - 5$
$= 4x + 8 + (-3x) + 12 - 5$
$= 4x + (-3x) + 8 + 12 - 5$
$= x + 8 + 12 - 5$
$= x + 15$

113. $-0.2(2 - x) + 4(y + 0.2)$
$= -0.2[2 + (-x)] + 4(y + 0.2)$
$= -0.2(2) + (-0.2)(-x) + 4(y) + 4(0.2)$
$= -0.4 + 0.2x + 4y + 0.8$
$= 0.2x + 4y + (-0.4) + (0.8)$
$= 0.2x + 4y + 0.4$

115. $-6 + 3y - (6 + x) + (x + 3)$
$= -6x + 3y + (-1)(6 + x) + (x + 3)$
$= -6x + 3y + (-1)(6) + (-1)x + x + 3$
$= -6x + 3y + (-6) + (-x) + x + 3$
$= -6x + 3y + (-6) + 0x + 3$
$= -6x + 3y + (-6) + 3$
$= -6x + 3y + (-3)$
$= -6x + 3y - 3$

117. $-(x + 3) + (2x + 4) - 6$
$= (-1)(x + 3) + (2x + 4) + (-6)$
$= (-1)x + (-1)3 + 2x + 4 + (-6)$
$= -x + (-3) + 2x + 4 + (-6)$
$= -x + (-3) + 2x + (-2)$
$= -x + 2x + (-3) + (-2)$
$= x + (-3) + (-2)$
$= x + (-5)$
$= x - 5$

119. $\frac{2}{3}(x - 2) - \frac{1}{2}(x + 4)$

$= \frac{2}{3}(x) + \frac{2}{3}(-2) - \frac{1}{2}(x) - \frac{1}{2}(4)$

$= \frac{2}{3}x - \frac{4}{3} - \frac{1}{2}x - 2$

$= \frac{2}{3}x - \frac{1}{2}x - \frac{4}{3} - 2$

$= \frac{4}{6}x - \frac{3}{6}x - \frac{4}{3} - \frac{6}{3}$

$= \frac{1}{6}x - \frac{10}{3}$

121. The signs of all the terms inside the parentheses are when the parentheses are removed.

123. a) $2x^2$, $3x$, -5;

The terms are the parts of the expression that are added or subtracted.

b) $1, 2, x, 2x, x^2, 2x^2$

Note that $1 \cdot 2x^2 = 2x^2$,

$2 \cdot x^2 = 2x^2$, and $x \cdot 2x = 2x^2$. Expressions that are multiplied are factors of the product.

Cumulative Review Exercises

125. $-|-16| = -16$

127. $-x^2 + 5x - 6$

$= -(-1)^2 + 5(-1) - 6$
$= -1 - 5 - 6$
$= -12$

Just for fun

1. $4x + 5y + 6(3x - 5y) - 4x + 3$
$= 4x + 5y + 6(3x) + 6(-5y) - 4x + 3$
$= 4x + 5y + 18x - 30y - 4x + 3$
$= 4x + 18x - 4x + 5y - 30y + 3$
$= 18x - 25y + 3$

2. $2x^2 - 4x + 8x^2 - 3(x + 2) - x^2 - 2$

$= 2x^2 - 4x + 8x^2 - 3(x) - 3(2) - x^2 - 2$

$= 2x^2 - 4x + 8x^2 - 3x - 6 - x^2 - 2$

$= 2x^2 + 8x^2 - x^2 - 4x - 3x - 6 - 2$

$= 9x^2 - 7x - 8$

3. $x^2 + 2y - y^2 + 3x + 5x^2 + 6y^2 + 5y$

$= x^2 + 5x^2 - y^2 + 6y^2 + 3x + 2y + 5y$

$= 6x^2 + 5y^2 + 3x + 7y$

4. $2[3 + 4(x-5)] - [2 - (x-3)]$
$= 2[3 + 4x - 20] - [2 - x + 3]$
$= 2[4x - 17] - [-x + 5]$
$= 8x - 34 + x - 5$
$= 8x + x - 34 - 5$
$= 9x - 39$

Exercise Set 2.2

1. $x = 4$
$3x - 1 = 11$
$3(4) - 1 = 11$
$12 - 1 = 11$
$11 = 11$ True
Since we obtain a true statement, 4 is a solution to the equation.

3. $x = -3$
$2x - 5 = 5(x + 2)$
$2(-3) - 5 = 5(-3 + 2)$
$-6 - 5 = 5(-1)$
$-11 = -5$ Not true
Since the value -3 does not check, -3 is not a solution to the equation.

5. $x = 0$
$3x - 5 = 2(x + 3) - 11$
$3(0) - 5 = 2(0 + 3) - 11$
$0 - 5 = 2(3) - 11$
$-5 = 6 - 11$

$-5 = -5$ True

Since the value 0 checks, 0 is a solution to the equation.

7. $x = 2.3$

$$\begin{aligned} 5(x + 2) - 3(x - 1) &= 4 \\ 5(2.3 + 2) - 3(2.3 - 1) &= 4 \\ 5(4.3) - 3(1.3) &= 4 \\ 21.5 - 3.9 &= 4 \\ 17.6 &= 4 \end{aligned}$$

Not true

Since the value 2.3 does not check, 2.3 is not a solution to the equation.

9. $x = \frac{1}{2}$

$$\begin{aligned} 4x - 4 &= 2x - 3 \\ 4(\tfrac{1}{2}) - 4 &= 2(\tfrac{1}{2}) - 3 \\ 2 - 4 &= 1 - 3 \\ -2 &= -2 \end{aligned}$$

True

Since the value $\frac{1}{2}$ checks, $\frac{1}{2}$ is a solution to the equation.

11. $x = \frac{11}{2}$

$$\begin{aligned} 3(x + 2) &= 5(x - 1) \\ 3(\tfrac{11}{2} + 2) &= 5(\tfrac{11}{2} - 1) \\ 3(\tfrac{11 + 4}{2}) &= 5(\tfrac{11 - 2}{2}) \\ 3(\tfrac{15}{2}) &= 5(\tfrac{9}{2}) \\ \tfrac{45}{2} &= \tfrac{45}{2} \end{aligned}$$

True

Since the value $\frac{11}{2}$ checks, $\frac{11}{2}$ is a solution to the the equation.

13.
$$\begin{aligned} x + 3 &= 8 \\ x + 3 + (-3) &= 8 + (-3) \\ x + 0 &= 5 \\ x &= 5 \end{aligned}$$

15.
$$\begin{aligned} x + 7 &= -3 \\ x + 7 + (-7) &= -3 + (-7) \\ x + 0 &= -10 \\ x &= -10 \end{aligned}$$

17.
$$\begin{aligned} x + 3 &= -4 \\ x + 3 + (-3) &= -4 + (-3) \\ x + 0 &= -7 \\ x &= -7 \end{aligned}$$

19.
$$\begin{aligned} x + 43 &= -18 \\ x + 43 + (-43) &= -18 + (-43) \\ x + 0 &= -61 \\ x &= -61 \end{aligned}$$

21.
$$\begin{aligned} -6 + x &= 12 \\ -6 + 6 + x &= 12 + 6 \\ 0 + x &= 18 \\ x &= 18 \end{aligned}$$

23.
$$\begin{aligned} 27 &= x - 16 \\ 27 + 16 &= x - 16 + 16 \\ 43 &= x + 0 \\ 43 &= x \end{aligned}$$

25.
$$\begin{aligned} -13 &= x - 1 \\ -13 + 1 &= x - 1 + 1 \\ -12 &= x + 0 \\ -12 &= x \end{aligned}$$

27.
$$\begin{aligned} 29 &= -43 + x \\ 29 + 43 &= -43 + 43 + x \\ 72 &= 0 + x \\ 72 &= x \end{aligned}$$

29.
$$\begin{aligned} 7 + x &= -19 \\ 7 + (-7) + x &= -19 + (-7) \\ 0 + x &= -26 \\ x &= -26 \end{aligned}$$

31.
$$\begin{aligned} x + 29 &= -29 \\ x + 29 + (-29) &= -29 + (-29) \\ x + 0 &= -58 \end{aligned}$$

$x = -58$

33. $6 = x - 4$
$6 + 4 = x - 4 + 4$
$10 = x + 0$
$10 = x$

35. $x + 7 = -5$
$x + 7 + (-7) = -5 + (-7)$
$x + 0 = -12$
$x = -12$

37. $9 + x = 12$
$9 + (-9) + x = 12 + (-9)$
$0 + x = 3$
$x = 3$

39. $-5 = 4 + x$
$-5 + (-4) = 4 + (-4) + x$
$-9 = 0 + x$
$-9 = x$

41. $30 = x - 19$
$30 + 19 = x + (-19) + 19$
$49 = x + 0$
$49 = x$

43. $x - 12 = -9$
$x + (-12) + 12 = -9 + 12$
$x + 0 = 3$
$x = 3$

45. $4 + x = 9$
$4 + (-4) + x = 9 + (-4)$
$0 + x = 5$
$x = 5$

47. $-8 = -9 + x$
$-8 + 9 = -9 + 9 + x$
$1 = 0 + x$
$1 = x$

49. $2 = x - 9$
$2 + 9 = x + (-9) + 9$
$11 = x + 0$
$11 = x$

51. $-50 = x - 24$
$-50 + 24 = x - 24 + 24$
$-26 = x$

53. $16 + x = -20$
$16 + (-16) + x = -20$
$+ (-16)$
$0 + x = -36$
$x = -36$

55. $40.2 + x = -7.3$
$40.2 + 40.2 + x = -7.3$
$- 40.2$
$x = -47.5$

57. $-37 + x = 9.5$
$-37 + 37 + x = 9.5 + 37$
$x = 46.5$

59. $x - 8.42 = -30$
$x - 8.42 + 8.42 = -30 + 8.42$
$x = -21.58$

61. $9.75 = x + 9.75$
$9.75 - 9.75 = x + 9.75 - 9.75$
$0 = x$

63. $600 = x - 120$
$600 + 120 = x - 120 + 120$
$720 = x$

65. Equations are said to be equivalent if they have the same solution.

67. It is necessary to subtract 3 from both sides of the equation. To solve an equation, you must get the variable alone on one side of the equal sign. Subtracting 5 does not accomplish this.

$5 - 5 = x + 3 - 5$
$0 = x - 2$

Cumulative Review Exercises

69. $6x - 2(2x + 1)$
$= 6(-3) - 2[2(-3) + 1]$
$= -18 - 2(-6 + 1)$
$= -18 - 2(-5)$
$= -18 + 10$
$= -8$

71. $-(x - 3) + 7(2x - 5) - 3x$
$= -x + 3 + 14x - 35 - 3x$
$= -x + 14x - 3x + 3 - 35$
$= 10x - 32$

Just for fun

1. $2(x + 3) = 2x + 6$

a) $2(-1 + 3) = 2(-1) + 6$
$2(2) = -2 + 6$
$4 = 4$ True

Since the value -1 checks, -1 is a solution to the equation.

b) $2(5 + 3) = 2(5) + 6$
$2(8) = 10 + 6$
$16 = 16$ True

Since the value 5 checks, 5 is a solution to the equation.

c) $2(\frac{1}{2} + 3) = 2(\frac{1}{2}) + 6$

$2(\frac{1 + 6}{2}) = 1 + 6$

$2(\frac{7}{2}) = 7$

$7 = 7$ True

Since the value $\frac{1}{2}$ checks, $\frac{1}{2}$ is a solution to the equation.

d) All

2. $2x^2 - 7x + 3 = 0$

a) $2(3)^2 - 7(3) + 3 = 0$
$2(9) - 21 + 3 = 0$
$18 - 21 + 3 = 0$
$0 = 0$
True

Since the value 3 checks, 3 is a solution to the equation.

b) $2(2)^2 - 7(2) + 3 = 0$
$2(4) - 14 + 3 = 0$
$8 - 14 + 3 = 0$
$-3 = 0$
False

Since the value 2 does not check, 2 is not a solution to the equation.

c) $2(\frac{1}{2})^2 - 7(\frac{1}{2}) + 3 = 0$

$2(\frac{1}{4}) - \frac{7}{2} + 3 = 0$

$\frac{1}{2} - \frac{7}{2} + 3 = 0$

$-\frac{6}{2} + 3 = 0$

$-3 + 3 = 0$
$0 = 0$
True

Since the value $\frac{1}{2}$ checks, $\frac{1}{2}$ is a solution to the equation.

Exercise Set 2.3

1. $3x = 9$

$\frac{\overset{1}{\cancel{3}}x}{\underset{1}{\cancel{3}}} = \frac{\overset{3}{\cancel{9}}}{\underset{3}{\cancel{3}}}$

$x = 3$

3. $\frac{x}{2} = 4$

$$\frac{\overset{1}{\cancel{2}}}{1} \cdot \frac{x}{\underset{1}{\cancel{2}}} = \frac{2}{1} \cdot \frac{4}{1}$$

$$x = 8$$

5. $-4x = 8$

$$\frac{\overset{1}{-\cancel{4}x}}{\underset{1}{-\cancel{4}}} = \frac{\overset{-2}{\cancel{8}}}{\underset{1}{-\cancel{4}}}$$

$$x = -2$$

7. $\dfrac{x}{6} = -2$

$$\frac{\overset{1}{\cancel{6}}}{1} \cdot \frac{x}{\underset{1}{\cancel{6}}} = \frac{6}{1} \cdot \frac{-2}{1}$$

$$x = -12$$

9. $\dfrac{x}{5} = 1$

$$\frac{\overset{1}{\cancel{5}}}{1} \cdot \frac{x}{\underset{1}{\cancel{5}}} = \frac{5}{1} \cdot \frac{1}{1}$$

$$x = 5$$

11. $-32x = -96$

$$\frac{\overset{1}{-\cancel{32}x}}{\underset{1}{-\cancel{32}}} = \frac{\overset{3}{-\cancel{96}}}{\underset{1}{-\cancel{32}}}$$

$$x = 3$$

13. $-6 = 4z$

$$\frac{\overset{-3}{-\cancel{6}}}{\underset{2}{\cancel{4}}} = \frac{\overset{1}{\cancel{4}z}}{\underset{1}{\cancel{4}}}$$

$$\frac{-3}{2} = z$$

15. $-x = -4$

$$-1x = -4$$

$$(-1)(-1x) = (-1)(-4)$$

$$x = 4$$

17. $-2 = -y$

$$-2 = -1y$$

$$(-1)(-2) = (-1)(-1y)$$

$$2 = y$$

19. $\dfrac{-x}{7} = -7$

$$\frac{x}{-7} = -7$$

$$\frac{\overset{1}{-\cancel{7}}}{1} \cdot \frac{x}{\underset{1}{-\cancel{7}}} = \frac{-7}{1} \cdot \frac{-7}{1}$$

$$x = 49$$

21. $9 = -18x$

$$\frac{1}{\underset{-2}{-\cancel{18}}} \cdot \frac{\overset{1}{\cancel{9}}}{1} = \frac{1}{\underset{1}{-\cancel{18}}} \cdot \frac{\overset{1}{-\cancel{18}x}}{1}$$

$$\frac{-1}{2} = x$$

23. $\dfrac{-x}{3} = -2$

$$\frac{x}{-3} = -2$$

$$\frac{\overset{1}{-\cancel{3}}}{1} \cdot \frac{x}{\underset{1}{-\cancel{3}}} = \frac{-3}{1} \cdot \frac{-2}{1}$$

$$x = 6$$

25. $19x = 35$

$$\frac{\overset{1}{\cancel{19}x}}{\underset{1}{\cancel{19}}} = \frac{35}{19}$$

$$x = \frac{35}{19}$$

27. $-4.2x = -8.4$

$$\frac{-4.2x}{-4.2} = \frac{-8.4}{-4.2}$$

$$x = 2$$

29. $7x = -7$

$$\frac{\overset{1}{\cancel{7}}x}{\underset{1}{\cancel{7}}} = \frac{\overset{-1}{\cancel{-7}}}{\underset{1}{\cancel{7}}}$$

$$x = -1$$

31. $5x = \frac{-3}{8}$

$$\frac{1}{\underset{1}{\cancel{5}}} \cdot \frac{\overset{1}{\cancel{5}}x}{1} = \frac{1}{5} \cdot \frac{-3}{8}$$

$$x = \frac{-3}{40}$$

33. $15 = \frac{-x}{5}$

$$15 = \frac{x}{-5}$$

$$\frac{-5}{1} \cdot \frac{15}{1} = \frac{\overset{1}{\cancel{-5}}}{1} \cdot \frac{x}{\underset{1}{\cancel{-5}}}$$

$$-75 = x$$

35. $\frac{-x}{5} = -25$

$$\frac{x}{-5} = -25$$

$$\frac{\overset{1}{\cancel{-5}}}{1} \cdot \frac{x}{\underset{1}{\cancel{-5}}} = \frac{-5}{1} \cdot \frac{-25}{1}$$

$$x = 125$$

37. $\frac{x}{5} = -7$

$$\frac{\overset{1}{\cancel{5}}}{1} \cdot \frac{x}{\underset{1}{\cancel{5}}} = \frac{5}{1} \cdot \frac{-7}{1}$$

$$x = -35$$

39. $6 = \frac{x}{4}$

$$\frac{4}{1} \cdot \frac{6}{1} = \frac{\overset{1}{\cancel{4}}}{1} \cdot \frac{x}{\underset{1}{\cancel{4}}}$$

$$24 = x$$

41. $6c = -30$

$$\frac{\overset{1}{\cancel{6}}c}{\underset{1}{\cancel{6}}} = \frac{\overset{-5}{\cancel{-30}}}{\underset{1}{\cancel{6}}}$$

$$c = -5$$

43. $\frac{y}{-2} = -6$

$$\frac{\overset{1}{\cancel{-2}}}{1} \cdot \frac{y}{\underset{1}{\cancel{-2}}} = \frac{-2}{1} \cdot \frac{-6}{1}$$

$$y = 12$$

45. $\frac{-3}{8} \cdot x = 6$

$$\frac{\overset{1}{\cancel{8}}}{\underset{1}{\cancel{-3}}} \cdot \frac{\overset{1}{\cancel{-3}}}{\underset{1}{\cancel{8}}} \cdot x = \frac{8}{\underset{1}{\cancel{-3}}} \cdot \frac{\overset{-2}{\cancel{6}}}{1}$$

$$x = -16$$

47. $\frac{1}{3} \cdot x = -12$

$$\frac{\overset{1}{\cancel{3}}}{1} \cdot \frac{1}{\underset{1}{\cancel{3}}} \cdot x = \frac{3}{1} \cdot \frac{-12}{1}$$

$$x = -36$$

49. $-4 = \frac{-2}{3} \cdot z$

$$\frac{3}{-\cancel{2}_{1}} \cdot \frac{-\cancel{4}^{2}}{1} = \frac{\cancel{3}^{1}}{-\cancel{2}_{1}} \cdot \frac{-\cancel{2}^{1}}{\cancel{3}_{1}} \cdot z$$

$$6 = z$$

51. $-1.4x = 28.28$

$$\frac{-1.4x}{-1.4} = \frac{28.28}{-1.4}$$

$$x = -20.2$$

53. $2x = \frac{-5}{2}$

$$\frac{1}{\cancel{2}_{1}} \cdot \frac{\cancel{2}^{1}x}{1} = \frac{1}{2} \cdot \frac{-5}{2}$$

$$x = \frac{-5}{4}$$

55. $\frac{2}{3} \cdot x = 6$

$$\frac{\cancel{3}^{1}}{\cancel{2}_{1}} \cdot \frac{\cancel{2}^{1}}{\cancel{3}_{1}} \cdot x = \frac{3}{\cancel{2}_{1}} \cdot \frac{\cancel{6}^{3}}{1}$$

$$x = 9$$

57. Divide both sides by 3. To solve an equation, it is necessary to get the variable alone on one side of the equal sign. Dividing both sides by 5 does not do this.

$3x = 5$

$$\frac{3x}{5} = \frac{\cancel{5}^{1}}{\cancel{5}_{1}}$$

$$\frac{3}{5} \cdot x = 1$$

59. Multiply both sides by $\frac{3}{2}$.

$$\frac{2}{3} \cdot x = 4$$

$$\frac{\cancel{3}^{1}}{\cancel{2}_{1}} \cdot \frac{\cancel{2}^{1}}{\cancel{3}_{1}} \cdot x = \frac{3}{\cancel{2}_{1}} \cdot \frac{\cancel{4}^{2}}{1}$$

$$x = 6$$

61. Multiply both sides by $\frac{7}{3}$.

$$\frac{3}{7} \cdot x = \frac{4}{5}$$

$$\frac{\cancel{7}^{1}}{\cancel{3}_{1}} \cdot \frac{\cancel{3}^{1}}{\cancel{7}_{1}} \cdot x = \frac{7}{3} \cdot \frac{4}{5}$$

$$x = \frac{28}{15}$$

Cumulative Review

63. $6 - (-3) - 5 - 4$
$= 6 + 3 - 5 - 4$
$= 0$

65. $-48 = x + 9$
$-48 - 9 = x + 9 - 9$
$-57 = x$

Exercise Set 2.4

1. $2x + 3 = 7$
$2x + 3 - 3 = 7 - 3$
$2x = 4$
$\frac{2x}{2} = \frac{4}{2}$
$x = 2$

3. $-2x - 5 = 7$

$$-2x - 5 + 5 = 7 + 5$$
$$\frac{-2x}{-2} = \frac{12}{-2}$$
$$x = -6$$

5. $$5x - 6 = 19$$
$$5x - 6 + 6 = 19 + 6$$
$$5x = 25$$
$$\frac{5x}{5} = \frac{25}{5}$$
$$x = 5$$

7. $$5x - 2 = 10$$
$$5x - 2 + 2 = 10 + 2$$
$$5x = 12$$
$$\frac{5x}{5} = \frac{12}{5}$$
$$x = \frac{12}{5}$$

9. $$-x - 4 = 8$$
$$-x - 4 + 4 = 8 + 4$$
$$-x = 12$$
$$(-1)(-x) = (-1)(12)$$
$$x = -12$$

11. $$12 - x = 9$$
$$12 - 12 - x = 9 - 12$$
$$-x = -3$$
$$(-1)(-x) = (-1)(-3)$$
$$x = 3$$

13. $$9 + 2x = 24$$
$$9 - 9 + 2x = 24 - 9$$
$$2x = 15$$
$$\frac{2x}{2} = \frac{15}{2}$$
$$x = \frac{15}{2}$$

15. $$32x + 9 = -12$$
$$32x + 9 - 9 = -12 - 9$$
$$32x = -21$$
$$\frac{32x}{32} = \frac{-21}{32}$$
$$x = \frac{-21}{32}$$

17. $$-44 = 9x + 12$$
$$-44 - 12 = 9x + 12 - 12$$
$$-56 = 9x$$
$$\frac{-56}{9} = \frac{9x}{9}$$
$$\frac{-56}{9} = x$$

19. $$6x - 9 = 21$$
$$6x - 9 + 9 = 21 + 9$$
$$6x = 30$$
$$\frac{6x}{6} = \frac{30}{6}$$
$$x = 5$$

21. $$12 = -6x + 5$$
$$12 - 5 = -6x + 5 - 5$$
$$7 = -6x$$
$$\frac{7}{-6} = \frac{-6x}{-6}$$
$$\frac{-7}{6} = x$$

23. $$-2x - 7 = -13$$
$$-2x - 7 + 7 = -13 + 7$$
$$-2x = -6$$
$$\frac{-2x}{-2} = \frac{-6}{-2}$$
$$x = 3$$

25. $$x + 0.05x = 21$$
$$1.05x = 21$$
$$\frac{1.05x}{1.05} = \frac{21}{1.05}$$
$$x = 20$$

27. $$2.3x - 9.34 = 6.3$$
$$2.3x - 9.34 + 9.34 = 6.3 + 9.34$$
$$2.3x = 15.64$$

$$\frac{2.3x}{2.3} = \frac{15.64}{2.3}$$

$$x = 6.8$$

29. $28.8 = x - 0.10x$
$28.8 = 0.9x$

$$\frac{28.8}{0.9} = \frac{0.9x}{0.9}$$

$$32 = x$$

31. $2(x + 1) = 6$
$2x + 2 = 6$
$2x + 2 - 2 = 6 - 2$
$2x = 4$

$$\frac{2x}{2} = \frac{4}{2}$$

$$x = 2$$

33. $4(3 - x) = 12$
$12 - 4x = 12$
$12 - 12 - 4x = 12 - 12$
$-4x = 0$

$$\frac{-4x}{-4} = \frac{0}{-4}$$

$$x = 0$$

35. $-4 = -(x + 5)$
$-4 = -1(x + 5)$
$-4 = -x - 5$
$-4 + 5 = -x - 5 + 5$
$1 = -x$
$(-1)(1) = (-1)(-x)$
$-1 = x$

37. $12 = 4(x + 3)$
$12 = 4x + 12$
$12 - 12 = 4x + 12 - 12$
$0 = 4x$

$$\frac{0}{4} = \frac{4x}{4}$$

$$0 = x$$

39. $5 = 2(3x + 6)$
$5 = 6x + 12$
$5 - 12 = 6x + 12 - 1$
$-7 = 6x$

$$\frac{-7}{6} = \frac{7x}{6}$$

$$\frac{-7}{6} = x$$

41. $2x + 3(x + 2) = 11$
$2x + 3x + 6 = 11$
$5x + 6 = 11$
$5x + 6 - 6 = 11 - 6$
$5x = 5$

$$\frac{5x}{5} = \frac{5}{5}$$

$$x = 1$$

43. $x - 3(2x + 3) = 11$
$x - 6x - 9 = 11$
$-5x - 9 = 11$
$-5x - 9 + 9 = 11 + 9$
$-5x = 20$

$$\frac{-5x}{-5} = \frac{20}{-5}$$

$$x = -4$$

45. $5x + 3x - 4x - 7 = 9$
$4x - 7 = 9$
$4x - 7 + 7 = 9 + 7$
$4x = 16$

$$\frac{4x}{4} = \frac{16}{4}$$

$$x = 4$$

47. $0.7(x + 3) = 4.2$
$0.7x + 2.1 = 4.2$
$0.7x + 2.1 - 2.1 = 4.2 - 2.1$
$0.7x = 2.1$

$$\frac{0.7x}{0.7} = \frac{2.1}{0.7}$$

$$x = 3$$

49. $1.4(5x - 4) = -1.4$
$7x - 5.6 = -1.4$
$7x - 5.6 + 5.6 = -1.4 + 5.6$
$7x = 4.2$

$$\frac{7x}{7} = \frac{4.2}{7}$$

$$x = 0.6$$

51. $3 - 2(x + 3) + 2 = 1$
$$3 - 2x - 6 + 2 = 1$$
$$-1 - 2x = 1$$
$$-1 + 1 - 2x = 1 + 1$$
$$-2x = 2$$
$$\frac{-2x}{-2} = \frac{2}{-2}$$
$$x = -1$$

53. $1 - (x + 3) + 2x = 4$
$$1 + (-1)(x + 3) + 2x = 4$$
$$1 - x - 3 + 2x = 4$$
$$x - 2 = 4$$
$$x - 2 + 2 = 4 + 2$$
$$x = 6$$

55. $4 - 6x + 9 - 3 = -8$
$$10 - 6x = -8$$
$$10 - 10 - 6x = -8 - 10$$
$$-6x = -18$$
$$\frac{-6x}{-6} = \frac{-18}{-6}$$
$$x = 3$$

57. Use the addition property first. Using this property first will result in an equation of the form ax = b. Once the equation is in this form, the multiplication property may be used.

59. a)
1. Use distributive law.
2. Combine like terms.
3. Add 6 to both sides of equation.
4. Divide both sides of equation by 2.

b) $4x - 2(x + 3) = 4$
$$4x - 2x - 6 = 4$$
$$2x - 6 = 4$$
$$2x - 6 + 6 = 4 + 6$$
$$2x = 10$$
$$\frac{2x}{2} = \frac{10}{2}$$
$$x = 5$$

<u>Cumulative Review Exercises</u>

61. $[5(2 - 6) + 3(8 \div 4)^2]^2$
$$= [5(-4) + 3(2)^2]^2$$
$$= [-20 + 3(4)]^2$$
$$= [-20 + 12]^2$$
$$= [-8]^2$$
$$= 64$$

63. Divide both sides of the equation by -4.

<u>Just for fun</u>

1. $3(x - 2) - (x + 5) - 2(3 - 2x) = 18$
$$3x - 6 - x - 5 - 6 + 4x = 18$$
$$3x - x + 4x - 6 - 5 - 6 = 18$$
$$6x - 17 = 18$$
$$6x - 17 + 17 = 18 + 17$$
$$6x = 35$$
$$\frac{6x}{6} = \frac{35}{6}$$
$$x = \frac{35}{6}$$

2. $-6 = -(x - 5) - 3(5 + 2x) - 4(2x - 4)$
$$-6 = -x + 5 - 15 - 6x - 8x + 16$$
$$-6 = -x - 6x - 8x + 5 - 15 + 16$$
$$-6 = -15x + 6$$
$$-6 - 6 = -15x + 6 - 6$$
$$-12 = -15x$$
$$\frac{-12}{-15} = \frac{-15x}{-12}$$

$$\frac{4}{5} = x$$

3. $4[3 - 2(x + 4)] - (x + 3) = 13$
$4[3 - 2x - 8] - x - 3 = 13$
$4[-2x + 3 - 8] - x - 3 = 13$
$4[-2x - 5] - x - 3 = 13$
$-8x - 20 - x - 3 = 13$
$-8x - x - 20 - 3 = 13$
$-9x - 23 = 13$
$-9x - 23 + 23 = 13 + 23$
$-9x = 36$
$x = -4$

Exercise Set 2.5

1. $2x + 4 = 3x$
$2x - 2x + 4 = 3x - 2x$
$4 = x$

3. $-4x + 10 = 6x$
$-4x + 4x + 10 = 6x + 4x$
$10 = 10x$
$$\frac{10}{10} = \frac{10x}{10}$$
$1 = x$

5. $5x + 3 = 6$
$5x + 3 - 3 = 6 - 3$
$5x = 3$
$$\frac{5x}{5} = \frac{3}{5}$$
$$x = \frac{3}{5}$$

7. $15 - 3x = 4x - 2x$
$15 - 3x = 2x$
$15 - 3x + 3x = 2x + 3x$
$15 = 5x$
$$\frac{15}{5} = \frac{5x}{5}$$
$3 = x$

9. $x - 3 = 2x + 18$
$x - x - 3 = 2x - x + 18$
$-3 = x + 18$
$-3 - 18 = x + 18 - 18$
$-21 = x$

11. $3 - 2x = 9 - 8x$
$3 - 2x + 8x = 9 - 8x + 8x$
$3 + 6x = 9$
$3 - 3 + 6x = 9 - 3$
$6x = 6$
$$\frac{6x}{6} = \frac{6}{6}$$
$x = 1$

13. $4 - 0.6x = 2.4x - 8.48$
$4.06x - 2.4x = 2.4x - 2.4x - 8.48$
$4 - 3.0x = -8.48$
$4 - 4 - 3.0x = -8.48 - 4$
$-3x = -12.48$
$$\frac{-3x}{-3} = \frac{-12.48}{-3}$$
$x = 4.16$

15. $5x = 2(x + 6)$
$5x = 2x + 12$
$5x - 2x = 2x - 2x + 12$
$3x = 12$
$$\frac{3x}{3} = \frac{12}{3}$$
$x = 4$

17. $x - 25 = 12x + 9 + 3x$
$x - 25 = 15x + 9$
$x - x - 25 = 15x - x + 9$
$-25 = 14x + 9$
$-25 - 9 = 14x + 9 - 9$
$34 = 14x$
$$\frac{-34}{14} = \frac{14x}{14}$$

$$\frac{-17}{7} = x$$

19. $2(x + 2) = 4x + 1 - 2x$

$2x + 4 = 4x + 1 - 2x$

$2x + 4 = 2x + 1$

$2x - 2x + 4 = 2x - 2x + 1$

$4 = 1$ False

No solution

21. $-(x + 2) = -6x + 32$

$-x - 2 = -6x + 32$

$-x + 6x - 2 = -6x + 6x + 32$

$5x - 2 = 32$

$5x - 2 + 2 = 32 + 2$

$5x = 34$

$$\frac{5x}{5} = \frac{34}{5}$$

$$x = \frac{34}{5}$$

23. $4 - (2x + 5) = 6x + 31$

$4 - 2x - 5 = 6x + 31$

$-2x + 2x - 1 = 6x + 2x + 31$

$-1 = 8x + 31$

$-1 - 31 = 8x + 31 - 31$

$-32 = 8x$

$$\frac{-32}{8} = \frac{8x}{8}$$

$-4 = x$

25. $0.1(x + 10) = 0.3x - 4$

$0.1x + 1 = 0.3x - 4$

$0.1x - 0.3x + 1$

$= 0.3x - 0.3x - 4$

$-0.2x + 1 = -4$

$-0.2x + 1 - 1 = -4 - 1$

$-0.2x = -5$

$$\frac{-0.2x}{-0.2} = \frac{-5}{-0.2}$$

$x = 25$

27. $2(x + 4) = 4x + 3 - 2x + 5$

$2x + 8 = 4x + 3 - 2x + 5$

$2x + 8 = 2x + 3 + 5$

$2x + 8 = 2x + 8$

$2x - 2x + 8 = 2x - 2x + 8$

$8 = 8$ (true)

All real numbers

29. $9(-y + 3)$

$= -6y + 15 - 3y + 12$

$-9y + 27$

$= -6y + 15 - 3y + 12$

$-9y + 27 = -9y + 15 + 12$

$-9y + 27 = -9y + 27$

$-9y + 9y + 27$

$= -9y + 9y + 27$

$27 = 27$ (true)

All real numbers

31. $-(3 - x) = -(2x + 3)$

$-3 + x = -2x - 3$

$-3 + x + 2x = -2x + 2x - 3$

$-3 + 3x = -3$

$-3 + 3 + 3x = -3 + 3$

$3x = 0$

$$\frac{3x}{3} = \frac{0}{3}$$

$x = 0$

33. $-(x + 4) + 5 = 4x + 1 - 5x$

$-x + 1 = 4x + 1 - 5x$

$-x + 1 = -x + 1$

$-x + x + 1 = -x + x + 1$

$1 = 1$ (true)

All real numbers

35. $35(2x + 12) = 7(x - 4) + 3x$

$70x + 420 = 7x - 28 + 3x$

$70x + 420 = 10x - 28$

$70x - 10x + 420 = 10x - 10x - 28$

$60x + 420 = -28$

$60x + 420 - 420 = -28 - 420$

$60x = -448$

$$\frac{60x}{60} = \frac{-448}{60}$$

$$x = \frac{-112}{15}$$

37. $0.4(x + 0.7) = 0.6(x - 4.2)$

$0.4x + 0.28 = 0.6x - 2.52$

$0.4x - 0.6x + 0.28 = 0.6x - 0.6x - 2.52$

$-0.2x + 0.28 = -2.52$

$$-0.2x + 0.28 - 0.28 = -2.52 - 0.28$$
$$-0.2x = -2.8$$
$$\frac{-0.2x}{-0.2} = \frac{-2.8}{-0.2}$$
$$x = 14$$

39. $-(x - 5) + 2 = 3(4 - x) + 5x$
$$-x + 5 + 2 = 12 - 3x + 5x$$
$$-x + 7 = 12 + 2x$$
$$-x + x + 7 = 12 + 2x + x$$
$$7 = 12 + 3x$$
$$7 - 12 = 12 - 12 + 3x$$
$$-5 = 3x$$
$$\frac{-5}{3} = \frac{3x}{3}$$
$$\frac{-5}{3} = x$$

41. $2(x - 6) + 3(x + 1) = 4x + 3$
$$2x - 12 + 3x + 3 = 4x + 3$$
$$5x - 9 = 4x + 3$$
$$5x - 4x - 9 = 4x - 4x + 3$$
$$x - 9 = 3$$
$$x - 9 + 9 = 3 + 9$$
$$x = 12$$

43. $5 + 2x = 6(x + 1) - 5(x - 3)$
$$5 + 2x = 6x + 6 - 5x + 15$$
$$5 + 2x = x + 6 + 15$$
$$5 + 2x = x + 21$$
$$5 + 2x - x = x - x + 21$$
$$5 + x = 21$$
$$5 - 5 + x = 21 - 5$$
$$x = 16$$

45. $5 - (x - 5) = 2(x + 3) - 6(x + 1)$
$$5 - x + 5 = 2x + 6 - 6x - 6$$
$$-x + 10 = -4x$$
$$-x + 4x + 10 = -4x + 4x$$
$$3x + 10 = 0$$
$$3x + 10 - 10 = 0 - 10$$
$$3x = -10$$
$$\frac{3x}{3} = \frac{-10}{3}$$
$$x = \frac{-10}{3}$$

47. The same expression will appear on both sides of the equation.
Ex: $7 = 7$

49. a)
1. Use distributive law.
2. Subtract 6x from both sides of equation.
3. Subtract 12 from both sides of equation.
4. Divide both sides of equation by -2.

b) $4(x + 3) = 6(x - 5)$
$$4x + 12 = 6x - 30$$
$$4x - 6x + 12 = 6x - 6x - 30$$
$$-2x + 12 = -30$$
$$-2x + 12 - 12 = -30 - 12$$
$$-2x = -42$$
$$\frac{-2x}{-2} = \frac{-42}{-2}$$
$$x = 21$$

Cumulative Review

51. Numbers or letters multiplied together are factors, numbers or letters added or subtracted are terms.

53. $2(x - 3) + 4x - (4 - x) = 0$
$$2x - 6 + 4x - 4 + x = 0$$
$$2x + 4x + x - 6 - 4 = 0$$
$$7x - 10 = 0$$
$$7x - 10 + 10 = 10$$
$$7x = 10$$

$$\frac{7x}{7} = \frac{10}{7}$$

$$x = \frac{10}{7}$$

Just for fun

1. $-2(x + 3) + 5x = -3(5 - 2x) + 3(x + 2) + 6x$

$$-2x - 6 + 5x = -15 + 6x + 3x + 6 + 6x$$
$$3x - 6 = -15 + 15x + 6$$
$$3x - 6 = 15x - 9$$
$$3x - 3x - 6 = 15x - 3x - 9$$
$$-6 = 12x - 9$$
$$-6 + 9 = 12x - 9 + 9$$
$$3 = 12x$$
$$\frac{3}{12} = \frac{12x}{12}$$
$$\frac{1}{4} = x$$

2. $4(2x - 3) - (x + 7) - 4x + 6 = 5(x - 2) - 3x + 7(2x + 2)$

$$8x - 12 - x - 7 - 4x + 6 = 5x - 10 - 3x + 14x + 14$$
$$3x - 12 - 7 + 6 = 16x - 10 + 14$$
$$3x - 13 = 16x + 4$$
$$3x - 3x - 13 = 16x - 3x + 4$$
$$-13 = 13x + 4$$
$$-13 - 4 = 13x + 4 - 4$$
$$-17 = 13x$$
$$\frac{-17}{13} = \frac{13x}{13}$$
$$\frac{-17}{13} = x$$

3. $4 - [5 - 3(x + 2)] = x - 3$

$$4 - [5 - 3x - 6] = x - 3$$
$$4 - [-3x - 6 + 5] = x - 3$$
$$4 - [-3x - 1] = x - 3$$
$$4 + 3x + 1 = x - 3$$
$$3x + 5 = x - 3$$
$$3x - x + 5 = x - x - 3$$
$$2x + 5 = -3$$
$$2x + 5 - 5 = -3 - 5$$
$$2x = -8$$
$$x = -4$$

Exercise Set 2.6

1. Ratio of A's to C's: 5:8

3. Ratio of D's to F's:

$\frac{4}{2} = \frac{2}{1}$ or 2:1

5. Ratio of total grades to D's:

$\frac{5 + 6 + 8 + 4 + 2}{4} = \frac{25}{4}$

or 25:4

7. 5:3

9. 20:60 or 1:3

11. 4 hours = 240 min.
240:40 or 6:1

13. 4 pounds = 64 ounces
26:64 or 13:32

15.

$$\frac{4}{x} = \frac{5}{20}$$
$$4(20) = 5x$$
$$80 = 5x$$
$$\frac{80}{5} = \frac{5x}{5}$$
$$16 = x$$

17.
$$\frac{5}{3} = \frac{75}{x}$$
$$5x = 3(75)$$
$$5x = 225$$
$$\frac{5x}{5} = \frac{225}{5}$$
$$x = 45$$

19.
$$\frac{90}{x} = \frac{-9}{10}$$
$$90(10) = -9x$$
$$900 = -9x$$
$$\frac{900}{-9} = \frac{-9x}{-9}$$
$$-100 = x$$

21.
$$\frac{1}{9} = \frac{x}{45}$$
$$1(45) = 9x$$
$$45 = 9x$$
$$\frac{4x}{9} = \frac{9x}{9}$$
$$5 = x$$

23.
$$\frac{3}{z} = \frac{2}{-20}$$
$$3(-20) = 2z$$
$$-60 = 2z$$
$$\frac{-60}{2} = \frac{2z}{2}$$
$$-30 = z$$

25.
$$\frac{15}{20} = \frac{x}{8}$$
$$15(8) = 20x$$
$$120 = 20x$$
$$\frac{120}{20} = \frac{20x}{20}$$
$$6 = x$$

27. Let x = miles

$$\frac{\text{miles}}{\text{gallons}} = \frac{\text{miles}}{\text{gallons}}$$
$$\frac{32}{1} = \frac{x}{12}$$
$$32(12) = 1x$$
$$384 \text{ mi} = x$$

29. Let x = time.

$$\frac{\text{units}}{\text{time}} = \frac{\text{units}}{\text{time}}$$
$$\frac{12}{2.5} = \frac{60}{x}$$
$$12x = 2.5(60)$$
$$12x = 150$$
$$\frac{12x}{12} = \frac{150}{12}$$
$$x = 12.5 \text{ minutes}$$

31. Let x = amount of tax

$$\frac{\text{tax}}{\text{per 1000 assessed value}} = \frac{\text{tax}}{\text{per 1000 assessed value}}$$
$$\frac{8.235}{1000} = \frac{x}{122{,}000}$$
$$1000x = (8.235)(122{,}000)$$
$$1000x = 1{,}004{,}670$$
$$x = \frac{1{,}004{,}670}{1000}$$
$$= \$1{,}004.67$$

33. Let x = size of the model box car.

$$\frac{\text{size of model}}{\text{size of actual box car}} = \frac{\text{size of model}}{\text{size of actual box car}}$$
$$\frac{1}{87} = \frac{x}{12.2}$$
$$1(12.2) = 87x$$
$$12.2 = 87x$$

$$\frac{12.2}{87} = \frac{87x}{87}$$

$0.14 = x$

$x = 0.14$ meters (rounded)

35. Let x = amount of fertilizer.

$$\frac{\text{fertilizer}}{\text{sq. feet}} = \frac{\text{fertilizer}}{\text{sq. feet}}$$

$$\frac{40}{5{,}000} = \frac{x}{26{,}000}$$

$40(26{,}000) = 5{,}000x$

$1{,}040{,}000 = 5{,}000x$

$$\frac{1{,}040{,}000}{5{,}000} = \frac{5{,}000x}{5{,}000}$$

$208 = x$

$x = 208$ pounds

37. Let x = pounds of beef.

$$\frac{\text{beef}}{\text{servings}} = \frac{\text{beef}}{\text{servings}}$$

$$\frac{4.5}{20} = \frac{x}{12}$$

$4.5(12) = 20x$

$54 = 20x$

$$\frac{54}{20} = \frac{20x}{20}$$

$2.7 \text{ lbs} = x$

39. Let x = number on VCR counter.

$$\frac{\text{number}}{\text{minutes}} = \frac{\text{number}}{\text{minutes}}$$

$$\frac{5.5}{1} = \frac{x}{90}$$

$x = (90)(5.5)$

$x = 495$

41. $$\frac{12 \text{ inches}}{1 \text{ foot}} = \frac{57 \text{ inches}}{x \text{ foot}}$$

$$\frac{12}{1} = \frac{57}{x}$$

$12x = 57$

$$x = \frac{57}{12} = 4.75 \text{ feet}$$

43. $$\frac{5280 \text{ feet}}{1 \text{ mile}} = \frac{17{,}952 \text{ feet}}{x \text{ miles}}$$

$$\frac{5280}{1} = \frac{17{,}952}{x}$$

$5280x = 17{,}952$

$$x = \frac{17{,}952}{5280}$$

$= 3.4$ miles

45. $$\frac{16 \text{ ounces}}{1 \text{ pound}} = \frac{146.4 \text{ ounces}}{x \text{ pounds}}$$

$16x = 146.4$

$$x = \frac{146.4}{16} = 9.15 \text{ pounds}$$

47. $$\frac{1 \text{ liter}}{1.06 \text{ quarts}} = \frac{5 \text{ liters}}{x \text{ quarts}}$$

$$\frac{1}{1.06} = \frac{5}{x}$$

$x = (5)(1.06)$

$x = 5.3$ quarts

49. $$\frac{1 \text{ mile}}{1.6 \text{ kilometer}} = \frac{x \text{ miles}}{25 \text{ kilometers}}$$

$$\frac{1}{1.6} = \frac{x}{25}$$

$1.6x = 25$

$$x = \frac{25}{1.6}$$

$= 15.625$ miles

51. $$\frac{\$500}{480 \text{ grains}} = \frac{\$x}{1 \text{ grain}}$$

$$\frac{500}{480} = \frac{x}{1}$$

$480x = 500$

$$x = \frac{500}{480} = \$1.04$$

53. $\dfrac{16 \text{ points}}{3.2 \text{ standard deviations}}$

$= \dfrac{x \text{ points}}{1 \text{ standard deviation}}$

$\dfrac{16}{3.2} = \dfrac{x}{1}$

$3.2x = 16$

$x = \dfrac{16}{3.2} = 5 \text{ points}$

55. $\dfrac{1 \text{ Italian lira}}{0.00081 \text{ U.S. dollars}}$

$= \dfrac{x \text{ Italian lira}}{1200 \text{ U.S. dollars}}$

$\dfrac{1}{0.00081} = \dfrac{x}{1200}$

$0.00081x = 1200$

$x = \dfrac{1200}{0.00081}$

$x = 1{,}481{,}481.5$ lira

57. $\dfrac{3}{12} = \dfrac{8}{x}$

$3x = (12)(8)$

$3x = 96$

$x = 32$ inches

59. $\dfrac{5}{7} = \dfrac{8}{x}$

$5x = (7)(8)$

$5x = 56$

$x = 11.2$ ft.

61. $\dfrac{14}{x} = \dfrac{20}{8}$

$20x = (14)(8)$

$20x = 112$

$x = 5.6$ inches

63. $\dfrac{127 \text{ mg/dl}}{60 \text{ mg/dl}} = \dfrac{127}{60} = \dfrac{x}{1}$

$60x = 127$

$x = 2.12$

Her ratio is 2.12:1, which is less than 4:1.

65. a) Answers will vary.
b) Answers will vary.

Cumulative Review Exercises

67. Commutative property of addition.

69. Distributive property.

Just for fun

1. Let x = amount of flour:

$\dfrac{\text{apples}}{\text{flour}} = \dfrac{\text{apples}}{\text{flour}}$

$\dfrac{12}{0.5} = \dfrac{8}{x}$

$12x = 0.5(8)$

$12x = 4$

$\dfrac{12x}{12} = \dfrac{4}{12}$

$x = \dfrac{1}{3}$ cup of flour

Let x = amount of salt.

$\dfrac{\text{apples}}{\text{salt}} = \dfrac{\text{apples}}{\text{salt}}$

$\dfrac{12}{0.25} = \dfrac{8}{x}$

$12x = 8(0.25)$

$12x = 2$

$\dfrac{12x}{12} = \dfrac{2}{12}$

$x = \dfrac{1}{6}$ teaspoon salt

Let x = amount of butter.

$$\frac{\text{apples}}{\text{butter}} = \frac{\text{apples}}{\text{butter}}$$

$$\frac{12}{2} = \frac{8}{x}$$

$$12x = 8(12)$$
$$12x = 16$$

$$\frac{12x}{12} = \frac{16}{12}$$

$$x = \frac{4}{3}$$

$$x = 1\frac{1}{3} \text{ tablespoon butter}$$

Let x = amount of nutmeg.

$$\frac{\text{apples}}{\text{nutmeg}} = \frac{\text{apples}}{\text{nutmeg}}$$

$$\frac{12}{1} = \frac{8}{x}$$

$$12x = 8(1)$$
$$12x = 8$$

$$\frac{12x}{12} = \frac{8}{12}$$

$$x = \frac{2}{3} \text{ teaspoon nutmeg}$$

Let x = amount of cinnamon.

$$\frac{\text{apples}}{\text{cinnamon}} = \frac{\text{apples}}{\text{cinnamon}}$$

$$\frac{12}{1} = \frac{8}{x}$$

$$12x = 8(1)$$
$$12x = 8$$

$$\frac{12x}{12} = \frac{8}{12}$$

$$x = \frac{2}{3} \text{ teaspoon cinnamon}$$

Let x = amount of sugar.

$$\frac{\text{apples}}{\text{sugar}} = \frac{\text{apples}}{\text{sugar}}$$

$$\frac{12}{1.5} = \frac{8}{x}$$

$$12x = 1.5(8)$$
$$12x = 12$$

$$\frac{12x}{12} = \frac{12}{12}$$

$$x = 1 \text{ cup sugar}$$

2. Let x = number of cc of fluid.

$$\frac{\text{units of insulin}}{\text{cubic centimeters}} = \frac{\text{units of insulin}}{\text{cubic centimeters}}$$

$$\frac{40}{1} = \frac{25}{x}$$

$$40x = 25(1)$$
$$40x = 25$$

$$\frac{40x}{40} = \frac{25}{40}$$

$$x = 0.625 \text{ cc of insulin}$$

Exercise Set 2.7

1. $$x + 3 > 7$$
$$x + 3 - 3 > 7 - 3$$
$$x > 4$$

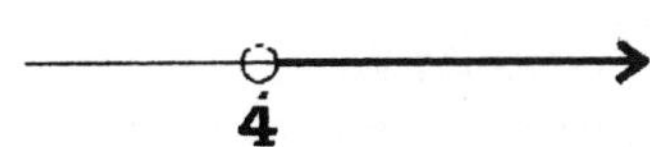

3. $$x + 5 \geq 3$$
$$x + 5 - 5 \geq 3 - 5$$
$$x \geq -2$$

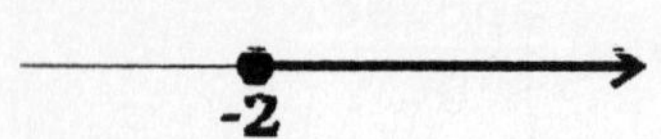

5. $-x + 3 < 8$

$-x + 3 - 3 < 8 - 3$

$-1x < 5$

$(-1)(-1x) > (-1)5$

$x > -5$

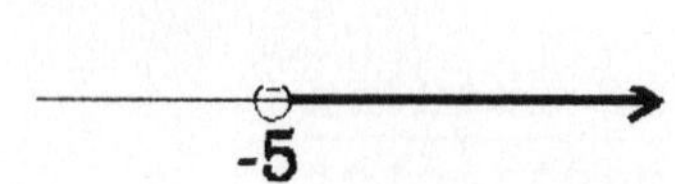

7. $6 > x - 4$

$6 + 4 > x - 4 + 4$

$10 > x$ or

$x < 10$

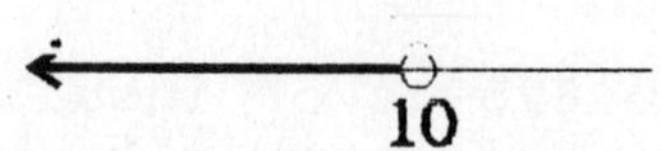

9. $8 \leq 4 - x$

$8 - 4 \leq 4 - 4 - x$

$4 \leq -1x$

$(-1)(4) \geq (-1)(-1x)$

$-4 \geq x$ or

$x \leq -4$

11. $-2x < 3$

$\frac{-2x}{-2} > \frac{3}{-2}$

$x > \frac{-3}{2}$

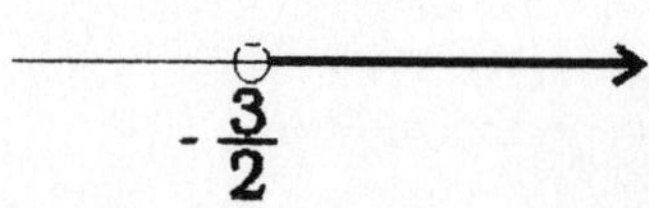

13. $2x + 3 \leq 5$

$2x + 3 - 3 \leq 5 - 3$

$2x \leq 2$

$\frac{2x}{2} \leq \frac{2}{2}$

$x \leq 1$

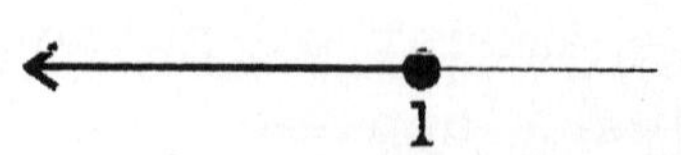

15. $12x + 24 < -12$

$12x + 24 - 24 < -12 - 24$

$12x < -36$

$\frac{-12x}{12} < \frac{-36}{12}$

$x < -3$

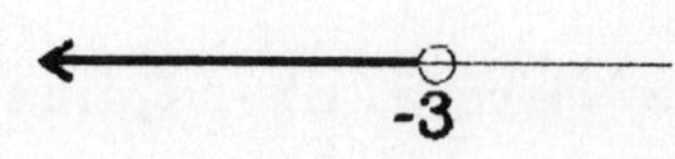

17. $4 - 6x > -5$

$4 - 4 - 6x > -5 - 4$

$-6x > -9$

$\frac{-6x}{-6} < \frac{-9}{-6}$

$x < \frac{3}{2}$

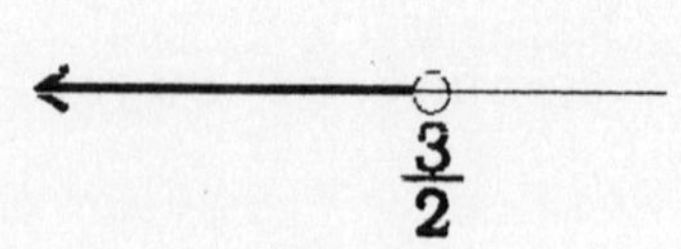

19.
$$\begin{aligned} 15 &> -9x + 50 \\ 15 - 50 &> -9x + 50 - 50 \\ -35 &> -9x \\ \frac{-35}{-9} &< \frac{-9x}{-9} \\ \frac{35}{9} &< x \text{ or} \\ x &> \frac{35}{9} \end{aligned}$$

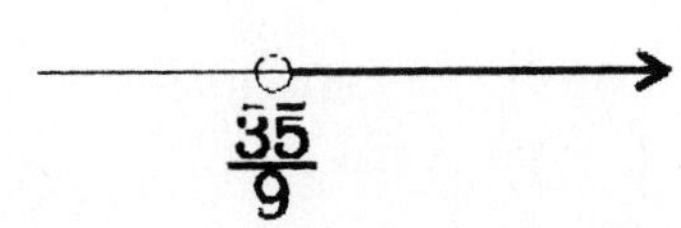

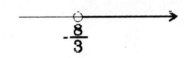

21.
$$\begin{aligned} 4 &< 3x + 12 \\ 4 - 12 &< 3x + 12 - 12 \\ -8 &< 3x \\ \frac{-8}{3} &< \frac{3x}{3} \\ \frac{-8}{3} &< x \text{ or} \\ x &> \frac{-8}{3} \end{aligned}$$

$-\frac{8}{3}$

23.
$$\begin{aligned} 6x + 2 &\le 3x - 9 \\ 6x - 3x + 2 &\le 3x - 3x - 9 \\ 3x + 2 &\le -9 \\ 3x + 2 - 2 &\le -9 - 2 \\ 3x &\le -11 \\ \frac{3x}{3} &\le \frac{-11}{3} \\ x &\le \frac{-11}{3} \end{aligned}$$

$-\frac{11}{3}$

25.
$$\begin{aligned} x - 4 &\le 3x + 8 \\ x - x - 4 &\le 3x - x + 8 \\ -4 &\le 2x + 8 \\ -4 - 8 &\le 2x + 8 - 8 \\ -12 &\le 2x \\ \frac{-12}{2} &\le \frac{2x}{2} \\ -6 &\le x \text{ or} \\ x &\ge -6 \end{aligned}$$

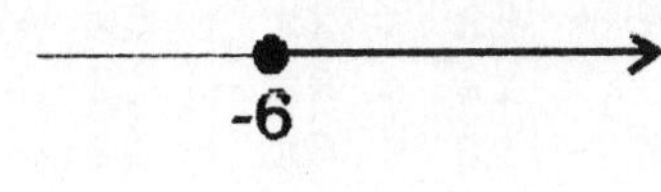

27.
$$\begin{aligned} -x + 4 &< -3x + 6 \\ -x + 3x + 4 &< -3x + 3x + 6 \\ 2x + 4 &< 6 \\ 2x + 4 - 4 &< 6 - 4 \\ 2x &< 2 \\ \frac{2x}{2} &< \frac{2}{2} \\ x &< 1 \end{aligned}$$

29.
$$\begin{aligned} -3(2x - 4) &> 2(6x - 12) \\ -6x + 12 &> 12x - 24 \\ -6x + 6x + 12 &> 12x + 6x \\ &\quad - 24 \\ 12 &> 18x - 24 \end{aligned}$$

$$\begin{aligned}
12 + 24 &> 18x - 24 + 24 \\
36 &> 18x \\
\frac{36}{18} &> \frac{18x}{18} \\
2 &> x \quad \text{or} \\
x &< 2
\end{aligned}$$

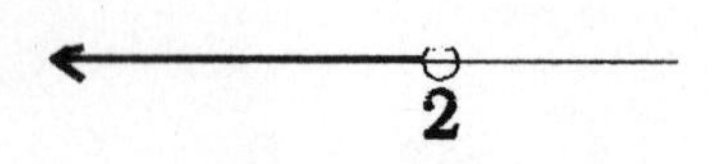

31.
$$\begin{aligned}
x + 3 &< x + 4 \\
x - x + 3 &< x - x + 4 \\
3 &< x
\end{aligned}$$

All real numbers.

33.
$$\begin{aligned}
6(3 - x) &< 2x + 12 \\
18 - 6x &< 2x + 12 \\
18 - 6x + 6x &< 2x + 6x + 12 \\
18 &< 8x + 12 \\
18 - 12 &< 8x + 12 - 12 \\
6 &< 8x \\
\frac{6}{8} &< \frac{8x}{8} \\
\frac{6}{8} &< x \\
\frac{3}{4} &< x \quad \text{or} \\
x &> \frac{3}{4}
\end{aligned}$$

35.
$$\begin{aligned}
-21(2 - x) + 3x &> 4x + 4 \\
-42 + 21x + 3x &> 4x + 4 \\
-42 + 24x &> 4x + 4
\end{aligned}$$

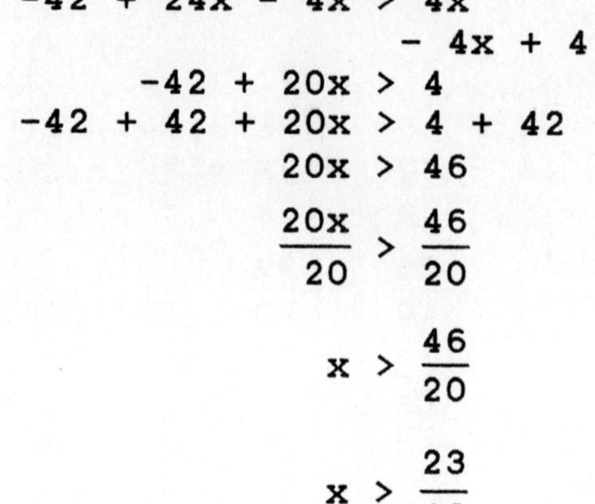

$$\begin{aligned}
-42 + 24x - 4x &> 4x - 4x + 4 \\
-42 + 20x &> 4 \\
-42 + 42 + 20x &> 4 + 42 \\
20x &> 46 \\
\frac{20x}{20} &> \frac{46}{20} \\
x &> \frac{46}{20} \\
x &> \frac{23}{10}
\end{aligned}$$

37.
$$\begin{aligned}
4x - 4 &< 4(x - 5) \\
4x - 4 &< 4x - 20 \\
4x - 4x - 4 &< 4x - 4x - 20 \\
-4 &< -20
\end{aligned}$$

No solution.

39.
$$\begin{aligned}
5(2x + 3) &\geq 6 + (x + 2) - 2x \\
10x + 15 &\geq 6 + x + 2 - 2x \\
10x + 15 &\geq 6 + 2 - x \\
10x + 15 &\geq 8 - x \\
10x + x + 15 &\geq 8 - x + x \\
11x + 15 &\geq 8 \\
11x + 15 - 15 &\geq 8 - 15 \\
11x &\geq -7 \\
\frac{11x}{11} &\geq \frac{-7}{11} \\
x &\geq \frac{-7}{11}
\end{aligned}$$

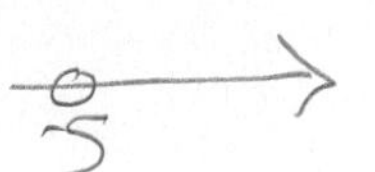

41. All real numbers.

43. No solution.

45. It is necessary to change the sense of the inequality when multiplying or dividing by a negative number.

47. $-x^2 = -(3)^2 = -9$

49.
$$
\begin{aligned}
4 - 3(2x - 4) &= 5 - (x + 3) \\
4 - 6x + 12 &= 5 - x - 3 \\
-6x + 16 &= -x + 2 \\
-6x + x + 16 &= -x + x + 2 \\
-5x + 16 &= 2 \\
-5x + 16 - 16 &= 2 - 16 \\
-5x &= -14 \\
x &= \frac{-14}{-5} = \frac{14}{5} \text{ or } 2\tfrac{4}{5}
\end{aligned}
$$

Just for fun

1.
$$
\begin{aligned}
&3(2 - x) - 4(2x - 3) \\
&\quad \le 6 + 2x - 6(x - 5) + 2x \\
&6 - 3x - 8x + 12 \\
&\quad \le 6 + 2x - 6x + 30 + 2x \\
&6 - 11x + 12 \le 6 - 2x + 30 \\
&-11x + 18 \le -2x + 36 \\
&-11x + 11x + 18 \\
&\quad \le -2x + 11x + 36 \\
&18 \le 9x + 36 \\
&18 - 36 \le 9x + 36 - 36 \\
&-18 \le 9x \\
&\frac{-18}{9} \le \frac{9x}{9} \\
&-2 \le x \text{ or } x \le -2
\end{aligned}
$$

2.
$$
\begin{aligned}
&-(x + 4) + 6x - 5 \\
&\quad > -4(x + 3) + 2(x + 6) - 5x \\
&-1(x + 4) + 6x - 5 \\
&\quad > -1[4(x + 3)] + 2(x + 6) - 5x \\
&-x - 4 + 6x - 5 \\
&\quad > -4x - 12 + 2x + 12 - 5x \\
&5x - 4 - 5 > -7x - 12 + 12 \\
&5x - 9 > -7x \\
&5x + 7x - 9 > -7x + 7x \\
&12x - 9 > 0 \\
&12x - 9 + 9 > 9 \\
&12x > 9 \\
&\frac{12x}{12} > \frac{9}{12} \\
&x > \frac{9}{12} \\
&x > \frac{3}{4}
\end{aligned}
$$

3. $\nleq$

Review Exercises

1.
$$
\begin{aligned}
&2(x + 4) \\
&= 2x + 2(4) \\
&= 2x + 8
\end{aligned}
$$

3.
$$
\begin{aligned}
&2(4x - 3) \\
&= 2(4x) + 2(-3) \\
&= 8x + (-6) \\
&= 8x - 6
\end{aligned}
$$

5.
$$
\begin{aligned}
&-(x + 2) \\
&= -1(x + 2) \\
&= -1(x) + (-1)(2) \\
&= -x + (-2) \\
&= -x - 2
\end{aligned}
$$

7.
$$
\begin{aligned}
&-4(4 - x) \\
&= -4(4) + (-4)(-x) \\
&= -16 + 4x
\end{aligned}
$$

9.
$$
\begin{aligned}
&4(5x - 6) \\
&= 4(5x) + 4(-6) \\
&= 20x + (-24) \\
&= 20x - 24
\end{aligned}
$$

11. $6(6x - 6)$

$= 6(6x) + 6(-6)$
$= 36x + (-36)$
$= 36x - 36$

13. $-3(x + y)$
$= -3(x) + (-3)(y)$
$= -3x + (-3y)$
$= -3x - 3y$

15. $-(3 + 2y)$
$= -1(3) + (-1)(2y)$
$= -3 + (-2y)$
$= -3 - 2y$

17. $3(x + 3y - 2z)$
$= 3(x) + 3(3y) + 3(-2x)$
$= 3x + 9y + (-6z)$
$= 3x + 9y - 6z$

19. $2x + 3x = 5x$

21. $4 - 2y + 3$
$= -2y + 4 + 3$
$= -2y + 7$

23. $6x + 2y + y = 6x + 3y$

25. $2x + 3y + 4x + 5y$
$= 2x + 4x + 3y + 5y$
$= 6x + 8y$

27. $2x - 3x - 1 = -x - 1$

29. $x + 8x - 9x + 3$
$= 9x - 9x + 3$
$= 0 + 3$
$= 3$

31. $3(x + 2) + 2x$
$= 3x + 6 + 2x$
$= 3x + 2x + 6$
$= 5x + 6$

33. $2x + 3(x + 4) - 5$
$= 2x + 3x + 12 - 5$
$= 5x + 7$

35. $6 - (-x + 3) + 4x$
$= 6 - 1(-x + 3) + 4x$
$= 6 + x - 3 + 4x$
$= 6 - 3 + x + 4x$
$= 3 + 5x$
$= 5x + 3$

37. $-6(4 - 3x) - 18 + 4x$
$= -24 + 18x - 18 + 4x$
$= -24 - 18 + 18x + 4x$
$= -42 + 22x$
$= 22x - 42$

39. $3(x + y) - 2(2x - y)$
$= 3x + 3y - 4x + 2y$
$= 3x - 4x + 3y + 2y$
$= -x + 5y$

41. $3 - (x - y) + (x - y)$
$= 3 - x + y + x - y$
$= 3 - x + x + y - y$
$= 3 + 0 + 0$
$= 3$

43. $2x = 4$
$\frac{2x}{2} = \frac{4}{2}$
$x = 2$

45. $x - 4 = 7$
$x - 4 + 4 = 7 + 4$
$x = 11$

47. $2x + 4 = 8$
$2x + 4 - 4 = 8 - 4$
$2x = 4$
$\frac{2x}{x} = \frac{4}{2}$
$x = 2$

49. $8x - 3 = -19$
$8x - 3 + 3 = -19 + 3$
$8x = -16$
$\frac{8x}{8} = \frac{-16}{8}$
$x = -2$

51. $-x = -12$
$-1x = -12$
$-1(-1x) = -1(-12)$
$x = 12$

53. $-3(2x - 8) = -12$
$-6x + 24 = -12$
$-6x + 24 - 24 = -12 - 24$
$-6x = -36$

$$\frac{-6x}{-6} = \frac{-36}{-6}$$
$$x = 6$$

55. $3x + 2x + 6 = -15$
$$5x + 6 = -15$$
$$5x + 6 - 6 = -15 - 6$$
$$5x = -21$$
$$\frac{5x}{5} = \frac{-21}{5}$$
$$x = \frac{-21}{5}$$

57. $27 = 46 + 2x - x$
$$27 = 46 + x$$
$$27 - 46 = 46 - 46 + x$$
$$-19 = x$$

59. $4 + 3(x + 2) = 12$
$$4 + 3x + 6 = 12$$
$$3x + 10 = 12$$
$$3x + 10 - 10 = 12 - 10$$
$$3x = 2$$
$$\frac{3x}{3} = \frac{2}{3}$$
$$x = \frac{2}{3}$$

61. $3x - 6 = -5x + 30$
$$3x + 5x - 6 = -5x + 5x + 30$$
$$8x - 6 = 30$$
$$8x - 6 + 6 = 30 + 6$$
$$8x = 36$$
$$\frac{8x}{8} = \frac{36}{8}$$
$$x = \frac{36}{8}$$
$$x = \frac{9}{2}$$

63. $2x + 6 = 3x + 9$
$$2x - 2x + 6 = 3x - 2x + 9$$
$$6 = x + 9$$
$$6 - 9 = x + 9 - 9$$
$$-3 = x$$

65. $3x - 12x = 24 - 6x$
$$-9x = 24 - 6x$$
$$-9x + 6x = 24 - 6x + 6x$$
$$-3x = 24$$
$$\frac{-3x}{-3} = \frac{24}{-3}$$
$$x = -8$$

67. $4(2x - 3) + 4 = 9x + 2$
$$8x - 12 + 4 = 9x + 2$$
$$8x - 8 = 9x + 2$$
$$8x - 8x - 8 = 9x - 8x + 2$$
$$-8 = x + 2$$
$$-8 - 2 = x + 2 - 2$$
$$-10 = x$$

69. $2(x + 7) = 6x + 9 - 4x$
$$2x + 14 = 2x + 9$$
$$2x - 2x + 14 = 2x - 2x + 9$$
$$14 = 9 \quad \text{False}$$
No solution

71. $4(x - 3) - (x + 5) = 0$
$$4x - 12 - x - 5 = 0$$
$$3x - 12 - 5 = 0$$
$$3x - 17 = 0$$
$$3x - 17 + 17 = 0 + 17$$
$$3x = 17$$
$$\frac{3x}{3} = \frac{17}{3}$$
$$x = \frac{17}{3}$$

73. $-3(2x - 5) + 5x = 4x - 7$
$$-6x + 15 + 5x = 4x - 7$$
$$-x + 15 = 4x - 7$$
$$-x + x + 15 = 4x + x - 7$$
$$15 = 5x - 7$$
$$15 + 7 = 5x - 7 + 7$$
$$22 = 5x$$
$$\frac{22}{5} = \frac{5x}{5}$$
$$\frac{22}{5} = x$$

75. 80 ounces = 5 pounds
5:12 is the ratio.

77. 32 ounces = 3 pounds
2:2 or 1:1 is the ratio.

79. $\frac{15}{100} = \frac{x}{20}$

$20(15) = 100x$
$300 = 100x$
$\frac{300}{100} = \frac{100x}{100}$
$3 = x$

81. $\frac{20}{45} = \frac{15}{x}$

$20x = 45(15)$
$20x = 675$
$\frac{20x}{20} = \frac{675}{20}$
$x = \frac{135}{4}$

83. $\frac{x}{9} = \frac{8}{-3}$

$-3x = 9(8)$
$-3x = 72$
$\frac{-3x}{-3} = \frac{72}{-3}$
$x = -24$

85. $\frac{x}{-15} = \frac{30}{-5}$

$-5x = 30(-15)$
$-5x = -450$
$\frac{-5x}{-5} = \frac{-450}{-5}$
$x = 90$

87. $\frac{2}{x} = \frac{7}{3.5}$

$7x = (2)(3.5)$
$7x = 7$
$x = 1$ ft.

89. $6 - 2x > 4x - 12$
$6 - 2x + 2x > 4x + 2x - 12$
$6 > 6x - 12$
$6 + 12 > 6x - 12 + 12$
$18 > 6x$
$\frac{18}{6} > \frac{6x}{6}$
$3 > x$
$x < 3$

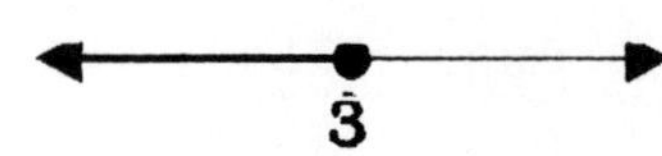

91. $2(x + 4) \le 2x - 5$
$2x + 8 \le 2x - 5$
$2x - 2x + 8 \le 2x - 2x - 5$
$8 \le -5$ False
No solution

93. $x + 6 > 9x + 30$
$x - x + 6 > 9x - x + 30$
$6 - 30 > 8x + 30 - 30$
$-24 > 8x$
$\frac{-24}{8} > \frac{8x}{8}$
$-3 > x$
$x < -3$

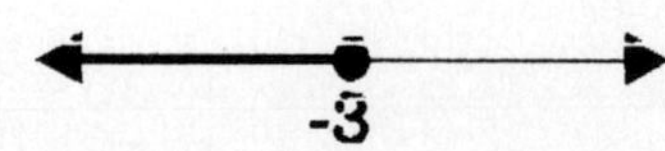

95. $-(x + 2) < -2(-2x + 5)$
$-x - 2 < 4x - 10$
$-x + x - 2 < 4x + x - 10$
$-2 < 5x - 10$
$-2 + 10 < 5x - 10 + 10$
$8 < 5x$
$\frac{8}{5} < \frac{5x}{5}$
$\frac{8}{5} < x$
$x > \frac{8}{5}$

$$x > \frac{8}{5}$$

$\frac{8}{5}$

97.

$$\begin{aligned}
-6x - 3 &\ge 2(x - 4) + 3x \\
-6x - 3 &\ge 2x - 8 + 3x \\
-6x - 3 &\ge 5x - 8 \\
-6x + 6x - 3 &\ge 5x + 6x - 8 \\
-3 &\ge 11x - 8 \\
-3 + 8 &\ge 11x - 8 + 8 \\
5 &\ge 11x \\
\frac{5}{11} &\ge \frac{11x}{11} \\
\frac{5}{11} &\ge x \\
x &\le \frac{5}{11}
\end{aligned}$$

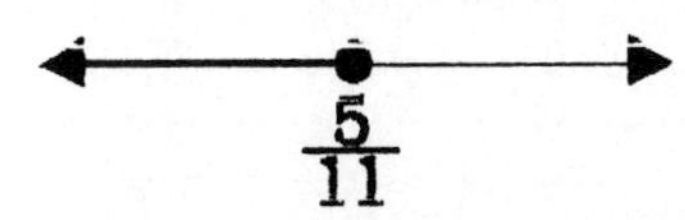

99.

$$\begin{aligned}
2(2x + 4) &> 4(x + 2) - 6 \\
4x + 8 &> 4x + 8 - 6 \\
4x + 8 &> 4x + 2 \\
4x - 4x + 8 &> 4x - 4x + 2 \\
8 &> 2 \quad \text{True}
\end{aligned}$$

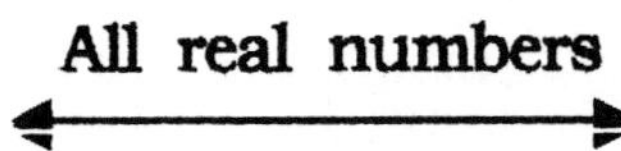

101. Let x = miles.

$$\frac{\text{miles}}{\text{minutes}} = \frac{\text{miles}}{\text{minutes}}$$

$$\begin{aligned}
\frac{45}{60} &= \frac{x}{90} \\
45(90) &= 60x \\
4050 &= 60x \\
\frac{4050}{60} &= \frac{60x}{60} \\
x &= 67.5 \text{ miles}
\end{aligned}$$

103. Let x = feet.

$$\frac{\text{inches}}{\text{feet}} = \frac{\text{inches}}{\text{feet}}$$

$$\begin{aligned}
\frac{1}{0.9} &= \frac{10.5}{x} \\
1x &= 0.9(10.5) \\
x &= 9.45 \text{ feet}
\end{aligned}$$

105.

$$\frac{1 \text{ U.S. dollar}}{2788 \text{ Mexican pesos}} = \frac{x \text{ U.S. dollars}}{1 \text{ Mexican peso}}$$

$$\begin{aligned}
\frac{1}{2788} &= \frac{x}{1} \\
2788x &= 1 \\
x &= \frac{1}{2788} = \$0.00036
\end{aligned}$$

107.

$$\frac{3 \text{ slugs}}{96.6 \text{ pounds}} = \frac{x \text{ slugs}}{1 \text{ pound}}$$

$$\begin{aligned}
\frac{3}{96.6} &= \frac{x}{1} \\
96.6x &= 3 \\
x &= \frac{3}{96.6} = 0.03 \text{ slugs}
\end{aligned}$$

Practice Test

1.

$$\begin{aligned}
&-2(4 - 2x) \\
&= -2(4) - (-2)(2x) \\
&= -8 + 4x \\
&= 4x - 8
\end{aligned}$$

2. $-(x + 3y - 4)$
$= -1[x + 3y + (-4)]$
$= -1x + (-1)(3y) + (-1)(-4)$
$= -x - 3y + 4$

3. $3x - x + 4 = 2x + 4$

4. $4 + 2x - 3x + 6$
$= 4 - 1x + 6$
$= -x + 6 + 4$
$= -x + 10$

5. $y - 2x - 4x - 6$
$= y - 6x - 6$
$= -6x + y - 6$

6. $x - 4y + 6x - y + 3$
$= x + 6x - 4y - y + 3$
$= 7x - 5y + 3$

7. $2x + 3 + 2(3x - 2)$
$= 2x + 3 + 6x - 4$
$= 2x + 6x + 3 - 4$
$= 8x - 1$

8.
$$
\begin{aligned}
2x + 4 &= 12 \\
2x + 4 - 4 &= 12 - 4 \\
2x &= 8 \\
\frac{2x}{2} &= \frac{8}{2} \\
x &= 4
\end{aligned}
$$

9.
$$
\begin{aligned}
-x - 3x + 4 &= 12 \\
-4x + 4 &= 12 \\
-4x + 4 - 4 &= 12 - 4 \\
-4x &= 8 \\
\frac{-4x}{-4} &= \frac{8}{-4} \\
x &= -2
\end{aligned}
$$

10.
$$
\begin{aligned}
4x - 2 &= x + 4 \\
4x - x - 2 &= x - x + 4 \\
3x - 2 &= 4 \\
3x - 2 + 2 &= 4 + 2 \\
3x &= 6 \\
\frac{3x}{3} &= \frac{6}{3} \\
x &= 2
\end{aligned}
$$

11.
$$
\begin{aligned}
3(x - 2) &= -(5 - 4x) \\
3x - 6 &= -5 + 4x \\
3x - 3x - 6 &= -5 + 4x - 3x \\
-6 &= -5 + x \\
-6 + 5 &= -5 + 5 + x \\
-1 &= x
\end{aligned}
$$

12.
$$
\begin{aligned}
2x - 3(-2x + 4) &= -13 + x \\
2x + 6x - 12 &= -13 + x \\
8x - 12 &= -13 + x \\
8x - x - 12 &= -13 + x - x \\
7x - 12 &= -13 \\
7x - 12 + 12 &= -13 + 12 \\
7x &= -1 \\
\frac{7x}{7} &= \frac{-1}{7} \\
x &= \frac{-1}{7}
\end{aligned}
$$

13.
$$
\begin{aligned}
3x - 4 - x &= 2(x + 5) \\
2x - 4 &= 2x + 10 \\
2x - 2x - 4 &= 2x - 2x + 10 \\
-4 &= 10 \quad \text{False}
\end{aligned}
$$
No solution

14.
$$
\begin{aligned}
-3(2x + 3) &= -2(3x + 1) - 7 \\
-6x - 9 &= -6x - 2 - 7 \\
-6x - 9 &= -6x - 9 \\
-6x + 6x - 9 &= -6x + 6x - 9 \\
-9 &= -9 \quad \text{True}
\end{aligned}
$$
All real numbers

15.
$$
\begin{aligned}
\frac{9}{x} &= \frac{3}{-15} \\
9(-15) &= 3(x) \\
-135 &= 3x \\
\frac{-135}{3} &= \frac{3x}{3} \\
-45 &= x
\end{aligned}
$$

16.
$$
\begin{aligned}
2x - 4 &< 4x + 10 \\
2x - 2x - 4 &< 4x - 2x + 10 \\
-4 &< 2x + 10 \\
-4 - 10 &< 2x + 10 - 10 \\
-14 &< 2x \\
\frac{-14}{2} &< \frac{2x}{2} \\
-7 &< x \text{ or}
\end{aligned}
$$

$x > -7$

17. $$3(x + 4) \geq 5x - 12$$
$$3x + 12 \geq 5x - 12$$
$$3x - 3x + 12 \geq 5x - 3x - 12$$
$$12 \geq 2x - 12$$
$$12 + 12 \geq 2x - 12 + 12$$
$$24 \geq 2x$$
$$\frac{24}{2} \geq \frac{2x}{2}$$
$$12 \geq x \quad \text{or} \quad x \leq 12$$

12

18. $$4(x + 3) + 2x < 6x - 3$$
$$4x + 12 + 2x < 6x - 3$$
$$6x + 12 < 6x - 3$$
$$6x - 6x + 12 < 6x - 6x - 3$$
$$12 < -3$$
False

No solution

19. $$\frac{x}{8} = \frac{4}{3}$$
$$3x = (8)(4)$$
$$3x = 32$$
$$x = \frac{32}{3} = 10\frac{2}{3} \text{ ft.}$$

20. Let x = number of gallons of insecticide.

$$\frac{\text{insecticide}}{\text{acres}} = \frac{\text{insecticide}}{\text{acres}}$$
$$\frac{6}{3} = \frac{x}{75}$$
$$6(75) = 3x$$
$$450 = 3x$$
$$\frac{450}{3} = \frac{3x}{3}$$
$$150 = x$$
$$x = 150 \text{ gallons}$$

<u>Cumulative Review Test</u>

1. $$\frac{16}{\underset{5}{\cancel{20}}} \cdot \frac{\overset{1}{\cancel{4}}}{5} = \frac{16}{25}$$

2. $$\frac{8}{24} \div \frac{2}{3} = \frac{\overset{4}{\cancel{8}}}{\underset{8}{\cancel{24}}} \cdot \frac{\overset{1}{\cancel{3}}}{\underset{1}{\cancel{2}}} = \frac{4}{8} = \frac{1}{2}$$

3. $>$

4. $$-6 - (-3) + 5 - 8$$
$$= -6 + 3 + 5 - 8$$
$$= -6$$

5. $$-12 - (-4) = -12 + 4$$
$$= -8$$

6. $$16 - 6 \div 2 \cdot 3$$
$$= 16 - 3 \cdot 3$$
$$= 16 - 9$$
$$= 7$$

7. $$3[6 - (4 - 3^2)] - 30$$
$$= 3[6 - (4 - 9)] - 30$$
$$= 3[6 - (-5)] - 30$$
$$= 3[6 + 5] - 30$$
$$= 3[11] - 30$$
$$= 33 - 30$$
$$= 3$$

8. $$-3x^2 - 4x + 5$$
$$= -3(-2)^2 - 4(-2) + 5$$
$$= -3(4) + 8 + 5$$

$= -12 + 8 + 5$
$= 1$

9. Associative property of addition.

10. $6x + 2y + 4x - y$
$= 6x + 4x + 2y - y$
$= 10x + y$

11. $3x - 2x + 16 + 2x$
$= 3x - 2x + 2x + 16$
$= 3x + 16$

12. $4x - 2 = 10$
$4x - 2 + 2 = 10 + 2$
$4x = 12$
$x = \frac{12}{4} = 3$

13. $\frac{1}{4}x = -10$
$x = (-10)(4)$
$x = -40$

14. $6x + 5x + 6 = 28$
$11x + 6 = 28$
$11x + 6 - 6 = 28 - 6$
$11x = 22$
$x = \frac{22}{11} = 2$

15. $3(x - 2) = 5(x - 1) + 3x + 4$
$3x - 6 = 5x - 5 + 3x + 4$
$3x - 6 = 5x + 3x - 5 + 4$
$3x - 6 = 8x - 1$
$3x - 8x - 6 = 8x - 8x - 1$
$-5x - 6 = -1$
$-5x - 6 + 6 = -1 + 6$
$-5x = 5$
$x = \frac{5}{-5} = -1$

16. $\frac{15}{30} = \frac{3}{x}$
$15x = (3)(30)$
$15x = 90$
$x = \frac{90}{15} = 6$

17. $x - 4 > 6$
$x - 4 + 4 > 6 + 4$
$x > 10$

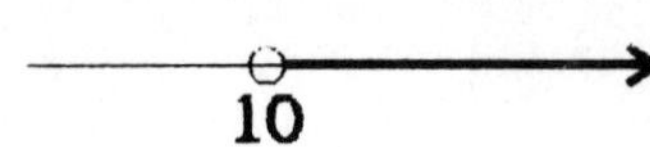

18. $2x - 7 \leq 3x + 5$
$2x - 3x - 7 \leq 3x - 3x + 5$
$-x - 7 \leq 5$
$-x - 7 + 7 \leq 5 + 7$
$-x \leq 12$
$x \geq -12$

-12

19. $\frac{36 \text{ pounds}}{5000 \text{ square feet}} = \frac{x \text{ pounds}}{22{,}000 \text{ square feet}}$
$\frac{36}{5000} = \frac{x}{22{,}000}$
$5000x = (36)(22{,}000)$
$5000x = 792{,}000$
$x = \frac{792{,}000}{5000} = 158.4$ pounds

20. $\frac{\$10.50}{2 \text{ hours}} = \frac{\$x}{8 \text{ hours}}$
$\frac{10.5}{2} = \frac{x}{8}$
$2x = (8)(10.5)$
$2x = 84$
$x = \$42$

CHAPTER 3

Exercise Set 3.1

1. $A = s^2$

$A = (4)^2$
$A = 16$

3. $P = 2l + 2w$
$P = 2(8) + 2(5)$
$P = 16 + 10$
$P = 26$

5. $A = \frac{1}{2}h(b + d)$

$A = \frac{1}{2}6(18 + 24)$

$A = 3(42)$
$A = 126$

7. $C = 2\pi r$
$C = 2(3.14)2$
$C = 12.56$

9. $A = \frac{1}{2} \cdot b \cdot h$

$20 = \frac{1}{2} \cdot 4h$

$20 = 2h$

$\frac{20}{2} = \frac{2h}{2}$

$10 = h$

11. $V = lwh$
$18 = l \cdot 1 \cdot 3$

$\frac{18}{3} = \frac{3l}{3}$

$6 = l$

13. $A = P(1 + rt)$
$A = 1000(1 + 0.08 \cdot 1)$
$A = 1000(1 + 0.08)$
$A = 1000(1.08)$
$A = 1080$

15. $M = \frac{a + b}{2}$

$36 = \frac{16 + b}{2}$

$2(36) = [\frac{16 + b}{2}] \cdot 2$

$72 = 16 + b$
$72 - 16 = 16 - 16 + b$
$56 = b$

17. $C = \frac{5}{9}(F - 32)$

$C = \frac{5}{9}(41 - 32)$

$C = \frac{5}{9} \cdot 9$

$C = 5$

19. $Z = \frac{x - m}{s}$

$2 = \frac{x - 50}{5}$

$2(5) = [\frac{x - 50}{5}] \cdot 5$

$10 = x - 50$
$10 + 50 = x - 50 + 50$
$60 = x$

21. $K = \frac{1}{2} mv^2$

$288 = \frac{1}{2} m(6)^2$

$288 = \frac{1}{2} m(36)$

$$288 = 18m$$

$$\frac{288}{18} = \frac{18m}{18}$$

$$m = 16$$

23. $V = \pi r^2 h$

$$678.24 = (3.14)(6)^2 h$$
$$678.24 = (3.14)(36)h$$
$$678.24 = 113.04h$$

$$\frac{678.24}{113.04} = \frac{113.04h}{113.04}$$

$$6 = h$$

25. $2x + y = 8$

$$2x - 2x + y = 8 - 2x$$
$$y = 8 - 2x$$

Now solve when $x = 2$.

$$y = 8 - 2(2)$$
$$y = 8 - 4$$
$$y = 4$$

27. $2x = 6y - 4$

$$2x + 4 = 6y - 4 + 4$$
$$2x + 4 = 6y$$

$$\frac{2x + 4}{6} = \frac{6y}{6}$$

$$\frac{2x + 4}{6} = y$$

Now solve when $x = 10$.

$$\frac{2(10) + 4}{6} = y$$

$$\frac{20 + 4}{6} = y$$

$$\frac{24}{6} = y$$

$$4 = y$$

29. $2y = 6 - 3x$

$$\frac{2y}{2} = \frac{6 - 3x}{2}$$

$$y = \frac{6 - 3x}{2}$$

Now solve when $x = 2$.

$$y = \frac{6 - 3(2)}{2}$$

$$y = \frac{6 - 6}{2}$$

$$y = \frac{0}{2}$$

$$y = 0$$

31. $-4x + 5y = -20$

$$-4x + 4x + 5y = 4x - 20$$
$$5y = 4x - 20$$

$$\frac{5y}{5} = \frac{4x - 20}{5}$$

$$y = \frac{4x - 20}{5}$$

Now solve when $x = 4$.

$$y = \frac{4(4) - 20}{5}$$

$$y = \frac{16 - 20}{5}$$

$$y = \frac{-4}{5}$$

33. $-3x = 18 - 6y$

$$-3x + 6y = 18 - 6y + 6y$$
$$-3x + 6y = 18$$
$$-3x + 3x + 6y = 3x + 18$$
$$6y = 3x + 18$$

$$\frac{6y}{6} = \frac{3x + 18}{6}$$

$$y = \frac{3x + 18}{6}$$

Now solve when $x = 0$.

$$y = \frac{3(0) + 18}{6}$$

$$y = \frac{0 + 18}{6}$$

$$y = \frac{18}{6}$$

$$y = 3$$

35. $$-8 = -x - 2y$$
$$-8 + 2y = -x - 2y + 2y$$
$$-8 + 2y = -x$$
$$-8 + 8 + 2y = -x + 8$$
$$2y = -x + 8$$
$$\frac{2y}{2} = \frac{-x + 8}{2}$$

$$y = \frac{-x + 8}{2}$$

Now solve when $x = -4$.

$$y = \frac{-(-4) + 8}{2}$$

$$y = \frac{4 + 8}{2}$$

$$y = \frac{12}{2}$$

$$y = 6$$

37. $$d = rt$$
$$\frac{d}{r} = \frac{rt}{r}$$

$$\frac{d}{r} = t$$

39. $$i = prt$$
$$\frac{i}{rt} = \frac{prt}{rt}$$

$$\frac{i}{rt} = p$$

41. $$C = \pi d$$
$$\frac{C}{\pi} = \frac{\pi d}{\pi}$$

$$\frac{C}{\pi} = d$$

43. $$A = \frac{1}{2} bh$$

$$2A = [\tfrac{1}{2}] \cdot 2bh$$
$$2A = bh$$
$$\frac{2A}{h} = \frac{bh}{h}$$

$$\frac{2A}{h} = b$$

45. $$P = 2l + 2w$$
$$P - 2l = 2l - 2l + 2w$$
$$P - 2l = 2w$$
$$\frac{P - 2l}{2} = \frac{2w}{2}$$

$$\frac{P - 2l}{2} = w$$

47. $$4n + 3 = m$$
$$4n + 3 - 3 = m - 3$$
$$4n = m - 3$$
$$\frac{4n}{4} = \frac{m - 3}{4}$$

$$n = \frac{m - 3}{4}$$

49. $$y = mx + b$$
$$y - mx = mx - mx + b$$
$$y - mx = b$$

51. $$I = P + Prt$$
$$I - P = P - P + Prt$$
$$I - P = Prt$$
$$\frac{I - P}{Pt} = \frac{Prt}{Pt}$$

$$\frac{I - P}{Pt} = r$$

53. $$A = \frac{m + 2d}{3}$$

$$3A = \frac{m + 2d}{3} \cdot 3$$
$$3A = m + 2d$$
$$3A - m = m - m + 2d$$
$$3A - m = 2d$$
$$\frac{3A - m}{2} = \frac{2d}{2}$$
$$\frac{3A - m}{2} = d$$

55. $$d = a + b + c$$
$$d - a - c = a - a + b + c - c$$
$$d - a - c = b$$

57. $$ax + by = c$$
$$ax - ax + by = c - ax$$
$$by = c - ax$$
$$\frac{by}{b} = \frac{c - ax}{b}$$
$$y = \frac{c - ax}{b}$$

59. $$V = \pi r^2 h$$
$$\frac{V}{\pi r^2} = \frac{\pi r^2 h}{\pi r^2}$$
$$\frac{V}{\pi r^2} = h$$

61. $$d = \frac{1}{2} n^2 - \frac{3}{2} n$$
$$n = 10$$
$$d = \frac{1}{2}(10)^2 - \frac{3}{2}(10)$$
$$d = \frac{1}{2}(100) - \frac{3}{2}(10)$$
$$d = 50 - 15$$
$$d = 35$$

63. $$C = \frac{5}{9}(F - 32)$$
$$F = 50^0$$
$$C = \frac{5}{9}(50 - 32)$$
$$C = \frac{5}{9}(18)$$
$$C = 10^0$$

65. $$F = \frac{9}{5} C + 32$$
$$C = 35^0$$
$$F = \frac{9}{5}(35) + 32$$
$$F = 63 + 32$$
$$F = 95^0$$

67. $$P = \frac{KT}{V}$$
$$T = 10,\ K = 1,\ V = 1$$
$$P = \frac{(1)(10)}{1}$$
$$P = 10$$

69. $$P = \frac{KT}{V}$$
$$P = 80,\ T = 100,\ V = 5$$
$$80 = \frac{100K}{5}$$
$$80 = 20K$$
$$\frac{80}{20} = \frac{20K}{20}$$
$$4 = K$$

71. $$S = n^2 + n$$
$$n = 5$$
$$S = (5)^2 + (5)$$
$$S = 25 + 5$$
$$S = 30$$

73. $$i = prt$$
$$i = (4000)(0.12)(3)$$
$$i = \$1440$$

75. $$i = prt$$
$$1050 = p(0.07)(3)$$
$$1050 = p(0.21)$$
$$\frac{1050}{0.21} = \frac{0.21\ p}{0.21}$$
$$\$5000 = p$$

77. $p = a + b + c$
$p = 5 + 12 + 13$
$p = 30$ inches

79. $A = \frac{1}{2}bh$

$A = \frac{1}{2}(6)(8)$

$A = 24$ sq. cm.

81. $A = \pi r^2$

$A = (3.14)(4)^2$
$A = (3.14)(16)$
$A = 50.24$ sq. in.

83. radius = 4 inches
$C = 2\pi r$
$C = 2(3.14)(4)$
$C = 25.12$ in

85. $A = lw$
$48 = (6)w$
$\frac{48}{6} = \frac{6w}{6}$
$8 = w$
$w = 8$ ft.

87. a) $C = 2\pi r$
$390 = 2(3.14)r$
$390 = 6.28r$
$\frac{390}{6.28} = \frac{6.28r}{6.28}$
$62.1 = r$
$r = 62.1$ ft.

b) $d = 2r$
$d = 2(62.1)$
$d = 124.2$ ft.

89. radius = 11 in.
4 ft. = 48 in.

$V = \pi r^2 h$

$V = (3.14)(11)^2(48)$
$V = (3.14)(121)(48)$
$V = 18{,}237.12$ cu. in.

91. When you multiply a unit by the same unit, you get a square unit.

93. a) $C = 2\pi r$
$\frac{C}{2r} = \frac{2\pi r}{2r}$
$\frac{C}{2r} = \pi$

b) π or about 3.14.
From (a) we solved for π and got $\frac{C}{2r}$. Since
$2r = d$, $\frac{C}{d} = \pi$.

c) π or about 3.14.

Cumulative Review Exercises

95. $\frac{\text{Arabians}}{\text{Morgans}} = \frac{6}{4} = 6:4$ or $3:2$

97. $2(x - 4) \geq 3x + 9$
$2x - 8 \geq 3x + 9$
$2x - 3x - 8 \geq 3x - 3x + 9$
$-x - 8 \geq 9$
$-x - 8 + 8 \geq 9 + 8$
$-x \geq 17$
$x \leq -17$

Just for fun

1. a) $d = 2r$
$\frac{d}{2} = r$

$A = \pi r^2$

$$A = \pi(\frac{d}{2})^2$$

Area of shaded region:

$$A = d^2 - \pi(\frac{d}{2})^2$$

b) $A = d^2 - \pi(\frac{d}{2})^2$

$A = (4)^2 - (3.14)(\frac{4}{2})^2$

$A = 16 - (3.14)(2)^2$
$A = 16 - (3.14)(4)$
$A = 16 - 12.56$
$A = 3.44$ sq. ft.

c) $A = d^2 - \pi(\frac{d}{2})^2$

$A = (6)^2 - (3.14)(\frac{6}{2})^2$

$A = 36 - (3.14)(3)^2$
$A = 36 - (3.14)(9)$
$A = 36 - 28.26$
$A = 7.74$ sq. ft.

2. a) $V = lwh$
$V = (6x - 1)(3x)(x)$

$V = (6x - 1)(3x^2)$

$V = 18x^3 - 3x^2$

b) $V = 18x^3 - 3x^2$

$V = 18(7)^3 - 3(7)^2$
$V = 18(343) - 3(49)$
$V = 6174 - 147$
$V = 6027$ cu. cm.

c) $S = 2(x)(6x - 1)$
$+ 2(x)(3x)$
$+ 2(3x)(6x - 1)$
$S = 2x(6x - 1)$

$+ 6x^2 + 6x(6x - 1)$

$S = 12x^2 - 2x + 6x^2$

$+ 36x^2 - 6x$

$S = 54x^2 - 8x$

d) $S = 54x^2 - 8x$

$S = 54(7)^2 - 8(7)$
$S = 54(49) - 56$
$S = 2646 - 56$
$S = 2590$ sq. cm.

Exercise Set 3.2

1. $x + 5$

3. $4x$

5. $0.70x$

7. $0.10c$

9. $6x - 3$

11. $7 + \frac{3}{4}x$

13. $2(x + 8)$

15. $25x$

17. $12x$

19. $16a + b$

21. Three more than a number.

23. Three times a number, decreased by four.

25. The difference of twice a number and three.

27. The difference of five and

a number.

29. The sum of four and six times a number.

31. Three times the sum of a number and two.

33. Brother's age: x
Boy's age: $x + 12$

35. First integer: x
Next consecutive integer: $x + 1$

37. One person's share: x
Second person's share: $100 - x$

39. A number: x
Second number: $4x - 5$

41. Cost of Ford: x
Cost of Cadillac: $1.7x$

43. Piece of tree: x
Another piece: $80 - x$

45. A number: x
The number increased by 12%: $x + 0.12x$

47. Cost of an item: x
Cost plus 7% sales tax: $x + 0.07x$

49. Number of times: x
Cost: $4x$ dollars

51. Number of miles: x
Total cost: \$0.23x

53. Number of hours: x
Total cost: \$15x

55. Number of days: x
Total employees hired: $10x$

57. Number of years: n
Total population growth: $300n$

59. Number of dollars: x
Total sales tax: $0.075x$ dollars

61. Number of dimes: a
Total number of cents: $10a$ cents

63. Number of \$5 bills: p
Total number of dollars: $5p$ dollars

65. A number: x
Another number: $5x$
Equation: $x + 5x = 18$

67. An integer: x
Next consecutive integer: $x + 1$
Equation: $x + (x + 1) = 47$

69. A number: x
Equation: $2x - 8 = 12$

71. A number: x
Equation: $\frac{1}{5}(x + 10) = 150$

73. Distance one train travels: x
Distance another travels: $2x - 8$
Equation: $x + (2x - 8) = 1000$

75. A number: x
Equation: $x + 0.08x = 92$

77. Cost of jacket: x
Cost of jacket minus 25%: $x - 0.25x$
Equation: $x - 0.25x = 65$

79. Cost of video cassette recorder: x
Cost of recorder reduced by 20%: $x - 0.20\%$
Equation: $x - 0.20x = 215$

81. A number: x
Another number: $2x - 3$
Equation:
$x + (2x - 3) = 21$

83. Number of hours: t
Equation: $40t = 180$

85. Number of French Fries: y
Equation: $15y = 215$

87. Number of quarters: q
Equation: $25q = 150$

89. Three more than a number is six.

91. Three times a number, decreased by one, is four more than twice the number.

93. Four times the difference of a number and one is six.

95. Six more than five times a number is the difference of six times the number and one.

97. The sum of a number and the number increased by four is eight.

99. The sum of twice a number and the number increased by three is five.

101. The cost of purchasing x items at 6 dollars each is the product of x and 6, which is 6x.

Cumulative Review Exercises

103. $$\frac{1/2 \text{ teaspoon}}{1 \text{ pound}} = \frac{x \text{ teaspoon}}{6.7 \text{ pounds}}$$

$$\frac{0.5}{1} = \frac{x}{6.7}$$

$$x = (0.5)(6.7)$$
$$x = 3.35 \text{ teaspoons}$$

105. $$P = 2l + 2w$$

$$40 = 2l + 2(5)$$
$$40 = 2l + 10$$
$$40 - 10 = 2l + 10 - 10$$
$$30 = 2l$$
$$\frac{30}{2} = \frac{2l}{2}$$
$$15 = l$$

Just for fun

1. a) $P = 100(\frac{9f}{c})$

b) $P = 100[\frac{9(8)}{150}]$

$P = 100[\frac{72}{150}]$

$P = \frac{7200}{150} = 48\%$

c) $P = 100[\frac{9(24)}{510}]$

$P = 100[\frac{216}{510}]$

$P = \frac{21,600}{510} = 42.4\%$

2. a) Seconds in d days:
$d \cdot 24 \cdot 60 \cdot 60 = 86,400d$
Seconds in h hours:
$h \cdot 60 \cdot 60 = 3600h$
Seconds in m minutes:
$m \cdot 60 = 60m$
s seconds: s
Expression: $86,400d + 3600h + 60m + s$

b) $86,400(4) + 3600(6) + 60(15) + 25$
$= 345,600 + 21,600 + 900 + 25$
$= 368,125$ seconds

Exercise Set 3.3

1. Let x = 1st integer.
   $x + 1$ = next consecutive integer

$$x + (x + 1) = 71$$
$$2x + 1 = 71$$
$$2x = 70$$
$$x = 35$$

One integer is 35 and the next consecutive integer is 36.

3. Let x = an odd integer.
   $x + 2$ = next consecutive odd integer.

$$x + (x + 2) = 76$$
$$2x + 2 = 76$$
$$2x + 2 - 2 = 76 - 2$$
$$2x = 74$$
$$\frac{2x}{2} = \frac{74}{2}$$
$$x = 37$$

First odd integer: 37
Next consecutive odd integer is:
$x + 2 = (37) + 2 = 39$

5. Let x = a number.
   $3x - 5$ = another number.

$$x + (3x - 5) = 43$$
$$4x - 5 = 43$$
$$4x - 5 + 5 = 43 + 5$$
$$4x = 48$$
$$\frac{4x}{4} = \frac{48}{4}$$
$$x = 12$$

One number is 12, and other number is $3x - 5 = 3(12) - 5 = 31$.

7. x = an integer.
   $x + 2$ = next consecutive odd integer.
   $x + 4$ = third consecutive odd integer.

$$x + (x + 2) + (x + 4) = 87$$
$$3x + 6 = 87$$
$$3x + 6 - 6 = 87 - 6$$
$$3x = 81$$
$$\frac{3x}{3} = \frac{81}{3}$$
$$x = 27$$

An odd integer: 27
Second odd integer:
$x + 2 = (27) + 2 = 29$
Third odd integer:
$x + 4 = (27) + 4 = 31$

9. x = smaller integer.
   $3x - 4$ = larger integer.

$$(3x - 4) - x = 12$$
$$2x - 4 = 12$$
$$2x - 4 + 4 = 12 + 4$$
$$2x = 16$$
$$\frac{2x}{2} = \frac{16}{2}$$
$$x = 8$$

Smaller integer: 8
Larger integer:
$3x - 4 = 3(8) - 4 = 20$

11. Let x = number of years.

$$5200 + 300x = 9400$$
$$300x = 4200$$
$$\frac{300x}{300} = \frac{4200}{300}$$
$$x = 14 \text{ years}$$

13. Let x = number of months.

$$3.50x = 14.00$$

$$\frac{3.50x}{3.50} = \frac{14.00}{3.50}$$

$x = 4$ months

15. Let x = number of miles.

$$15 + 0.20x = 55$$
$$15 - 15 + 0.20x = 55 - 15$$
$$0.20x = 40$$
$$\frac{0.20x}{0.20} = \frac{40}{0.20}$$
$x = 200$ miles

17. Let x = amount Mary spends.

$$0.08x = 60$$
$$\frac{0.08x}{0.08} = \frac{60}{0.08}$$
$$x = \$750$$

19. Let x = Paul's present salary.

$$x + 0.08x = 31,320$$
$$1.08x = 31,320$$
$$\frac{1.08x}{1.08} = \frac{31,320}{1.08}$$
$$x = \$29,000$$

21. Let x = cost of hot dog.

$$x + 0.05x = 1.25$$
$$1.05x = 1.25$$
$$\frac{1.05x}{1.05} = \frac{1.25}{1.05}$$
$$x = \$1.19$$

23. Let x = number of French Fries.

$$300 + 20x = 500$$
$$300 - 300 + 20x = 500 - 300$$
$$20x = 200$$
$$\frac{20x}{20} = \frac{200}{20}$$
$x = 10$ French Fries

25. x = number of hours each younger worker worked overtime.
$2x$ = number of hours 3rd worker worked overtime.
$3x$ = number of hours 4th worker worked overtime.

$$x + x + 2x + 3x = 91$$
$$7x = 91$$
$$x = 13$$

Each younger worker worked 13 hours.
Third worker:
$2x = 2(13) = 26$ hours.
Fourth worker:
$3x = 3(13) = 39$ hours.

27. Let x = price of meal.

$$x + 0.07x + 0.15x = 20.00$$
$$1.22x = 20.00$$
$$\frac{1.22x}{1.22} = \frac{20.00}{1.22}$$
$$x = \$16.39$$

29. Let x = number of hours.

$$7.25x = 35$$
$$\frac{7.25x}{7.25} = \frac{35}{7.25}$$
$$x = 4.827$$

5 hours in full hour intervals.

31. Let x = regular price of ski.

$$x - 0.20x - 25 = 231$$
$$0.80x - 25 = 231$$
$$0.80x = 256$$
$$\frac{0.80x}{0.80} = \frac{256}{0.80}$$
$$x = \$320$$

33. Let x = number of terminals.

$$65,000 + 1500x$$

$= 20{,}000 + 3000x$
$65{,}000 = 20{,}000 + 1500x$
$45{,}000 = 1500x$
$x = 30$ terminals

35. Answers will vary.

Cumulative Review Exercises

37. Associative property of addition.

39. Distributive property

41.
$$M = \frac{a + b}{2}$$
$$2M = \frac{a + b}{2} \cdot 2$$
$$2M = a + b$$
$$2M - a = a - a + b$$
$$2m - a = b$$

Just for fun

1. a) Let x = grade on fourth exam.

$$\frac{x + 74 + 88 + 76}{4} = 80$$

b) $$\frac{x + 238}{4} = 80$$
$$x + 238 = 320$$
$$x = 82$$

2. Let x = number of 3-point field goals.
4x = number of 2-point field goals.

$$12 + 3(x) + 2(4x) = 78$$
$$12 + 3x + 8x = 78$$
$$12 + 11x = 78$$
$$11x = 66$$
$x = 6$ 3-pointers
$4x = 4(6) = 24$ 2-pointers

3. Let n = given number.

$$[(4n + 6) \div 2] - 3 = 2n$$
$$\frac{4n + 6}{2} - 3 = 2n$$
$$\frac{4n}{2} + \frac{6}{2} - 3 = 2n$$
$$2n + 3 - 3 = 2n$$
$$2n = 2n$$
Therefore, n can be any number.

Exercise Set 3.4

1. Let x = length of each side.

$$3x = 28.5$$
$$x = \frac{28.5}{3} = 9.5 \text{ inches}$$

3. Let x = measure of angle A.
3x - 8 = measure of angle B.

$$x + (3x - 8) = 180$$
$$4x - 8 = 180$$
$$4x = 188$$
$$x = 47^{\circ} \text{ (angle A)}$$
$$3x - 8 = 3(47) - 8$$
$$= 141 - 8$$
$$= 133^{\circ} \text{ (angle B)}$$

5. Let x = smallest angle
x + 10 = 2nd angle.
2x - 30 = 3rd angle.

$$x + (x+10) + (2x-30) = 180$$
$$4x - 20 = 180$$
$$4x = 200$$
$$x = 50^{\circ}$$

Smallest angle: 50°
Second angle:
$x + 10 = 50 + 10 = 60^\circ$
Third angle: $2x - 30$
$= 2(50) - 30 = 100 - 30$
$= 70^\circ$

7. Let x = first side.
x = second side.
x - 2 = third side.
a + b + c = P

$$x + x + x - 2 = 10$$
$$3x - 2 = 10$$
$$3x - 2 + 2 = 10 + 2$$
$$3x = 12$$
$$\frac{3x}{3} = \frac{12}{3}$$
$$x = 4$$

First side: x = 4m
Second side: x = 4m
Third side: x - 2 = 4 - 2
= 2m

9. Let w = width of basement.
2 to -24 = length of basement.

$$P = 21 + 2w$$
$$240 = 2(2w - 24) + 2(w)$$
$$240 = 4w - 48 + 2w$$
$$240 = 6w - 48$$
$$288 = 6w$$
$$48 = w$$

width: 48 ft.
length: $2w - 24$
$= 2(48) - 24$
$= 96 - 24$
= 72 feet

11. Let x = measure of each smaller angle.

2x - 27 = measure of each larger angle.

$$2(x) + 2(2x - 27) = 360$$
$$2x + 4x - 54 = 360$$
$$6x - 54 = 360$$
$$6x = 414$$
$$x = 69^\circ$$

Each smaller angle: 69°
Each larger angle:
$2x - 27 = 2(69) - 27$
$= 138 - 27$
$= 111^\circ$

13. Let x = width of bookcase.
x + 2 = height of bookcase.
4x = width of four shelves.
2(x + 2) = height of two sides.

$$4x + 2(x + 2) = 20$$
$$4x + 2x + 4 = 20$$
$$6x + 4 = 20$$
$$6x + 4 - 4 = 20 - 4$$
$$6x = 16$$
$$x = \frac{8}{3}$$

Width: $x = 2\frac{2}{3}$ feet

Height: $x + 2 = 2\frac{2}{3} + 2$

$= 4\frac{2}{3}$ feet

15. Let x = height of shelves.
3x = length of shelves.

$$3(x) + 4(3x) = 45$$
$$3x + 12x = 45$$
$$15x = 45$$
$$x = 3 \text{ feet}$$

height: 3 ft.
length: 3x = 3(3) = 9 ft.

17. The area remains the same.

19. The volume becomes eight times as great.

21. Answers will vary.

Cumulative Review Exercises

23. $>$

25. $-6y + x - 3(x - 2) + 2y$
$= -6y + x - 3x + 6 + 2y$
$= x - 3x - 6y + 2y + 6$
$= -2x - 4y + 6$

Just for fun

1. a) $A = S^2 - s^2$

b) $A = (9)^2 - (6)^2$
$A = 81 - 36$
$A = 45$ sq. in.

2. $A = ac + ad + bc + bd$

3. a)
$$P = 14.70 + 0.43x$$
$$162 = 14.70 + 0.43x$$
$$147.3 = 0.43x$$
$$\frac{147.3}{0.43} = \frac{0.43x}{0.43}$$
$$342.56 = x$$
$$x = 342.56 \text{ ft.}$$

b)
$$P = 14.70 + 0.43x$$
$$97.26 = 14.70 + 0.43x$$
$$82.56 = 0.43x$$
$$\frac{82.56}{0.43} = \frac{0.43x}{0.43}$$
$$192 = x$$
$$x = 192 \text{ ft.}$$

Review Exercises

1. $c = \pi d$
$c = (3.14)(4)$
$c = 12.56$

3. $p = 2l + 2w$
$p = 2(6) + 2(4)$
$p = 12 + 8$
$p = 20$

5. $E = IR$
$E = 0.12(2000)$
$E = 240$

7.
$$V = \frac{4}{3}\pi r^3$$
$$V = \frac{4}{3}(3.14)(3)^3$$
$$V = \frac{4}{3}(3.14)(27)$$
$$V = 113.04$$

9.
$$y = mx + b$$
$$15 = 3(-2) + b$$
$$15 = -6 + b$$
$$15 + 6 = -6 + 6 + b$$
$$21 = b$$

11.
$$4x - 3y = 15 + x$$
$$4(-3) - 3y = 15 + (-3)$$
$$-12 - 3y = 12$$
$$-12 + 12 - 3y = 12 + 12$$
$$-3y = 24$$
$$\frac{-3y}{-3} = \frac{24}{-3}$$
$$y = -8$$

13.
$$IR = E + Rr$$
$$5.0(200) = 100 + 200r$$
$$1000 = 100 + 200r$$
$$1000 - 100 = 100 - 100 + 200r$$
$$900 = 200r$$
$$\frac{900}{200} = \frac{200r}{200}$$
$$4.5 = r$$

15.
$$3x - 2y = -4$$
$$3x - 3x - 2y = -4 - 3x$$
$$-2y = -3x - 4$$
$$\frac{-2y}{-2} = \frac{-3x - 4}{-2}$$
$$y = \frac{3x + 4}{2}$$

Now solve when $x = 2$.

$$y = \frac{3}{2}(2) + 2$$

$$y = 3 + 2$$
$$y = 5$$

17. $$-6x - 2y = 20$$
$$-6x + 6x - 2y = 6x + 20$$
$$-2y = 6x + 20$$
$$\frac{-2y}{-2} = \frac{6x + 20}{-2}$$
$$y = \frac{6x}{-2} + \frac{20}{-2}$$
$$y = -3x - 10$$

Now solve when $x = 0$.

$$y = -3(0) - 10$$
$$y = -10$$

19. $$3y - 4x = -3$$
$$3y - 4x + 4x = -3 + 4x$$
$$3y = -3 + 4x$$
$$\frac{3y}{3} = \frac{-3 + 4x}{3}$$
$$y = \frac{-3 + 4x}{3}$$

Now solve when $x = 2$.

$$y = \frac{-3 + 4(2)}{3}$$
$$y = \frac{-3 + 8}{3}$$
$$y = \frac{5}{3}$$

21. $$A = \frac{1}{2}bh$$

$$2A = 2(\frac{1}{2}) \cdot bh$$
$$2A = bh$$
$$\frac{2A}{b} = \frac{bh}{b}$$
$$\frac{2A}{b} = h$$

23. $$P = 2l + 2w$$
$$P - 2l = 2l - 2l + 2w$$
$$P - 2l = 2w$$
$$\frac{P - 2l}{2} = \frac{2w}{2}$$

$$\frac{P - 2l}{2} = w$$

25. $$A = \frac{B + C}{2}$$

$$2A = \frac{B + C}{2} \cdot 2$$
$$2A = B + C$$
$$2A - C = B + C - C$$
$$2A - C = B$$

27. $$I = prt$$
$$I = 600(0.15)(2)$$
$$I = \$180$$

29. Let x = one number.
$x + 4$ = other number.

$$x + x + 4 = 62$$
$$2x + 4 = 62$$
$$2x + 4 - 4 = 62 - 4$$
$$2x = 58$$
$$\frac{2x}{2} = \frac{58}{2}$$
$$x = 29$$

One number: $x = 29$
Other number:
$x + 4 = (29) + 4 = 33$

31. Let x = smaller number.
$5x + 3$ = larger number.

$$5x + 3 - x = 31$$

$$4x + 3 = 31$$
$$4x + 3 - 3 = 31 - 3$$
$$4x = 28$$
$$\frac{4x}{4} = \frac{28}{4}$$
$$x = 7$$

Smaller number: $x = 7$

Larger number:

$5x + 3 = 5(7) + 3 = 38$

33. Let x = amount of Paul's sales.

$$500 + 0.03x = 400 + 0.08x$$
$$500 = 400 + 0.05x$$
$$100 = 0.05x$$
$$\frac{100}{0.05} = \frac{0.05x}{0.05}$$
$$2000 = x$$
$$x = \$2000$$

35. Let x = smallest angle.
$x + 10$ = second angle.
$2x - 10$ = third angle.

$$x + (x + 10) + (2x - 10) = 180$$
$$4x = 180$$
$$x = 45^0$$

Smallest angle: 45^0

Second angle:

$x + 10 = 55^0$

Third angle:

$$2x - 10 = 2(45) - 10 = 90 - 10 = 80^0$$

37. Let $x + 4$ = length.
x = width.

$$2l + 2w = \text{Perimeter}$$
$$2(x + 4) + 2(x) = 70$$
$$2x + 8 + 2x = 70$$
$$4x + 8 = 70$$
$$4x + 8 - 8 = 70 - 8$$
$$4x = 62$$
$$\frac{4x}{4} = \frac{62}{4}$$
$$x = 15.5$$

Width: $x = 15.5$ ft.

Length: $x + 4 = 15.5 + 4 = 19.5$ ft.

39. Let x = number of miles.

$$2(18) + 0.16x = 100$$
$$36 + 0.16x = 100$$
$$36 - 36 + 0.16x = 100 - 36$$
$$0.16x = 64$$
$$\frac{0.16x}{0.16} = \frac{64}{0.16}$$
$$x = 400 \text{ miles}$$

41. x = amount of sales.

$$300 + 0.05x = 900$$
$$300 - 300 + 0.05x = 900 - 300$$
$$0.05x = 600$$
$$\frac{0.05x}{0.05} = \frac{600}{0.05}$$
$$x = \$12,000$$

43. Let x = number of years.

$$427 + 25x = 627$$
$$25x = 200$$
$$x = 8 \text{ years}$$

Practice Test

1. $P = 2l + 2w$
$P = 2(6) + 2(3)$
$P = 12 + 6$
$P = 18$ feet

2. $A = P + Prt$
$A = 100 + 100(0.15)3$
$A = 100 + 45$
$A = 145$

3. $V = \frac{1}{3}\pi r^2 h$

$V = \frac{1}{3}(3.14)(4)^2(6)$

$V = \frac{1}{3}(3.14)(16)(6)$

$V = 100.48$

4. $P = IR$

$\frac{P}{I} = \frac{IR}{I}$

$\frac{P}{I} = R$

5.
$$\begin{aligned} 3x - 2y &= 6 \\ 3x - 3x - 2y &= -3x + 6 \\ -2y &= -3x + 6 \\ \frac{-2y}{-2} &= \frac{-3x + 6}{-2} \\ y &= \frac{3x - 6}{2} \end{aligned}$$

6.
$$\begin{aligned} A &= \frac{a + b}{3} \\ 3A &= 3[\frac{a + b}{3}] \\ 3A &= a + b \\ 3A - b &= a + b - b \\ 3A - b &= a \end{aligned}$$

7.
$$\begin{aligned} D &= R(c + a) \\ D &= Rc + Ra \\ D - Ra &= Rc + Ra - Ra \\ D - Ra &= Rc \\ \frac{D - Ra}{R} &= \frac{Rc}{R} \\ \frac{D - Ra}{R} &= c \end{aligned}$$

8. Let x = an integer.
$2x - 10$ = larger integer.

$$\begin{aligned} x + 2x - 10 &= 158 \\ 3x - 10 &= 158 \\ 3x - 10 + 10 &= 158 + 10 \\ 3x &= 168 \\ \frac{3x}{3} &= \frac{168}{3} \\ x &= 56 \end{aligned}$$

An integer: $x = 56$
Larger integer:
$2x - 10 = 2(56) - 10$
$= 102$

9. Let x = first integer.
$x + 1$ = second consecutive integer.
$x + 2$ = third consecutive integer.

$$\begin{aligned} x + x + 1 + x + 2 &= 42 \\ 3x + 3 &= 42 \\ 3x + 3 - 3 &= 42 - 3 \\ 3x &= 39 \\ \frac{3x}{3} &= \frac{39}{3} \\ x &= 13 \end{aligned}$$

First integer: $x = 13$
Second consecutive integer:
$x + 1 = 13 + 1 = 14$
Third consecutive integer:
$x + 2 = 13 + 2 = 15$

10. Let x = cost of meal.

$$\begin{aligned} x + 0.15x + 0.07x &= 20 \\ 1.22x &= 20 \\ \frac{1.22x}{1.22} &= \frac{20}{1.22} \\ x &= \$16.39 \end{aligned}$$

The most expensive meal that she could order would cost $16.39.

11. Let x = smallest side.
$x + 15$ = second side.
$2x$ = third side.

$$a + b + c = p$$
$$x + x + 15 + 2x = 75$$
$$4x + 15 = 75$$
$$4x + 15 - 15 = 75 - 15$$
$$4x = 60$$
$$\frac{4x}{4} = \frac{60}{4}$$
$$x = 15 \text{ in.}$$

Smallest side: $x = 15$ in.
Second side:
$x + 15 = 15 + 15 = 30$ in.
Third side:
$2x = 2(15) = 30$ in.

12. Let x = smaller angles.
$2x + 30$ = larger angles.

$$2(x) + 2(2x + 30) = 360$$
$$2x + 4x + 60 = 360$$
$$6x + 60 = 360$$
$$6x = 300$$
$$x = 50^{0}$$

Smaller angles: 50^{0}
Larger angles:
$$2x + 30 = 2(50) + 30$$
$$= 100 + 30$$
$$= 130^{0}$$

CHAPTER FOUR

Exercise Set 4.1

1. $x^4 \cdot x^3 = x^{4+3} = x^7$

3. $y^2 \cdot y = y^{2+1} = y^3$

5. $3^2 \cdot 3^3 = 3^{2+3} = 3^5 = 243$

7. $y^3 \cdot y^2 = y^{3+2} = y^5$

9. $y^4 \cdot y = y^{4+1} = y^5$

11. $\dfrac{x^{15}}{x^7} = x^{15-7} = x^8$

13. $\dfrac{5^4}{5^2} = 5^{4-2} = 5^2 = 25$

15. $\dfrac{x^9}{x^5} = x^{9-5} = x^4$

17. $\dfrac{y^2}{y} = y^{2-1} = y$

19. $\dfrac{x^2}{x^2} = x^{2-2} = x^0 = 1$

21. $x^0 = 1$

23. $3x^0 = 3 \cdot 1 = 3$

25. $(3x)^0 = 1$

27. $(-4x)^0 = 1$

29. $(x^5)^2 = x^{5 \cdot 2} = x^{10}$

31. $(x^5)^5 = x^{5 \cdot 5} = x^{25}$

33. $(x^3)^1 = x^{3 \cdot 1} = x^3$

35. $(x^3)^4 = x^{3 \cdot 4} = x^{12}$

37. $(x^4)^2 = x^{4 \cdot 2} = x^8$

39. $(1.3x)^2 = (1.3)^2x^2 = 1.69x^2$

41. $(-x)^2 = (-1)^2x^2 = 1x^2 = x^2$

43. $(4x^2)^3 = 4^3(x^2)^3 = 64x^6$

45. $(-3x^3)^3 = (-3)^3(x^3)^3 = -27x^9$

47. $(2x^2y)^3 = 2^3(x^2)^3y^3 = 8x^6y^3$

49. $(8.6x^2y^5)^2 = (8.6)^2(x^2)^2(y^5)^2$

$= 73.96x^4y^{10}$

51. $(-6x^3y^2)^3 = (-6)^3(x^3)^3(y^2)^3$

$= -216x^9y^6$

53. $(-x^4y^5z^6)^3 = (-x^4)^3(y^5)^3(z^6)^3$

$= -x^{12}y^{15}z^{18}$

55. $(\dfrac{x}{y})^2 = \dfrac{x^2}{y^2}$

57. $(\dfrac{x}{5})^3 = \dfrac{x^3}{5^3} = \dfrac{x^3}{125}$

59. $(\dfrac{y}{x})^5 = \dfrac{y^5}{x^5}$

61. $(\dfrac{6}{x})^3 = \dfrac{6^3}{x^3} = \dfrac{216}{x^3}$

63. $(\dfrac{3x}{y})^3 = \dfrac{3^3x^3}{y^3} = \dfrac{27x^3}{y^3}$

65. $(\frac{2x}{5})^2 = \frac{2^2x^2}{5^2} = \frac{4x^2}{25}$

67. $(\frac{4y^3}{x})^3 = \frac{4^3(y^3)^3}{x^3} = \frac{64y^9}{x^3}$

69. $(\frac{-3x^3}{4})^3 = \frac{(-1)^3 3^3(x^3)^3}{4^3}$

$= \frac{-27x^9}{64}$

71. $\frac{x^3y^2}{xy^5} = \frac{x \cdot x^2 \cdot y^2}{x \cdot y^3 \cdot y^2} = \frac{x^2}{y^3}$

73. $\frac{x^5y^7}{x^{12}y^3} = \frac{y^{7-3}}{x^{12-5}} = \frac{y^4}{x^7}$

75. $\frac{10x^3y^8}{2xy^{10}} = \frac{5x^{3-1}}{y^{10-8}} = \frac{5x^2}{y^2}$

77. $\frac{4xy}{16x^3y^2} = \frac{1}{4x^{3-1}y^{2-1}} = \frac{1}{4x^2y}$

79. $\frac{35x^4y^7}{10x^9y^{12}} = \frac{7}{2x^{9-4}y^{12-7}} = \frac{7}{2x^5y^5}$

81. $\frac{-36xy^9z}{12x^4y^5z^2} = \frac{-3 \cdot 12 \cdot x \cdot y^5 \cdot y^4 \cdot z}{12 \cdot x \cdot x^3 \cdot y^5 \cdot z \cdot z}$

$= \frac{-3y^4}{x^3z}$

83. $\frac{-6x^2y^7z^5}{2x^5y^9z^6} = \frac{-3 \cdot 2 \cdot x^2 \cdot y^7 \cdot z^5}{2 \cdot x^3 \cdot x^2 \cdot y^7 \cdot y^2 \cdot z^5 \cdot z}$

$= \frac{-3}{x^3y^2z}$

85. $(\frac{4x^4}{2x^6})^3 = (\frac{2 \cdot 2 \cdot x^4}{2 \cdot x^2 \cdot x^4})^3$

$= (\frac{2}{x^2})^3 = \frac{2^3}{(x^2)^3} = \frac{8}{x^6}$

87. $(\frac{8y^7}{2y^3})^3 = (\frac{4 \cdot 2 \cdot y^3 \cdot y^4}{2 \cdot y^3})^3$

$= (4y^4)^3 = 4^3(y^4)^3 = 64y^{12}$

89. $(\frac{27x^9}{30x^5})^2 = (\frac{3 \cdot 9 \cdot x^5 \cdot x^4}{3 \cdot 10 \cdot x^5})^2$

$= (\frac{9x^4}{10})^2 = \frac{9^2(x^4)^2}{10^2} = \frac{81x^8}{100}$

91. $(\frac{x^4y^3}{x^2y^5})^2 = (\frac{x^2 \cdot x^2 \cdot y^3}{x^2 \cdot y^3 \cdot y^2})^2$

$= (\frac{x^2}{y^2})^2 = \frac{(x^2)^2}{(y^2)^2} = \frac{x^4}{y^4}$

93. $(\frac{9y^2z^7}{18y^7z})^4 = (\frac{9 \cdot y^2 \cdot z \cdot z^6}{2 \cdot 9 \cdot y^2 \cdot y^5 \cdot z})^4$

$= (\frac{z^6}{2y^5})^4 = \frac{(z^6)^4}{2^4(y^5)^4} = \frac{z^{24}}{16y^{20}}$

95. $(\frac{3x^2y^5}{y^2})^3 = (\frac{3 \cdot x^2 \cdot y^2 \cdot y^3}{y^2})^3$

$= (3x^2y^3)^3 = 3^3(x^2)^3(y^3)^3$

$= 27x^6y^9$

97. $(\frac{-x^4y^6}{x^2})^2 = (\frac{-x^2 \cdot x^2 \cdot y^6}{x^2})^2$

$= (-x^2y^6)^2 = (-x^2)^2(y^6)^2$

$= x^4y^{12}$

99. $\left(\frac{-12x}{16x^7y^2}\right)^2 = \left(\frac{-3 \cdot 4 \cdot x}{4 \cdot 4 \cdot x \cdot x^6 \cdot y^2}\right)^2$

$= \left(\frac{-3}{4x^6y^2}\right)^2 = \frac{(-3)^2}{4^2(x^6)^2(y^2)^2}$

$= \frac{9}{16x^{12}y^4}$

101. $x^6(3xy^4) = 3 \cdot x^6 \cdot x \cdot y^4$

$= 3x^7y^4$

103. $(-6xy^5)(3x^2y^4)$

$= -6 \cdot 3 \cdot x \cdot x^2 \cdot y^5 \cdot y^4$

$= -18x^3y^9$

105. $(3x^4y^2)(4xy^6)$

$= 3 \cdot 4 \cdot x^4 \cdot y^2 \cdot y^6$

$= 12x^5y^8$

107. $(5xy)(2xy^6)$

$= 5 \cdot 2 \cdot x \cdot x \cdot y \cdot y^6$

$= 10x^2y^7$

109. $(2xy)^2(3xy^2) = (2^2x^2y^2)(3xy^2)$

$= 4 \cdot 3 \cdot x^2 \cdot x \cdot y^2 \cdot y^2$

$= 12x^3y^4$

111. $(x^4y^6)^3(3x^2y^5)$

$= (x^4)^3(y^6)^3(3x^2y^5)$

$= x^{12}y^{18}(3x^2y^5)$

$= 3 \cdot x^{12} \cdot x^2 \cdot y^{18} \cdot y^5$

$= 3x^{14}y^{23}$

113. $(2x^2y^5)(3x^5y^4)^3$

$= (2x^2y^5)(27x^{15}y^{12})$

$= 2 \cdot 27 \cdot x^2 \cdot x^{15}$

$\cdot y^5 \cdot y^{12}$

$= 54x^{17}y^{17}$

115. $(x^7y^5)(xy^2)^4$

$= (x^7y^5)(x^4y^8)$

$= x^7 \cdot x^4 \cdot y^5 \cdot y^8$

$= x^{11}y^{13}$

117. $(3x^4y^{10})^2(2x^2y^8)$

$= (9x^8y^{20})(2x^2y^8)$

$= 9 \cdot 2 \cdot x^8 \cdot x^2$

$\cdot y^{20} \cdot y^8$

$= 18x^{10}y^{28}$

119. $\frac{x + y}{x}$ cannot be simplified.

121. $\frac{x^2 + 2}{x}$ cannot be simplified.

123. $\frac{x + 4}{2}$ cannot be simplified.

125. $\frac{x^2y^2}{x^2} = \frac{\cancel{x^2} \cdot y^2}{\cancel{x^2}} = y^2$

127. $\frac{x}{x + 1}$ cannot be simplified.

129. $\frac{x^4}{x^2y} = \frac{x^2 \cdot \cancel{x^2}}{\cancel{x^2} \cdot y} = \frac{x^2}{y}$

131. $x^0 \neq 1$ when $x = 0$

133. The (sign of the simplified) expression is negative, because an expression to an odd power is negative.

135. The sign of the simplified expression is positive, because an expression to an even power is positive.

Cumulative Review Exercises

137. A linear equation is an equation of the form $ax + b = c$.

139. An identity is an equation that is true for any value of the variable.

141.

$$2x - 5y = 6$$
$$2x - 2x - 5y = 6 - 2x$$
$$-5y = 6 - 2x$$
$$\frac{-5y}{-5} = \frac{6 - 2x}{-5}$$
$$y = \frac{6 - 2x}{-5}$$

Just for fun

1. $$\left(\frac{3x^4y^5}{6x^6y^8}\right)^3\left(\frac{9x^7y^8}{3x^3y^5}\right)^2$$

$$= \left(\frac{3\cdot x^4\cdot y^5}{2\cdot 3\cdot x^2\cdot x^4\cdot y^5\cdot y^3}\right)^3 \cdot \left(\frac{3\cdot 3\cdot x^3\cdot x^4\cdot y^3\cdot y^5}{3\cdot x^3\cdot y^5}\right)^2$$

$$= \left(\frac{1}{2x^2y^3}\right)^3(3x^4y^3)^2$$

$$= \left(\frac{1}{8x^6y^9}\right)(9x^8y^6)$$

$$= \frac{9x^8y^6}{8x^6y^9}$$

$$= \frac{9\cdot x^2\cdot x^6\cdot y^6}{8\cdot x^6\cdot y^6\cdot y^3} = \frac{9x^2}{8y^3}$$

2. $$(2xy^4)^3\left(\frac{6x^2y^5}{3x^3y^4}\right)^3(3x^2y^4)^2$$

$$= (8x^3y^{12})\left(\frac{2\cdot 3\cdot x^2\cdot y\cdot y^4}{3\cdot x\cdot x^2\cdot y^4}\right)^3(9x^4y^8)$$

$$= (8x^3y^{12})\left(\frac{2y}{x}\right)^3(9x^4y^8)$$

$$= (8x^3y^{12})\left(\frac{8y^3}{x^3}\right)(9x^4y^8)$$

$$= \frac{8\cdot 8\cdot 9\cdot x^3\cdot x^4\cdot y^{12}\cdot y^3\cdot y^8}{x^3}$$

$$= \frac{576x^7y^{23}}{x^3}$$

$$= \frac{576\cdot x^3\cdot x^4\cdot y^{23}}{x^3}$$

$$= 576x^4y^{23}$$

Exercise Set 4.2

1. $x^{-2} = \frac{1}{x^2}$

3. $5^{-1} = \frac{1}{5}$

5. $\frac{1}{x^{-4}} = x^4$

7. $\frac{1}{x^{-1}} = x$

9. $\frac{1}{5^{-2}} = 5^2 = 25$

11. $(x^{-2})^3 = x^{(-2)(3)} = x^{-6}$
$= \frac{1}{x^6}$

13. $(y^{-7})^3 = y^{(-7)(3)}$
$= y^{-21} = \frac{1}{y^{21}}$

15. $(x^5)^{-2} = x^{(5)(-2)}$
$= x^{-10} = \frac{1}{x^{10}}$

17. $(2^{-3})^{-2} = 2^{(-3)(-2)} = 2^6 = 64$

19. $x^4 \cdot x^{-1} = x^{4+(-1)} = x^3$

21. $x^7 \cdot x^{-5} = x^{7+(-5)} = x^2$

23. $3^{-2} \cdot 3^4 = 3^{-2+4} = 3^2 = 9$

25. $\frac{x^9}{x^{12}} = x^{9-12} = x^{-3} = \frac{1}{x^3}$

27. $\frac{y^6}{y^{-3}} = y^{6-(-3)} = y^{6+3} = y^9$

29. $\frac{x^{-7}}{x^{-3}} = x^{-7-(-3)} = x^{-7+3}$
$= x^{-4} = \frac{1}{x^4}$

31. $\frac{3^2}{3^{-1}} = 3^{2-(-1)} = 3^{2+1} = 3^3 = 27$

33. $3^{-3} = \frac{1}{3^3} = \frac{1}{27}$

35. $\frac{1}{z^{-9}} = z^9$

37. $(x^5)^{-5} = x^{-25} = \frac{1}{x^{25}}$

39. $(y^{-2})^{-3} = y^{(-2)(-3)} = y^6$

41. $x^5 \cdot x^{-9} = x^{5+(-9)}$
$= x^{-4} = \frac{1}{x^4}$

43. $x^{-12} \cdot x^{-7} = x^{-12+(-7)} = x^{-19}$
$= \frac{1}{x^{19}}$

45. $\frac{x^{-3}}{x^5} = x^{-3-5} = x^{-8} = \frac{1}{x^8}$

47. $\frac{y^9}{y^{-1}} = y^{9-(-1)} = y^{9+1} = y^{10}$

49. $\frac{2^{-3}}{2^{-3}} = 2^{-3-(-3)} = 2^{-3+3} = 2^0 = 1$

51. $z^{-7} = \frac{1}{z^7}$

53. $\frac{1}{1^{-7}} = 1^7 = 1$

55. $(x^{-4})^{-1} = x^{(-4)(-1)} = x^4$

57. $(x^0)^{-3} = x^{(0)(-3)} = x^0 = 1$

59. $2^{-3} \cdot 2 = 2^{-3+(1)} = 2^{-2}$

$= \frac{1}{2^2} = \frac{1}{4}$

61. $6^{-4} \cdot 6^{2} = 6^{-4+2} = 6^{-2}$

$= \frac{1}{6^2} = \frac{1}{36}$

63. $\frac{x^{-1}}{x^{-4}} = x^{-1-(-4)} = x^{-1+4} = x^{3}$

65. $(3^2)^{-1} = 3^{(2)(-1)} = 3^{-2}$

$= \frac{1}{3^2} = \frac{1}{9}$

67. $\frac{5}{5^{-2}} = 5^{1-(-2)} = 5^{1+2} = 5^3 = 125$

69. $\frac{2^{-4}}{2^{-2}} = 2^{-4-(-2)} = 2^{-4+2} = 2^{-2}$

$= \frac{1}{2^2} = \frac{1}{4}$

71. $\frac{7^{-1}}{7^{-1}} = 7^{-1-(-1)} = 7^{-1+1} = 7^0 = 1$

73. $5x^{-1}y = 5 \cdot \frac{1}{x} \cdot y = \frac{5y}{x}$

75. $(3x^3)^{-1} = 3^{-1}x^{(3)(-1)} = 3^{-1}x^{-3}$

$= \frac{1}{3x^3}$

77. $5x^4y^{-1} = 5x^4 \cdot \frac{1}{y} = \frac{5x^4}{y}$

79. $(3x^2y^3)^{-2} = 3^{-2}x^{(2)(-2)}y^{(3)(-2)}$

$= 3^{-2}x^{-4}y^{-6} = \frac{1}{3^2x^4y^6}$

$= \frac{1}{9x^4y^6}$

81. $(x^5y^{-3})^{-3} = x^{(5)(-3)}y^{(-3)(-3)}$

$= x^{-15}y^{9} = \frac{1}{x^{15}} \cdot y^{9} = \frac{y^9}{x^{15}}$

83. $3x(5x^{-4}) = 3 \cdot 5 \cdot x \cdot x^{-4}$

$= 15x^{1+(-4)} = 15x^{-3}$

$= 15 \cdot \frac{1}{x^3}$

$= \frac{15}{x^3}$

85. $2x^5(3x^{-6}) = 2 \cdot 3 \cdot x^5 \cdot x^{-6}$

$= 6x^{5-6} = 6x^{-1} = 6 \cdot \frac{1}{x}$

$= \frac{6}{x}$

87. $(9x^5)(-3x^{-7}) = 9 \cdot -3 \cdot x^5 \cdot x^{-7}$

$= -27x^{5+(-7)} = -27x^{-2}$

$= -27 \cdot \frac{1}{x^2} = \frac{-27}{x^2}$

89. $(2x^{-3}y^{-2})(x^4y)$

$= 2 \cdot x^{-3} \cdot x^{4} \cdot y^{-2} \cdot y$

$= 2x^{-3+4}y^{-2+1} = 2x^1y^{-1}$

$= 2x \cdot \frac{1}{y} = \frac{2x}{y}$

91. $(3y^{-2})(5x^{-1}y^3) = 3 \cdot 5 \cdot x^{-1} \cdot y^{-2} \cdot y^3$

$= 15 \cdot x^{-1} \cdot y^{-2+3} = 15x^{-1}y^{1}$

$= 15 \cdot \frac{1}{x} \cdot y = \frac{15y}{x}$

93. $\frac{3x^5}{6x^{-2}} = \frac{3}{6} \cdot \frac{x^5}{x^{-2}} = \frac{1}{2} \cdot x^{5-(-2)}$

$= \frac{1}{2} \cdot x^{5+2} = \frac{1}{2} \cdot x^7 = \frac{x^7}{2}$

95. $$\frac{2y^{-6}}{6y^4} = \frac{2}{6}\cdot\frac{y^{-6}}{y^4} = \frac{1}{3}\cdot y^{-6-4}$$

$$= \frac{1}{3}\cdot y^{-10} = \frac{1}{3}\cdot\frac{1}{y^{10}} = \frac{1}{3y^{10}}$$

97. $$\frac{36x^{-4}}{9x^{-2}} = \frac{36}{9}\cdot\frac{x^{-4}}{x^{-2}} = 4x^{-4-(-2)}$$

$$= 4x^{-4+2} = 4x^{-2}$$

$$= 4 \cdot \frac{1}{x^2} = \frac{4}{x^2}$$

99. $$\frac{3x^4y^{-2}}{6y^3} = \frac{3}{6}\cdot x^4\cdot\frac{y^{-2}}{y^3}$$

$$= \frac{1}{2}\, x^4y^{-2-3}$$

$$= \frac{1}{2}\, x^4y^{-5}$$

$$= \frac{1}{2}\cdot x^4\cdot\frac{1}{y^5} = \frac{x^4}{2y^5}$$

101. $$\frac{32x^4y^{-2}}{4x^{-2}y^{-3}} = \frac{32}{4}\cdot\frac{x^4}{x^{-2}}\cdot\frac{y^{-2}}{y^{-3}}$$

$$= 8x^{4-(-2)}y^{-2-(-3)}$$

$$= 8x^{4+2}y^{-2+3} = 8x^6y$$

103. a) yes $a^{-1}b^{-1} = \frac{1}{a}\cdot\frac{1}{b} = \frac{1}{ab}$

b) no $a^{-1} + b^{-1} = \frac{1}{a} + \frac{1}{b}$

$$\neq \frac{1}{a + b}$$

105. An expression raised to a negative exponent may be rewritten as 1 divided by the expression to that positive exponent.

Cumulative Review Exercises

107. $$\frac{-3^2\cdot 4 \div 2}{\sqrt{9} - 2^2}$$

$$= \frac{-9\cdot 4 + 2}{3 - 4}$$

$$= \frac{-36 \div 2}{-1}$$

$$= \frac{-18}{-1} = 18$$

109. Let x = smaller integer.
3x + 1 = larger integer.

$$x + (3x + 1) = 37$$
$$4x + 1 = 37$$
$$4x = 36$$
$$x = 9$$ – smaller integer

$$3x + 1 = 28$$ – larger integer

Just for fun

1. a) $$\left(\frac{3x^2y^3}{z}\right)^{-2} = \frac{3^{-2}x^{-4}y^{-6}}{z^{-2}}$$

$$= \frac{1}{3^2}\cdot\frac{1}{x^4}\cdot\frac{1}{y^6}\cdot\frac{1}{z^{-2}}$$

$$= \frac{z^2}{9x^4y^6}$$

b) $$\left(\frac{3x^2y^3}{z}\right)^{-2} = \frac{1}{\left(\frac{3x^2y3}{z}\right)^2}$$

$$= \left(\frac{z}{3x^2y^3}\right)^2 = \frac{z^2}{9x^4y^6}$$

Exercise Set 4.3

1. $42,000 = 4.2 \times 10^4$

3. $900 = 9.0 \times 10^2$

5. $0.053 = 5.3 \times 10^{-2}$

7. $19,000 = 1.9 \times 10^4$

9. $0.00000186 = 1.86 \times 10^{-6}$

11. $0.00000914 = 9.14 \times 10^{-6}$

13. $107 = 1.07 \times 10^2$

15. $0.153 = 1.53 \times 10^{-1}$

17. $4.2 \times 10^3 = 4,200$

19. $4 \times 10^7 = 40,000,000$

21. $2.13 \times 10^{-5} = 0.0000213$

23. $3.12 \times 10^{-1} = 0.312$

25. $9 \times 10^6 = 9,000,000$

27. $5.35 \times 10^2 = 535$

29. $3.5 \times 10^4 = 35,000$

31. $1 \times 10^4 = 10,000$

33. $(4 \times 10^2)(3 \times 10^5)$

$= (4 \times 3)(10^2 \times 10^5)$

$= 12 \times 10^7$
$= 120,000,000$

35. $(5.1 \times 10^1)(3 \times 10^{-4})$

$= (5.1 \times 3)(10^{1-4})$

$= 15.3 \times 10^{-3}$

$= 0.0153$

37. $\dfrac{6.4 \times 10^5}{2 \times 10^3} = (\dfrac{6.4}{2})(\dfrac{10^5}{10^3})$

$= 3.2 \times 10^2$
$= 320$

39. $\dfrac{8.4 \times 10^{-6}}{4 \times 10^{-3}} = (\dfrac{8.4}{4})(\dfrac{10^{-6}}{10^{-3}})$

$= 2.1 \times 10^{-3}$
$= 0.0021$

41. $\dfrac{4 \times 10^5}{2 \times 10^4} = (\dfrac{4}{2})(\dfrac{10^5}{10^4})$

$= 2 \cdot 10 = 20$

43. $(700,000)(6,000,000)$

$= (7.0 \times 10^5)(6.0 \times 10^6)$

$= (7.0 \times 6.0)(10^5 \times 10^6)$

$= 42 \times 10^{11}$

$= 4.2 \times 10^{12}$

45. $(0.003)(0.00015)$

$= (3.0 \times 10^{-3})(1.5 \times 10^{-4})$

$= (3.0 \times 1.5)(10^{-3} \times 10^{-4})$

$= 4.5 \times 10^{-7}$

47. $\dfrac{1,400,000}{700} = \dfrac{1.4 \times 10^6}{7.0 \times 10^2}$

$= (\dfrac{1.4}{7.0})(\dfrac{10^6}{10^2})$

$= 0.2 \times 10^4$

$= 2.0 \times 10^3$

49. $\dfrac{0.00004}{200} = \dfrac{4.0 \times 10^{-5}}{2.0 \times 10^{2}}$

$= 2.0 \times 10^{-7}$

51. $\dfrac{150{,}000}{0.0005} = \dfrac{1.5 \times 10^{5}}{5.0 \times 10^{-4}}$

$= 0.3 \times 10^{9}$

$= 3.0 \times 10^{8}$

53. 9.2×10^{-5}, 1.3×10^{-1}, 8.4×10^{3}, 6.2×10^{4}

55. $(0.000004)(8{,}000{,}000{,}000{,}000)$

$= (4.0 \times 10^{-6})(8.0 \times 10^{12})$

$= 32.0 \times 10^{6}$

$= 3.2 \times 10^{7}$ seconds

57. $(100{,}000 \text{ cu. ft.})(60)(60)(24)$

$= 1.0 \times 10^{5}(86{,}400)$

$= 1.0 \times 10^{5}(8.64 \times 10^{4})$

$= 8.64 \times 10^{9}$
$= 8{,}640{,}000{,}000$ cu. ft.

59. $\dfrac{(4 \cdot 10^{3})(6 \cdot 10^{x})}{24 \cdot 10^{-5}} = 1$

$\dfrac{4 \cdot 6 \cdot 10^{3} \cdot 10^{x}}{24 \cdot 10^{-5}} = 1$

$\dfrac{24 \cdot 10^{3+x}}{24 \cdot 10^{-5}} = 1$

$\dfrac{10^{3+x}}{10^{-5}} = 1$

Since $10^{0} = 1$,
$(3 + x) - (-5) = 0$

$3 + x + 5 = 0$
$x + 8 = 0$
$x = -8$

Cumulative Review Exercises

61. $4x^{2} + 3x + \frac{x}{2}$, $x = 0$

$4(0)^{2} + 3(0) + \frac{0}{2}$

$= 0 + 0 + 0$
$= 0$

63. $2x - 3(x - 2) = x + 2$
$2x - 3x + 6 = x + 2$
$-x + 6 = x + 2$
$-x - x + 6 = x - x + 2$
$-2x + 6 = 2$
$-2x + 6 - 6 = 2 - 6$
$-2x = -4$
$\dfrac{-2x}{-2} = \dfrac{-4}{-2}$
$x = 2$

Just for fun

1. a) $\dfrac{1.86 \times 10^{5} \text{ mi}}{\cancel{\text{sec}}} \times \dfrac{3600 \cancel{\text{sec}}}{\cancel{\text{hr}}} \times \dfrac{365 \cancel{\text{days}}}{\text{yr}} \times \dfrac{24 \cancel{\text{hr}}}{\cancel{\text{day}}}$

$= \dfrac{1.86 \times 10^{5} \text{ mi}}{\cancel{\text{sec}}} \times \dfrac{3600 \cancel{\text{sec}}}{\cancel{\text{hr}}} \times \dfrac{24 \cancel{\text{hr}}}{\cancel{\text{day}}} \times \dfrac{365 \cancel{\text{days}}}{\text{yr}}$

$= 5.8656960 \times 10^{12}$ miles/year

$= 5.87 \times 10^{12}$ miles/year

b) $\dfrac{9.3 \times 10^7 \text{ miles}}{1.86 \times 10^5 \text{ mi/sec}}$

$= 5.0 \times 10^2$ sec.
$= 500$ sec.

or $8\frac{1}{3}$ min.

Exercise Set 4.4

1. Monomial

3. Monomial

5. Binomial

7. Trinomial

9. Not polynomial

11. Binomial

13. Monomial

15. Polynomial

17. Trinomial

19. Not polynomial

21. In descending order first degree

23. $2x^2 + x - 6$, second degree

25. $-x^2 - 4x - 8$, second degree

27. In descending order, third degree

29. In descending order, second degree

31. $-6x^3 + x^2 - 3x + 4$, third degree

33. $5x^2 - 2x - 4$, second degree

35. $-2x^3 + 3x^2 + 5x - 6$, third degree

37. $(2x + 3) + (4x - 2)$
$= 2x + 3 + 4x - 2$
$= 2x + 4x + 3 - 2$
$= 6x + 1$

39. $(-4x + 8) + (2x + 3)$
$= -4x + 8 + 2x + 3$
$= -4x + 2x + 8 + 3$
$= -2x + 11$

41. $(5x + 8) + (-6x - 10)$
$= 5x + 8 - 6x - 10$
$= 5x - 6x + 8 - 10$
$= -x - 2$

43. $(9x - 12) + (12x - 9)$
$= 9x - 12 + 12x - 9$
$= 9x + 12x - 12 - 9$
$= 21x - 21$

45. $(x^2 + 2x - 3) + (4x + 3.8)$

$= x^2 + 2x - 3 + 4x + 3.8$

$= x^2 + 2x + 4x - 3 + 3.8$

$= x^2 + 6x + 0.8$

47. $(5x - 7) + (2x^2 + 3x + 12)$

$= 5x - 7 + 2x^2 + 3x + 12$

$= 2x^2 + 5x + 3x - 7 + 12$

$= 2x^2 + 8x + 5$

49. $(3x^2 - 4x + 8)$

$+ (2x^2 + 5x + 12)$

$= 3x^2 - 4x + 8 + 2x^2$
$+ 5x + 12$

$= 3x^2 + 2x^2 - 4x + 5x$
$+ 8 + 12$

$= 5x^2 + x + 20$

51. $(-3x^2 - 4x + 8)$

$+ (5x - 2x^2 + \frac{1}{2})$

$= -3x^2 - 4x + 8 + 5x$

$- 2x^2 + \frac{1}{2}$

$= -3x^2 - 2x^2 - 4x + 5x$

$+ 8 + \frac{1}{2}$

$= 5x^2 + x + \frac{17}{2}$

53. $(8x^2 + 4) + (-2.6x^2 - 5x)$

$= 8x^2 + 4 - 2.6x^2 - 5x$

$= 8x^2 - 2.6x^2 - 5x + 4$

$= 5.4x^2 - 5x + 4$

55. $(-7x^3 - 3x^2 + 4)$

$+ (5x^3 + 4x - 7)$

$= -7x^3 - 3x^2 + 4 + 5x^3$
$+ 4x - 7$

$= -7x^3 + 5x^3 - 3x^2$
$+ 4x + 4 - 7$

$= -2x^3 - 3x^2 + 4x - 3$

57. $(x^2 + xy - y^2)$

$+ (2x^2 - 3xy + y^2)$

$= x^2 + xy - y^2 + 2x^2$

$- 3xy + y^2$

$= x^2 + 2x^2 + xy - 3xy$

$- y^2 + y^2$

$= 3x^2 - 2xy$

59. $(4x^2y + 2x - 3)$

$+ (3x^2y - 5x + 5)$

$= 4x^2y + 2x - 3 + 3x^2y$
$- 5x + 5$

$= 4x^2y + 3x^2y + 2x$

$- 5x - 3 + 5$

$= 7x^2y - 3x + 2$

61.
$$\begin{array}{l} 3x - 6 \\ 4x + 5 \\ \hline 7x - 1 \end{array}$$

63.
$$\begin{array}{r} x^2 - 2x + 4 \\ 3x + 12 \\ \hline x^2 + x + 16 \end{array}$$

65.
$$\begin{array}{l} -2x^2 + 4x - 12 \\ \;\; -x^2 - 2x \\ \hline -3x^2 + 2x - 12 \end{array}$$

67.
$$\begin{array}{l} 3x^2 + 4x - 5 \\ 4x^2 + 3x - 8 \\ \hline 7x^2 + 7x - 13 \end{array}$$

69. $(2x^2 + 3x^2 + 6x - 9)$

$+ (7 - 4x^2)$

$= 2x^3 + 3x^2 + 6x - 9$

$+ 7 - 4x^2$

$= 2x^3 + 3x^2 - 4x^2$
$+ 6x - 9 + 7$

$= 2x^3 - x^2 + 6x - 2$

71.
$$\begin{array}{l} 6x^3 - 4x^2 + x - 9 \\ -x^3 - 3x^2 - x + 7 \\ \hline 5x^3 - 7x^2 \qquad - 2 \end{array}$$

73.
$$\begin{array}{l} \;\; xy + 6x + 4 \\ 2xy - 3x - 1 \\ \hline 3xy + 3x + 3 \end{array}$$

75. $(3x - 4) - (2x + 2)$
$= 3x - 4 - 2x - 2$
$= 3x - 2x - 4 - 2$
$= x - 6$

77. $(-2x - 3) - (-5x - 7)$
$= -2x - 3 + 5x + 7$
$= -2x + 5x - 3 + 7$
$= 3x + 4$

79. $(-x + 4) - (-x + 9)$
$= -x + 4 + x - 9$
$= -x + x + 4 - 9$
$= -5$

81. $(6 - 12x) - (3 - 5x)$
$= 6 - 12x - 3 + 5x$
$= -12x + 5x + 6 - 3$
$= -7x + 3$

83. $(9x^2 + 7x - 5) - (3x^2 + 3.5)$

$= 9x^2 + 7x - 5 - 3x^2 - 3.5$

$= 9x^2 - 3x^2 + 7x - 5 - 3.5$

$= 6x^2 + 7x - 8.5$

85. $(5x^2 - x - 1)$

$- (-3x^2 - 2x - 5)$

$= 5x^2 - x - 1 + 3x^2$
$+ 2x + 5$

$= 5x^2 + 3x^2 - x + 2x$
$- 1 + 5$

$= 8x^2 + x + 4$

87. $(5x^2 - x + 12) - (5 + x)$

$= 5x^2 - x + 12 - 5 - x$

$= 5x^2 - x - x + 12 - 5$

$= 5x^2 - 2x + 7$

89. $(9x - 6) - (-2x^2 + 4x - 8)$

$= 9x - 6 + 2x^2 - 4x + 8$

$= 2x^2 + 9x - 4x - 6 + 8$

$= 2x^2 + 5x + 2$

91. $(4x^3 - 6x^2 + 5x - 7)$

$- (6x + \frac{2}{3}x^2 - 3)$

$= 4x^3 - 6x^2 + 5x - 7$

$- 6x - \frac{2}{3}x^2 + 3$

$= 4x^3 - 6x^2 - \frac{2}{3}x^2$

$+ 5x - 6x - 7 + 3$

$= 4x^3 - \frac{20}{3}x^2 - x - 4$

93. $(9x^3 - \frac{1}{5} - (x^2 + 5x)$

$= 9x^3 - \frac{1}{5} - x^2 - 5x$

$= 9x^3 - x^2 - 5x - \frac{1}{5}$

95. $(3x + 5) - (4x - 6)$
$= 3x + 5 - 4x + 6$
$= 3x - 4x + 5 + 6$
$= -x + 11$

97. $(2x^2 - 4x + 8) - (5x - 6)$

$= 2x^2 - 4x + 8 - 5x + 6$

$= 2x^2 - 4x - 5x + 8 + 6$

$= 2x^2 - 9x + 14$

99. $(3x^3 + 5x^2 + 9x - 7)$

$- (4x^3 - 6x^2)$

$= 3x^3 + 5x^2 + 9x - 7$

$- 4x^3 + 6x^2$

$= 3x^3 - 4x^3 + 5x^2 + 6x^2$
$+ 9x - 7$

$= -x^3 + 11x^2 + 9x - 7$

101. $\begin{array}{r} 5x + 10 \\ -2x + 7 \\ \hline 3x + 17 \end{array}$

103. $\begin{array}{r} -5x + 3 \\ 9x + 4 \\ \hline 4x + 7 \end{array}$

105. $\begin{array}{r} 9x^2 + 7x - 9 \\ -4x^2 \quad\quad + 7 \\ \hline 5x^2 + 7x - 2 \end{array}$

107. $\begin{array}{r} x - 6 \\ 4x^2 - 6x \quad\quad \\ \hline 4x^2 - 5x - 6 \end{array}$

109. $\begin{array}{r} 4x^3 - 6x^2 + 7x - 9 \\ -x^2 - 6x + 7 \\ \hline 4x^3 - 7x^2 + x - 2 \end{array}$

111. A polynomial is an expression that contains the sum of a finite number of terms of the form ax^n, where a is a real number and n is a whole number.

113. a) The degree of a term in one variable is the exponent on the variable.

b) The degree of a polynomial in one variable is the same as the degree of the highest powered term in the polynomial.

115. To write a poynomial in descending order of the variable, write the polynomial with exponents on the variable decreasing from left to right.

117. Answers will vary.

119. Answers will vary.

Cumulative Review Exercises

121. False

123. False

125. $\left(\frac{3x^4y^5}{6x^7y^4}\right)^3 = \left(\frac{y}{2x^3}\right)^3$

$= \frac{y^3}{8x^9}$

Just for fun

1. $(3x^2 - 6x + 3)$

$- (2x^2 - x - 6)$

$- (x^2 + 7x - 9)$

$= 3x^2 - 6x + 3 - 2x^2 + x$

$+ 6 - x^2 - 7x + 9$

$= 3x^2 - 2x^2 - x^2 - 6x + x$
$- 7x + 3 + 6 + 9$

$= -12x + 18$

2. $3x^2y - 6xy - 2xy + 9xy^2$
$- 5xy + 3x$

$= 3x^2y - 6xy - 2xy - 5xy$

$+ 9xy^2 + 3x$

$= 3x^2y - 13xy + 9xy^2 + 3x$

3. $4(x^2 + 2x - 3)$

$- 6(2 - 4x - x^2)$
$- 2x(x + 2)$

$= 4x^2 + 8x - 12 - 12 + 24x$

$+ 6x^2 - 2x^2 - 4x$

$= 4x^2 + 6x^2 - 2x^2 + 8x + 24x - 4x - 12 - 12$

$= 8x^2 + 28x - 24$

Exercise Set 4.5

1. $x^2 \cdot 3xy$

$= 3 \cdot x^2 \cdot x \cdot y$

$= 3x^{2+1}y = 3x^3y$

3. $5x^4y^5(6xy^2)$

$= 5 \cdot 6 \cdot x^4 \cdot x \cdot y^5 \cdot y^2$

$= 30x^{4+1}y^{5+2} = 30x^5y^7$

5. $4x^4y^6(-7x^2y^9)$

$= 4 \cdot -7 \cdot x^4 \cdot x^2 \cdot y^6 \cdot y^9$

$= -28x^{4+2}y^{6+9}$

$= -28x^6y^{15}$

7. $9xy^6 \cdot 6x^5y^8$

$= 9 \cdot 6 \cdot x \cdot x^5 \cdot y^6 \cdot y^8$

$= 54x^{1+5}y^{6+8}$

$= 54x^6y^{14}$

9. $(6x^2y)(\frac{1}{2}x^4)$

$= 6 \cdot \frac{1}{2} \cdot x^2 \cdot x^4 \cdot y$

$= 3x^{2+4}y = 3x^6y$

11. $3(x + 4) = (3)(x) + (3)(4)$
$= 3x + 12$

13. $2x(x - 3)$
$= (2x)(x) + 2x(-3)$

$= 2x^2 - 6x$

15. $-4x(-2x + 6)$
$= (-4x)(-2x) + (-4x)(6)$

$= 8x^2 - 24x$

17. $2x(x^2 + 3x - 1)$

$= (2x)(x^2) + (2x)(3x) + (2x)(-1)$

$= 2x^3 + 6x^2 - 2x$

19. $-2x(x^2 - 2x + 5)$

$= (-2x)(x^2) + (-2x)(-2x) + (-2x)(5)$

$= -2x^3 + 4x^2 - 10x$

21. $5x(-4x^2 + 6x - 4)$

$= (5x)(-4x^2) + (5x)(6x)$
$= (5x)(-4)$

$= -20x^3 + 30x^2 - 20x$

23. $(3x^2 + 4x - 5)8x$

$= (3x^2)(8x) + (4x)(8x) + (-5)(8x)$

$= 24x^3 + 32x^2 - 40x$

25. $0.3x(2xy + 5x - 6y)$

$= (0.3x)(2xy) + (0.3x)(5x) + (0.3x)(-6y)$

$= 0.6x^2y + 1.5x^2 - 1.8xy$

27. $(x - y - 3)y$
$= (x)(y) + (-y)(y) + (-3)(y)$

$= xy - y^2 - 3y$

29. $(x + 3)(x + 4)$
$= (x)(x) + (x)(4) + (3)(x) + (3)(4)$

$= x^2 + 4x + 3x + 12$

$= x^2 + 7x + 12$

31. $(2x + 5)(3x - 6)$
$= (2x)(3x) + (2x)(-6) + (5)(3x) + (5)(-6)$

$= 6x^2 - 12x + 15x - 30$

$= 6x^2 + 3x - 30$

33. $(2x - 4)(2x + 4)$
$= (2x)(2x) + (2x)(4) + (-4)(2x) + (-4)(4)$

$= 4x^2 + 8x - 8x - 16$

$= 4x^2 - 16$

35. $(5 - 3x)(6 + 2x)$
$= (5)(6) + 5(2x) + (-3x)(6) + (-3x)(2x)$

$= -6x^2 - 18x + 10x + 30$

$= -6x^2 - 8x + 30$

37. $(-x + 3)(2x + 5)$
$= (-x)(2x) + (-x)(5) + (3)(2x) + (3)(5)$

$= -2x^2 - 5x + 6x + 15$

$= -2x^2 + x + 15$

39. $(x + 4)(x + 3)$
$= (x)(x) + (x)(3) + (4)(x) + (4)(3)$

$= x^2 + 3x + 4x + 12$

$= x^2 + 7x + 12$

41. $(x + 4)(x - 2)$
$= (x)(x) + (x)(-2) + (4)(x) + (4)(-2)$

$= x^2 - 2x + 4x - 8$

$= x^2 + 2x - 8$

43. $(3x + 4)(2x + 5)$
$= (3x)(2x) + (3x)(5) + (4)(2x) + (4)(5)$

$= 6x^2 + 15x + 8x + 20$

$= 6x^2 + 23x + 20$

45. $(3x + 4)(2x - 3)$
$= (3x)(2x) + (3x)(-3) + (4)(2x) + (4)(-3)$

$= 6x^2 - 9x + 8x - 12$

$= 6x^2 - x - 12$

47. $(x - 1)(x + 1)$
$= (x)(x) + (x)(1) + (-1)(x) + (-1)(1)$

$= x^2 + x - x - 1$

$= x^2 - 1$

49. $(2x - 3)(2x - 3)$
$= (2x)(2x) + (2x)(-3) + (-3)(2x) + (-3)(-3)$

$= 4x^2 - 6x - 6x + 9$

$= 4x^2 - 12x + 9$

51. $(4 - x)(3 + 2x)$
$= (4)(3) + (4)(2x) + (-x)(3) + (-x)(2x)$

$= 12 + 8x - 3x - 2x^2$

$= -2x^2 + 5x + 12$

53. $(2x + 3)(4 - 2x)$
$= (2x)(4) + (2x)(-2x) + (3)(4) + (3)(-2x)$

$= 8x - 4x^2 + 12 - 6x$

$= -4x^2 + 8x - 6x + 12$

$= -4x^2 + 2x + 12$

55. $(x + y)(x - y)$
$= (x)(x) + (x)(-y)$

$+ (y)(x) + (y)(-y)$

$= x^2 - xy + xy - y^2$

$= x^2 - y^2$

57. $(2x - 3y)(3x + 2y)$
$= (2x)(3x) + (2x)(2y)$
$+ (-3y)(3x) + (-3y)(2y)$
$= 6x^2 + 4xy - 9xy - 6y^2$

$= 6x^2 - 5xy - 6y^2$

59. $(4x - 3y)(2y - 3)$
$= (4x)(2y) + (4x)(-3)$
$+ (-3y)(2y) + (-3y)(-3)$

$= 8xy - 12x - 6y^2 + 9y$

61. $(x + 0.6)(x + 0.3)$
$= (x)(x) + (x)(0.3)$
$+ (0.6)(x) + (0.6)(0.3)$

$= x^2 + 0.3x + 0.6x + 0.18$

$= x^2 + 0.9x + 0.18$

63. $(2x + 4)(x + \frac{1}{2})$

$= (2x)(x) + (2x)(\frac{1}{2})$

$+ (4)(x) + (4)(\frac{1}{2})$

$= 2x^2 + x + 4x + 2$

$= 2x^2 + 5x + 2$

65. $(x + 4)(x - 4)$

$= (x)^2 - (4)^2$

$= x^2 - 16$

67. $(2x - 1)(2x + 1)$

$= (2x)^2 - (1)^2$

$= 4x^2 - 1$

69. $(x + y)^2$

$= (x)^2 + 2(x)(y) + (y)^2$

$= x^2 + 2xy + y^2$

71. $(x - 0.2)^2$

$= x^2 + 2(-0.2x) + (-0.2)^2$

$= x^2 - 0.4x + 0.04$

73. $(3x + 5)(3x - 5)$

$= (3x)^2 - (5)^2$

$= 9x^2 - 25$

75.

$$\begin{array}{r} 2x^2 + 4x - 1 \\ x + 3 \\ \hline 6x^2 + 12x - 3 \\ 2x^3 + 4x^2 - x \\ \hline 2x^3 + 10x^2 + 11x - 3 \end{array}$$

77.

$$\begin{array}{r} x^2 - x + 4 \\ 5x + 4 \\ \hline 4x^2 - 4x + 16 \\ 5x^3 - 5x^2 + 20x \\ \hline 5x^3 - x^2 + 16x + 16 \end{array}$$

79.

$$\begin{array}{r} -2x^2 - 4x + 1 \\ 7x - 3 \\ \hline 6x^2 + 12x - 3 \\ -14x^3 - 28x^2 + 7x \\ \hline -14x^3 - 22x^2 + 19x - 3 \end{array}$$

81.

$$\begin{array}{r} -6x^2 + 5x - 3 \\ - 3x + 9 \\ \hline - 54x^2 + 45x - 27 \\ 18x^3 - 15x^2 + 9x \\ \hline 18x^3 - 69x^2 + 54x - 27 \end{array}$$

83.

$$\begin{array}{r} 3x^2 - 2x + 4 \\ 2x^2 + 3x + 1 \\ \hline 3x^2 - 2x + 4 \\ 9x^3 - 6x^2 + 12x \\ 6x^4 - 4x^3 + 8x^2 \\ \hline 6x^4 + 5x^3 + 5x^2 + 10x + 4 \end{array}$$

85.
$$\begin{array}{r} x^2 - x + 3 \\ x^2 - 2x \\ \hline -2x^3 + 2x^2 - 6x \\ x^4 - x^3 + 3x^2 \qquad \\ \hline x^4 - 3x^3 + 5x^2 - 6x \end{array}$$

87.
$$\begin{array}{r} 3x^3 + 2x^2 - x \\ x-3 \\ \hline -9x^3 - 6x^2 + 3x \\ 3x^4 + 2x^3 - x^2 \qquad \\ \hline 3x^4 - 7x^3 - 7x^2 + 3x \end{array}$$

89.
$$\begin{array}{r} a^2 - ab + b^2 \\ a + b \\ \hline a^2b - ab^2 + b3 \\ a^3 - a^2b + ab^2 \qquad \\ \hline a^3 \qquad\qquad + b^3 \end{array}$$

91. The product of a monomial and a monomial will always be a monomial. Look at Examples 1-5.

93. The product of two binomials will not always be a trinomial after like terms are combined. The product of the sum and difference of the same two terms will be a binomial.

Cumulative Review Exercises

95. Let x = time in minutes.
$$150x - 120x = 600$$
$$30x = 600$$
$$x = 20 \text{ min.}$$

97. a) $-6^3 = -216$

b) $6^{-3} = \dfrac{1}{6^3} = \dfrac{1}{216}$

Just for fun

1. $\sqrt{5}x(2x^2 + \sqrt{5}x - \frac{1}{2})$

$= \sqrt{5}x(2x^2) + \sqrt{5}x(\sqrt{5})$
$+ \sqrt{5}(-\frac{1}{2})$

$= 2\sqrt{5}x^3 + 5x^2 - \dfrac{\sqrt{5}x}{2}$

2. $(\frac{x}{2} + \frac{2}{3})(\frac{2x}{3} - \frac{2}{5})$

$= (\frac{x}{2})(\frac{2x}{3}) + (\frac{x}{2})(-\frac{2}{5})$

$+ (\frac{2}{3})(\frac{2x}{3}) + (\frac{2}{3})(-\frac{2}{5})$

$= \dfrac{2x^2}{6} - \dfrac{2x}{10} + \dfrac{4x}{9} - \dfrac{4}{15}$

$= \dfrac{x^2}{3} + \dfrac{-9x + 20x}{45} - \dfrac{4}{15}$

$= \dfrac{x^2}{3} + \dfrac{11x}{45} - \dfrac{4}{15}$

3.
$$\begin{array}{r} 2x^3 - 6x^2 + 5x - 3 \\ 3x^3 \qquad - 6x + 4 \\ \hline 8x^3 - 24x^2 + 20x - 12 \\ -12x^4 + 36x^3 - 30x^2 + 18x \qquad \\ 6x^6 - 18x^5 + 15x^4 - 9x^3 \qquad\qquad\qquad \\ \hline 6x^6 - 18x^5 + 3x^4 + 35x^3 - 54x^2 + 38x - 12 \end{array}$$

Exercise Set 4.6

1. $\dfrac{2x + 4}{2} = \dfrac{2x}{2} + \dfrac{4}{2}$

$= x + 2$

3. $\dfrac{2x + 6}{2} = \dfrac{2x}{2} + \dfrac{6}{2}$

$= x + 3$

5. $\dfrac{3x + 8}{2} = \dfrac{3x}{2} + \dfrac{8}{2}$

$= \dfrac{3x}{2} + 4$

7. $\dfrac{-6x + 4}{2} = \dfrac{-6x}{2} + \dfrac{4}{2}$

$= -3x + 2$

9. $\dfrac{-9x - 3}{-3} = \dfrac{-9x}{-3} + \dfrac{-3}{-3}$

$= 3x + 1$

11. $\dfrac{3x + 6}{x} = \dfrac{3x}{x} + \dfrac{6}{x}$

$= 3 + \dfrac{6}{x}$

13. $\dfrac{9 - 3x}{-3x} = \dfrac{9}{-3x} + \dfrac{-3x}{-3x}$

$= - \dfrac{3}{x} + 1$

15. $\dfrac{3x^2 + 6x - 9}{3}$

$= \dfrac{3x^2}{3} + \dfrac{6x}{3} + \dfrac{-9}{3}$

$= x^2 + 2x - 3$

17. $\dfrac{-4x^2 + 6x + 8}{2}$

$= \dfrac{-4x^2}{2} + \dfrac{6x}{2} + \dfrac{8}{2}$

$= -2x^2 + 3x + 4$

19. $\dfrac{x^2 + 4x - 3}{x}$

$= \dfrac{x^2}{x} + \dfrac{4x}{x} + \dfrac{-3}{x}$

$= x + 4 - \dfrac{3}{x}$

21. $\dfrac{6x^2 - 4x + 12}{2x}$

$= \dfrac{6x^2}{2x} + \dfrac{-4x}{2x} + \dfrac{12}{2x}$

$= 3x - 2 + \dfrac{6}{x}$

23. $\dfrac{4x^3 + 6x^2 - 8}{-4x}$

$= \dfrac{4x^3}{-4x} + \dfrac{6x^2}{-4x} + \dfrac{-8}{-4x}$

$= -x^2 - \dfrac{3x}{2} + \dfrac{2}{x}$

25. $\dfrac{9x^3 + 3x^2 - 12}{3x^2}$

$= \dfrac{9x^3}{3x^2} + \dfrac{3x^2}{3x^2} + \dfrac{-12}{3x^2}$

$= 3x + 1 - \dfrac{4}{x^2}$

27. $\dfrac{x^2 + 4x + 3}{x + 1}$

```
         x + 3
x+1/ x²+4x+3
     -x²∓ x
     -------
        3x+3
       -3x∓3
       -----
         -0-
```

29. $\dfrac{2x^2 + 13x + 15}{x + 5}$

$$\begin{array}{r} 2x+3 \\ \hline x+5\,/\; 2x^2+13x+15 \\ -2x^2\mp 10x \\ \hline 3x+15 \\ -3x\mp 15 \\ \hline -0- \end{array}$$

31\. $\dfrac{6x^2+16x+8}{3x+2}$

$$\begin{array}{r} 2x+4 \\ \hline 3x+2\,/\; 6x^2+16x+8 \\ -6x^2\mp 4x \\ \hline 12x+8 \\ -12x\mp 8 \\ \hline -0- \end{array}$$

33\. $\dfrac{2x^2+x-10}{2x+5}$

$$\begin{array}{r} x-2 \\ \hline 2x+5\,/\; 2x^2+x-10 \\ -2x^2\mp 5x \\ \hline -4x-10 \\ \pm 4x\mp 10 \\ \hline -0- \end{array}$$

35\. $\dfrac{2x^2+7x-18}{2x-3}$

$$\begin{array}{r} x+5-3/2x-3 \\ \hline 2x-3\,/\; 2x^2+7x-18 \\ -2x^2\pm 3x \\ \hline 10x-18 \\ -10x\pm 15 \\ \hline -3 \end{array}$$

37\. $\dfrac{4x^2-9}{2x-3}$

$$\begin{array}{r} 2x+3 \\ \hline 2x-3\,/\; 4x^2\quad -9 \\ -4x^2\pm 6x \\ \hline 6x-9 \\ -6x\pm 9 \\ \hline -0- \end{array}$$

39\. $\dfrac{6x+8x^2-25}{4x+9}$

$$\begin{array}{r} 2x-3+2/4x+9 \\ \hline 4x+9\,/\; 8x^2+6x-25 \\ -8x^2\mp 18x \\ \hline -12x-25 \\ -12x\pm 27 \\ \hline 2 \end{array}$$

41\. $\dfrac{8x^2+6x-12}{2x+3}$

$$\begin{array}{r} 4x-3-3/2x+3 \\ \hline 2x+3\,/\; 8x^2+6x-12 \\ -8x^2\mp 12x \\ \hline -6x-12 \\ \pm 6x\pm 9 \\ \hline -3 \end{array}$$

43\. $\dfrac{x^3+3x^2+5x+3}{x+1}$

$$\begin{array}{r} x^2+2x+3 \\ \hline x+1\,/\; x^3+3x^2+5x+3 \\ -x^3\mp x^2 \\ \hline 2x^2+5x \\ -2x^2\mp 2x \\ \hline 3x+3 \\ -3x\mp 3 \\ \hline -0- \end{array}$$

45\. $\dfrac{2x^3-3x^2-3x+6}{x-1}$

$2x^2-x-4+2/x-1$

$x-1/\ 2x^3-3x^2-3x+6$

$-2x^3\pm 2x^2$

$-\ x^2-3x$

$\pm\ x^2-\ x$

$-4x+6$

$-4x\mp 4$

2

47. $\dfrac{2x^3 + 6x - 4}{x + 4}$

$2x^2-8x+38-156/x+4$

$x+4/\ 2x^3\quad +6x-\ 4$

$-2x^3\mp 8x^2$

$-8x^2+\ 6x$

$\pm 8x^2\pm 32x$

$38x-\ 4$

$-38x\mp 152$

-156

49. $\dfrac{x^3 + 8}{x + 2}$

$x^2 - 2x + 4$

$x+2/\ x^3\quad +8$

$-x^3\mp 2x^2$

$-2x^2$

$-2x^2\pm 4x$

$4x+8$

$-4x\mp 8$

$-0-$

51. $\dfrac{x^3 + 27}{x + 3}$

$x^2 - 3x + 9$

$x+3/\ x^3\quad +27$

$-x^3\mp 3x^2$

$-3x^2$

$\pm 3x^2\pm 9x$

$9x+27$

$-9x\mp 27$

$-0-$

53. $\dfrac{9x^3 - x + 3}{3x - 2}$

$3x^2+2x+1+5/3x-2$

$3x-2/\ 9x^3\quad - x+3$

$-9x^3\pm 6x^2$

$6x^2-\ x$

$-6x^2\pm 4x$

$3x+3$

$-3x\pm 2$

5

55. $\dfrac{15x^{\square} + 25x^{\square} + 5x^{\square} + 10x^{\square}}{5x^2}$

$= 3x^5 + 5x^4 + x^2 + 2$

$\dfrac{15x^{\square}}{5x^2} = 3x^5 \qquad \dfrac{25x^{\square}}{5x^2} = 5x^4$

$3x^{\square-2} = 3x^5 \qquad 5x^{\square-2} = 5x^4$

$\square - 2 = 5 \qquad \square - 2 = 4$

$\square = 7 \qquad \square = 6$

$\dfrac{5x^{\square}}{5x^2} = x^2 \qquad \dfrac{10x^{\square}}{5x^2} = 2$

$x^{\square-2} = x^2 \qquad 2x^{\square-2} = 2$

$\square - 2 = 2 \qquad x^{\square-2} = 1$

$\square = 4$

$x^{\square-2} = x^{\square}$

$\square - 2 = 0$

$\square = 2$

Cumulative Review Exercises

57. a) $\frac{0}{1} = 0$

b) $\frac{1}{0}$ is undefined.

59.
$$\begin{aligned} 2(x+3)+2x &= x+4 \\ 2x+6+2x &= x+4 \\ 4x+6 &= x+4 \\ 4x-x+6 &= x-x+4 \\ 3x+6 &= 4 \\ 3x+6-6 &= 4-6 \\ 3x &= -2 \\ x &= -\frac{2}{3} \end{aligned}$$

Just for fun

1. $\dfrac{4x^3 - 4x + 6}{2x + 3}$

$$\begin{array}{r} 2x^2-3x+5/2-3/2(2x+3) \\ 2x+3 \,/\, 4x^3 \qquad -4x+6 \\ -4x^3 \mp 6x^2 \qquad\qquad \\ \hline -6x^2-4x \quad \\ \pm 6x^2 \pm 9x \quad \\ \hline 5x+6 \\ -5x \mp \frac{15}{2} \\ \hline -\frac{3}{2} \end{array}$$

2. $\dfrac{3x^3 - 5}{3x - 2}$

$$\begin{array}{r} x^2+2/3x+4/9-37/9(3x-2) \\ 3x-2 \,/\, 3x^3 \qquad\qquad -5 \\ -3x^3+2x^2 \qquad\qquad \\ \hline 2x^2 \qquad\quad \\ -2x^2+-\frac{4}{3}x \quad \\ \hline \frac{4}{3}x-5 \\ -\frac{4}{3}x+\frac{8}{9} \\ \hline -\frac{37}{9} \end{array}$$

Exercise Set 4.7

1. distance = rate·time
150 = r·3
$\frac{150}{3} = r$
50 mph = r

3. Amount = rate·time
546 = 42·t
$\frac{546}{42} = t$
13 hr. = t

5. thickness of door = rt
d = 0.2(12)
d = 2.4 cm.

7. flow·time = total intravenous fluid

F(6 hours) = 1500 cm^3

$F = \dfrac{1500\ cm^3}{6\ hours}$

$$F = \frac{250\ cm^3}{hour}$$

9. distance = rate · time

$57 = r \cdot 9.5$

$\frac{57}{9.5} = r$

6 meters/hour = r

11. distance = rate · time

$d = 3 \cdot \frac{1}{3}$

d = 1 mile

13.

Plane	Rate	Time	Distance
North	500	t	500t
South	650	t	650t

$500t + 650t = 4025$

$1150t = 4025$

$\frac{1150t}{1150} = \frac{4025}{1150}$

$t = 3.5$ hours

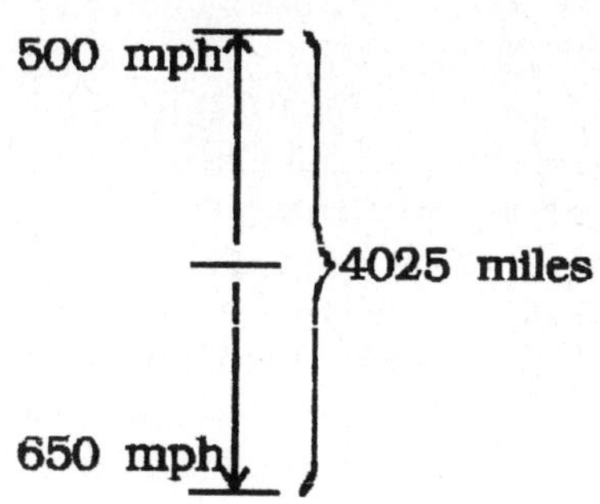

15.

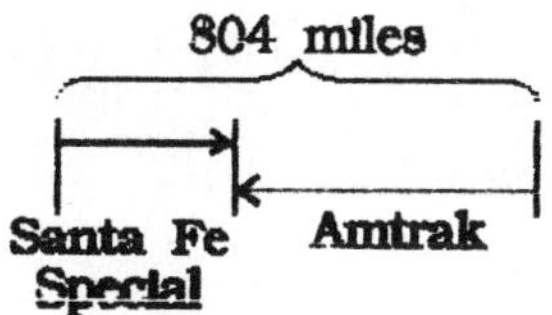

Train	Rate	Time	Distance
Santa Fe Special	x	6	6x
Amtrak	x+30	6	6(x+30)

$6x + 6(x + 30) = 804$

$6x + 6x + 180 = 804$

$12x + 180 = 804$

$12x + 180 - 180 = 804 - 180$

$12x = 624$

$\frac{12x}{12} = \frac{624}{12}$

$x = 52$ mph

Santa Fe Special train:
$x = 52$ mph
Amtrak train:
$x + 30 = 52 + 30$
$= 82$ mph

17.

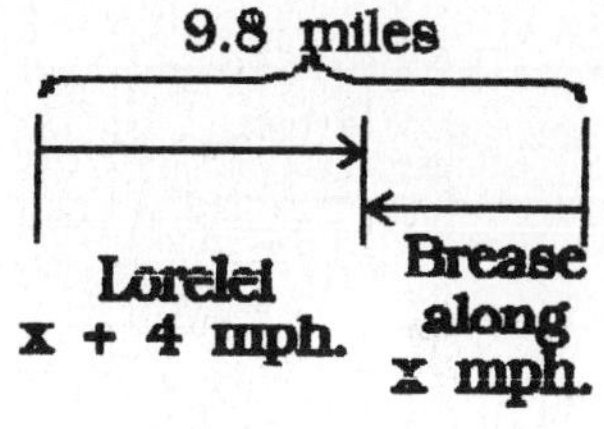

Boat	Rate	Time	Distance
Lorelei	x+4	0.7	0.7(x+4)
Brease Along	x	0.7	0.7x

$0.7x + 0.7(x+4) = 9.8$

$$0.7x + 0.7x + 2.8 = 9.8$$
$$1.4x + 2.8 = 9.8$$
$$1.4x = 7.0$$
$$x = 5 \text{ mph}$$
$$x + 4 = 9 \text{ mph}$$

Brease Along: 5 mph
Lorelei: 9 mph

19. length = rate · time
= 6.72 · 120
= 806.4
length = rate · time
806.4 = r · (6 · 60)
806.4 = r · 360
2.24 ft/min = r

21.

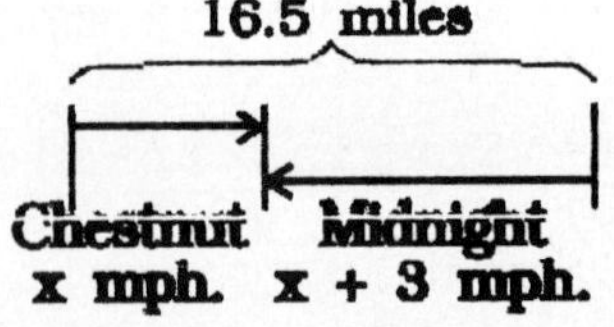

Horses	rate	time	distance
Chestnut	x	1.5	1.5x
Midnight	x+3	1.5	1.5(x+3)

$$1.5x + 1.5(x + 3) = 16.5$$
$$1.5x + 1.5x + 4.5 = 16.5$$
$$3.0x + 4.5 = 16.5$$
$$3x = 12$$
$$x = 4$$

Chestnut: 4 mph
Midnight: 7 mph

23.

	rate	time	distance
Serge	20	x	20x
Francine	18	x+30	18(x+30)

$$20x = 18(x + 30)$$
$$20x = 18x + 540$$
$$2x = 540$$
$$x = 270 \text{ min.}$$
$$20x = 5400 \text{ ft.}$$

25.

	rate	time	distance
walk	100	2.75	275
moving walkway	x	1.25	275

a) distance = rate · time
$275 = x \cdot 1.25$
$$\frac{275}{1.25} = x$$
$x = 220$ ft/min

b) $d = rt$
$d = 100 \cdot 2.75$
$d = 275$ ft.

27.

Amt.Invested	Rate	Interest
x	8%	0.08x
8900-x	11%	.11(8900-x)

$$100[0.08x + 0.11(8900-x)]$$
$$= 100(874)$$
$$8x + 11(8900 - x) = 87,400$$
$$8x + 97,900 - 11x = 87,400$$
$$-3x + 97,900 - 97,900$$
$$= 87,400 - 97,900$$
$$-3x = -10,500$$

$$\frac{-3x}{-3} = \frac{-10{,}500}{-3}$$

$$x = 3500$$

Amount invested at 8%:
$x = \$3500$
Amount invested at 11%:
$8900 - x = 8900 - 3500$
$= \$5400$

29.

Amt. Invested	Rate	Interest
x	10%	0.10x
5000 - x	15%	0.15(5000-x)

$$100(0.10x) = 100[0.15(5000-x)]$$
$$10x = 15(5000 - x)$$
$$10x = 75{,}000 - 15x$$
$$10x + 15x = 75{,}000 - 15x + 15x$$
$$25x = 75{,}000$$
$$\frac{25x}{25} = \frac{75{,}000}{25}$$
$$x = 3000$$

Amount invested at 10%:
$x = \$3000$
Amount invested at 15%:
$5000 - x = 5000 - 3000$
$= \$2000$

31.

Coin	Coin Value	Number of coins	Total Value
Dime	0.10	x	0.10x
Quarters	0.05	62-x	0.05 (62-x)

$$100[0.10x + 0.25(28-x)] = 100(3.55)$$
$$10x + 25(28 - x) = 355$$
$$10x + 700 - 25x = 355$$
$$-15x + 700 = 355$$
$$-15x + 700 - 700 = 355 - 700$$
$$-15x = -345$$
$$\frac{-15x}{-15} = \frac{-345}{-15}$$
$$x = 23$$

Number of dimes: $x = 23$
Number of quarters:
$28 - x = 28 - 23 = 5$

33.

Bills	Value of bills	Number of bills	Total Value
\$1	\$1	x	1x
\$10	\$10	12-x	20(12 -x)

$$x + 10(12 - x) = 39$$
$$x + 120 - 10x = 39$$
$$-9x + 120 = 39$$
$$-9x + 120 - 120 = 39 - 120$$
$$-9x = -81$$
$$\frac{-9x}{-9} = \frac{-81}{-9}$$
$$x = 9$$

Number of \$1: $x = 9$
Number of \$10:
$12 - x = 12 - 9 = 3$

35.

Part-time job	Rate	Number of hours	Total earnings
1	6.00	x	6.00x
2	6.50	18-x	6.50 (18-x)

$$10[6.00x + 6.50(18-x)] = 10(114.00)$$
$$60x + 65(18 - x) = 1140$$

$$60x + 1170 - 65x = 1140$$
$$-5x + 1170 = 1140$$
$$-5x + 1170 - 1170 = 1140 - 1170$$
$$-5x = -30$$
$$x = 6$$

Number of hours at $6.00:

x = 6 hours

Number of hours at $6.50:
18 - x = 18 - 6 = 12 hours

37.

Coffee	Cost	Number of pounds	Value of coffee
Less expensive	5.60	18	5.60(18)
More expensive	6.20	x	6.20(x)
Mixture	5.80	x+18	5.80 (x+18)

$$10[5.60(18) + 6.20(x)] = 10[5.80(x + 18)]$$

$$56(18) + 62x = 58(x + 18)$$
$$1008 + 62x = 58x + 1044$$
$$1008 + 62x - 58x = 58x - 58x + 1044$$
$$4x + 1008 = 1044$$
$$4x + 1008 - 1008 = 1044 - 1008$$
$$4x = 36$$
$$x = 9$$

Number of pounds of $6.20 coffee: x = 9 lbs.

39.

Strength of solution	Liters of solution	Amount of solution
20%	1	0.20(1)
12%	x	0.12x
15%	x+1	.15(x+1)

$$100[0.20(1) + 0.12x] = 100[0.15(x + 1)]$$
$$20 + 12x = 15(x + 1)$$
$$20 + 12x = 15x + 15$$
$$20 + 12x - 12x = 15x - 12x + 15$$
$$20 = 3x + 15$$
$$20 - 15 = 3x + 15 - 15$$
$$5 = 3x$$
$$\frac{5}{3} = \frac{3x}{3}$$
$$x = \frac{5}{3}$$
$$x = 1\frac{2}{3} \text{ liters}$$

Liters of 12% solution = $1\frac{2}{3}$ liters.

41.

Strength of solution	Number of quarts	Amount
12%	6	6(0.12)
0%	0.5	0.5(0)
x%	6.5	6.5(0.01x)

$$1000[0.12(6) + 0.5(0)] = 1000[6.5(0.01x)]$$
$$120(6) + 0 = 6500(0.01x)$$
$$720 = 65x$$

$11.1\% = x$

Percent of orange juice in solution: 11.1%

43.

Grass seed	Price per pound	number of pounds	total cost
Scott's Family	2.25	x	2.25x
Scott's spot filler	1.90	10-x	1.90(10-x)
Mixture	2.00	10	2.00(10)

$$100[2.25x + 1.90(10-x)] = [2.00(10)]100$$
$$225x + 190(10 - x) = 200(10)$$
$$225x + 1900 - 190x = 2000$$
$$35x + 1900 = 2000$$
$$35x = 100$$
$$x = 2.86 \text{ pounds}$$
$$10 - x = 7.14 \text{ pounds}$$

Pounds of Scott's Family: 2.86 pounds
Pounds of Scott's Spot Filler: 7.14 pounds

45.

Stock	Price	Number of shares	Cost of stock
United Airlines	140	4x	140(4x)
Getty Oil	22	x	22x

$$140(4x) + 22x = 8000$$
$$560x + 22x = 8000$$
$$582x = 8000$$
$$\frac{582x}{582} = \frac{8000}{582}$$
$$x = 13.746$$

a) Getty Oil:
$x = 13$ shares
United Airlines:
$4x = 4(13) = 52$ shares

b) $8000 - [(22)(13) + 52(140)]$ = money left over
$8000 - (286 + 7280)$
$= 8000 - 7566$
$= \$434$ left over

Cumulative Review Exercises

47. $6(x - 3) = 4x - 18 + 2x$
$6x - 18 = 6x - 18$
$6x = 6x$
Solution: all real numbers

49. $3x - 4 \le -4x + 3(x - 1)$
$3x - 4 \le - 4x + 3x - 3$
$3x - 4 \le - x - 3$
$4x - 4 \le - 3$
$4x \le 1$
$x \le \frac{1}{4}$

Just for fun

1. 20 min. - 1472 ft.
60 min. - 4416 ft.
10 miles = 52,800 ft.

a) $d = rt$
$52,800 = 4416\ t$
b) $t = 11.96$ hr.

2. 7 inches per line
66 lines per page

462 inches per page
@ 10 characters per inch
4620 characters per page
@ 100 characters per second

$\frac{7 \text{ in}}{\text{line}} \cdot \frac{66 \text{ line}}{1 \text{ pg.}} \cdot \frac{10 \text{ chr}}{\text{in}}$

$\cdot \frac{1 \text{ sec.}}{100 \text{ chr}}$

= 46.2 sec/page

3. 1.47 ft per sec.
= 1 mile per hour
5.86 ft per sec.
= 4 miles per hour
100 ft @ 5.86 ft per sec
= 17.06 sec. time to raise the door
17.06 seconds to raise the door to 6 feet:
6 ft. + 17.06 sec.
= 0.35 ft/sec (rounded)

4.

Solution of alcohol	Number of quarts	Total
20%	16-x	.20(16-x)
100%	x	1.00x
50%	16	0.50(16)

$100[0.20(16 - x) + 1.00x]$
$= 100[0.50(16)]$
$20(16-x) + 100x = 50(16)$
$320 - 20x + 100x = 800$
$320 + 80x = 800$
$320 - 320 + 80x = 800 - 320$
$80x = 480$
$\frac{80x}{80} = \frac{480}{80}$
$x = 6$

Amount of 20% solution that must be drained: x = 6 quarts

Review Exercises

1. $x^4 \cdot x^2 = x^{4+2} = x^6$

3. $3^2 \cdot 3^3 = 3^5 = 243$

5. $\frac{x^4}{x} = x^{4-1} = x^3$

7. $\frac{3^5}{3^3} = 3^{5-3} = 3^2 = 9$

9. $\frac{x^6}{x^8} = \frac{1}{x^{8-6}} = \frac{1}{x^2}$

11. $x^0 = 1$

13. $(3x)^0 = 1$

15. $(2x)^2 = 4x^2$

17. $(-2x)^2 = (-2)^2x^2 = 4x^2$

19. $(2x^2)^4 = 2^4(x^2)^4 = 16x^8$

21. $(-x^3)^4 = (-1)^4(x^3)^4 = x^{12}$

23. $[\frac{3x^4}{2y}]^3 = \frac{3^3x^{12}}{2^3y^3} = \frac{27x^{12}}{8y^3}$

25. $\frac{16x^2y}{4xy^2} = \frac{4x^{2-1}}{y^{2-1}} = \frac{4x}{y}$

27. $[\frac{9x^2y}{3xy}]^2 = (3x)^2 = 9x^2$

29. $4x^2y^3(2x^3y^4)^2$

$= 4x^2y^3(2^2x^6y^8)$

$= 4 \cdot 2^2 \cdot x^8y^{11}$

$= 16x^8y^{11}$

31. $(\frac{8x^4y^3}{2xy^5})^2 = (\frac{4x^3}{y^2})^2 = \frac{16x^6}{y^4}$

33. $x^{-3} = \frac{1}{x^3}$

35. $5^{-2} = \frac{1}{5^2} = \frac{1}{25}$

37. $\frac{1}{x^{-7}} = x^7$

39. $x^3 \cdot x^{-5} = \frac{x^3}{x^5} = \frac{1}{x^{5-3}} = \frac{1}{x^2}$

41. $x^4 \cdot x^{-7} = \frac{x^4}{x^7} = \frac{1}{x^{7-4}} = \frac{1}{x^3}$

43. $\frac{x^2}{x^{-3}} = x^2x^3 = x^{2+3} = x^5$

45. $\frac{x^{-3}}{x^3} = \frac{1}{x^3x^3} = \frac{1}{x^{3+3}} = \frac{1}{x^6}$

47. $(4x^{-3}y)^{-3} = 4^{-3}x^9y^{-3}$

$= \frac{x^9}{4^3y^3} = \frac{x^9}{64y^3}$

49. $-2x^3 \cdot 4x^5 = -8x^{3+5} = -8x^8$

51. $(4x^{-2}y^3)^{-2} = 4^{-2}x^4y^{-6}$

$= \frac{x^4}{16y^6}$

53. $(5x^{-2}y)(2x^4y)$

$= 5 \cdot 2 \cdot x^{-2+4}y^{1+1}$

$= 10x^2y^2$

55. $2x^{-4}(3x^{-2}y^{-1})$

$= 2 \cdot 3 \cdot x^{-4+(-2)}y^{-1}$

$= 6x^{-6}y^{-1} = \frac{6}{x^6y}$

57. $\frac{9x^{-2}y^3}{3xy^2} = 3x^{-2-1}y^{3-2}$

$= 3x^{-3}y = \frac{3y}{x^3}$

59. $\frac{36x^4y^7}{9x^5y^{-3}} = 4x^{4-5}y^{7-(-3)}$

$= 4x^{-1}y^{10} = \frac{4y^{10}}{x}$

61. $364{,}000 = 3.64 \times 10^5$

63. $0.00763 = 7.63 \times 10^{-3}$

65. $2080 = 2.08 \times 10^3$

67. $4.2 \times 10^{-3} = 0.0042$

69. $9.7 \times 10^5 = 970{,}000$

71. $9.14 \times 10^{-1} = 0.914$

73. $(2.3 \times 10^2)(2 \times 10^4)$

$= 4.6 \times 10^6$
$= 4{,}600{,}000$

75. $(6.4 \times 10^{-3})(3.1 \times 10^3)$

$= 19.84 \times 10^0 = 19.84$

77. $\frac{36 \times 10^4}{4 \times 10^6} = 9 \times 10^{-2} = 0.09$

79. $(60{,}000)(20{,}000)$

$= (6 \times 10^4)(2 \times 10^4)$

$= 12 \times 10^8$

$= 1.2 \times 10^9$

81. $(0.00023)(40{,}000)$

$= (2.3 \times 10^{-4})(4.0 \times 10^4)$

$= 9.2$

83. $\frac{0.000068}{0.02} = \frac{6.8 \times 10^{-5}}{2 \times 10^{-2}}$

$= 3.4 \times 10^{-3}$

85. $x + 3$, binomial, in descending order, 1st degree

87. $x^2 - 4 + 3x$
$x^2 + 3x - 4$, trinomial, 2nd

degree

89. $-5x^2 + 3$, binomial, in descending order, 2nd degree

91. $x - 4x^2$, $-4x^2 + x$, binomial, 2nd degree

93. $x^3 - 2x - 6 + 4x^2$

$x^3 + 4x^2 - 2x - 6$, 3rd degree

95. $(5x - 5) + (4x + 6)$
$= 5x - 5 + 4x + 6$
$= 5x + 4x - 5 + 6$
$= 9x + 1$

97. $4x^2 + 6x + 5 + (-6x + 9)$

$= 4x^2 + 6x + 5 - 6x + 9$

$= 4x^2 + 6x - 6x + 5 + 9$

$= 4x^2 + 14$

99. $(12x^2 + 4x - 8)$

$+ (-x^2 - 6x + 5)$

$= 12x^2 + 4x - 8 - x^2 - 6x + 5$

$= 12x^2 - x^2 + 4x - 6x - 8 + 5$

$= 11x^2 - 2x - 3$

101. $(-4x + 8) - (-2x + 6)$
$= -4x + 8 + 2x - 6$
$= -4x + 2x + 8 - 6$
$= -2x + 2$

103. $(6x^2 - 6x + 1) - (12x + 5)$

$= 6x^2 - 6x + 1 - 12x - 5$

$= 6x^2 - 6x - 12x + 1 - 5$

$= 6x^2 - 18x - 4$

105. $(x^2 + 7x - 3)$

$- (x^2 + 3x - 5)$

$= x^2 + 7x - 3 - x^2 - 3x + 5$

$= x^2 - x^2 + 7x - 3x - 3 + 5$
$= 4x + 2$

107. $x(2x - 4) = 2x^2 - 4x$

109. $3x(2x^2 - 4x + 7)$

$= 6x^3 - 12x^2 + 21x$

111. $-4x(-6x^2 + 4x - 2)$

$= 24x^3 - 16x^2 + 8x$

113. $(2x + 4)(x - 3)$

$= 2x^2 - 6x + 4x - 12$

$= 2x^2 - 2x - 12$

115. $(6 - 2x)(2 + 3x)$

$= 12 + 18x - 4x - 6x^2$

$= 12 + 14x - 6x^2$

$= -6x^2 + 14x + 12$

117. $(3x + 1)(x + 2x + 4)$

$$\begin{array}{r} = x + 2x + 4 \\ 3x + 1 \\ \hline x^2 + 2x + 4 \\ 3x^3 + 6x^2 + 12x \\ \hline 3x^3 + 7x^2 + 14x + 4 \end{array}$$

119. $(-5x + 2)(-2x^2 + 3x - 6)$

$$\begin{array}{r} = \quad -2x^2 + 3x - 6 \\ -5x + 2 \\ \hline -4x^2 + 6x - 12 \\ 10x^3 - 15x^2 + 30x \\ \hline 10x^3 - 19x^2 + 36x - 12 \end{array}$$

121. $\dfrac{4x - 8}{4} = \dfrac{4x}{4} - \dfrac{8}{4} = x - 2$

123. $\dfrac{6x^2 + 9x - 4}{3} = \dfrac{6x^2}{3} + \dfrac{9x}{3} - \dfrac{4}{3}$

$= 2x^2 + 3x - \dfrac{4}{3}$

125. $\dfrac{8x^2 - 4x}{2x} = \dfrac{8x^2}{2x} - \dfrac{4x}{2x}$

$= 4x - 2$

127. $\dfrac{12 + 6x}{-3} = \dfrac{12}{-3} + \dfrac{6x}{-3}$

$= -4 - 2x$

$= -2x - 4$

129. $\dfrac{x^2 + x - 12}{x - 3}$

$$\begin{array}{r} x + 4 \\ \hline x-3 /\; x^2 + x - 12 \\ -x^2 + 3x \\ \hline 4x - 12 \\ -4x + 12 \\ \hline -0- \end{array}$$

131. $\dfrac{5x^2 + 28x - 10}{x + 6}$

$$\begin{array}{r} 5x-2+2/x+6 \\ \hline x+6 /\; 5x^2 + 28x - 10 \\ -5x^2 - 30x \\ \hline -\; 2x - 10 \\ 2x + 12 \\ \hline 2 \end{array}$$

133. $\dfrac{4x^3 - 5x + 4}{2x - 1}$

$$\begin{array}{r} 2x^2+x-2+2/2x-1 \\ \hline 2x-1 /\; 4x^3 \quad -5x+4 \\ -4x^3 + 2x^2 \\ \hline 2x^2 - 5x \\ -2x^2 + x \\ \hline -4x + 4 \\ 4x - 2 \\ \hline 2 \end{array}$$

135. x = plane speed

$6.5x = 3500$

$\dfrac{6.5x}{6.5} = \dfrac{3500}{6.5}$

$x = 538.46$ mph (rounded)

137.

Amount Invested	Rate	Interest
x	8%	0.08x
12,000-x	$7\frac{1}{4}$%	0.0725(12,000-x)

$0.08x + 0.0725(12,000 - x) = 900$

$0.08x + 870 - 0.0725x = 900$

$0.0075x + 870 = 900$

$0.0075x + 870 - 870 = 900 - 870$

$0.0075x = 30$

$\dfrac{0.0075x}{0.0075} = \dfrac{30}{0.0075}$

$x = 4000$

Amount at 8%:
$x = \$4000$

Amount at $7\frac{1}{4}$%

$12,000 - x = 12,000 - 4000 = \$8000$

139. distance = rate' time

$26 = r \cdot 4$

$\frac{26}{4} = r$

$r = 6.5$ mph

141.

Cost of hamburger	Number of pounds	Total
3.50	x	3.50x
4.10	80-x	4.10(80-x)
3.65	80	3.65(80)

$$100[3.50x + 4.10(80-x)] = 100[3.65(80)]$$
$$350x + 410(80-x) = 365(80)$$
$$350x + 32{,}800 - 410x = 29{,}200$$
$$-60x = -3600$$
$$x = 60$$

Pounds of \$3.50 hamburger: 60 pounds
Pounds of \$4.10 hamburger: $80 - x = 80 - 60 = 20$ pounds

143.

Brothers	rate	time	distance
Younger	x+5	2	2(x+5)
Older	x	2	2x

$$2(x + 5) + 2x = 230$$
$$2x + 10 + 2x = 230$$
$$4x + 10 = 230$$
$$4x = 220$$
$$x = 55$$
$$x + 5 = 60$$

Rate of older brother: 55 mph
Rate of younger brother: 60 mph

Practice test

1. $2x^2 \cdot 3x^4 = 2 \cdot 3x^{2+4} = 6x^6$

2. $(3x^2)^3 = 3^3 \cdot x^6 = 27x^6$

3. $\frac{8x^4}{2x} = 4x^{4-1} = 4x^3$

4. $(\frac{3x^2y}{6xy^3})^3 = (\frac{x}{2y^2})^3 = \frac{x^3}{2^3y^6}$
$= \frac{x^3}{8y^6}$

5. $(2x^3y^{-2})^{-2} = 2^{-2}x^{-6}y^4$
$= \frac{y^4}{4x^6}$

6. $\frac{2x^4y^{-2}}{10x^7y^4} = \frac{1}{5x^{7-4}y^{4+2}} = \frac{1}{5x^3y^6}$

7. $x^2 - 4 + 6x$, trinomial

8. -3, monomial

9. $x^{-2} + 4$, not polynomial

10. $-5 + 6x^3 - 2x^2 + 5x$
$6x^3 - 2x^2 + 5x - 5$, third degree

11. $(2x + 4) + (3x^2 - 5x - 3)$
$= 2x + 4 + 3x^2 - 5x - 3$
$= 3x^2 + 2x - 5x + 4 - 3$
$= 3x^2 - 3x + 1$

12. $(x^2 - 4x + 7)$
$- (3x^2 - 8x + 7)$
$= x^2 - 4x + 7 - 3x^2 + 8x - 7$
$= x^2 - 3x^2 - 4x + 8x$

$+ 7 - 7$

$= -2x^2 + 4x$

13. $(4x^2 - 5) - (x^2 + x - 8)$

$= 4x^2 - 5 - x^2 - x + 8$

$= 4x^2 - x^2 - x + 8 - 5$

$= 3x^2 - x + 3$

14. $3x(4x^2 - 2x + 5)$

$= 12x^3 - 6x^2 + 15x$

15. $(4x + 7)(2x - 3)$

$= 8x^2 + 14x - 12x - 21$

$= 8x^2 + 2x - 21$

16. $(6 - 4x)(5 + 3x)$

$= 30 + 18x - 20x - 12x^2$

$= 30 - 2x - 12x^2$

$= -12x^2 - 2x + 30$

17. $(2x - 4)(3x^2 + 4x - 6)$

$$\begin{array}{r} = 3x^2 + 4x - 6 \\ 2x - 4 \\ \hline -12x^2 - 16x + 24 \\ 6x^3 + 8x^2 - 12x \qquad\quad \\ \hline 6x^3 - 4x^2 - 28x + 24 \end{array}$$

18. $\dfrac{16x^2 + 8x - 4}{4}$

$= \dfrac{16x^2}{4} + \dfrac{8x}{4} - \dfrac{4}{4}$

$= 4x^2 + 2x - 1$

19. $\dfrac{3x^2 - 6x + 5}{-3x}$

$= \dfrac{3x^2}{-3x} - \dfrac{6x}{-3x} + \dfrac{5}{-3x}$

$= -x + 2 - \dfrac{5}{3x}$

20. $\dfrac{8x^2 - 2x - 15}{2x - 3}$

$$\begin{array}{r} 4x + 5 \\ 2x-3 \overline{)\, 8x^2 - 2x - 15} \\ -8x^2 \pm 12x \qquad \\ \hline 10x - 15 \\ -10x \pm 15 \\ \hline -0- \end{array}$$

21. Length of time needed to fertilize: t

$0.7t = 40$

$\dfrac{0.7t}{0.7} = \dfrac{40}{0.7}$

$t = 57.14$ hours (rounded)

22.

Train	Rate	Time	Distance
A	60	4	240
B	x	3	3x

$3x = 240$

$\dfrac{3x}{3} = \dfrac{240}{3}$

$x = 80$ mph

Train B's rate:
$x = 80$ mph

23.

Salt Solution	Amount	Total
20%	x	0.20x
40%	60	0.40(60)
35%	x+60	0.35(x+60)

$$100[0.20x + 0.40(60)]$$
$$= 100[0.35(x + 60)]$$
$$20x + 40(60) = 35(x + 60)$$
$$20x + 2400 = 35x + 2100$$
$$20x - 35x + 2400$$
$$= 35x - 35x + 2100$$
$$-15x + 2400 = 2100$$
$$-15x + 2400 - 2400$$
$$= 2100 - 2400$$
$$-15x = -300$$
$$\frac{-15x}{-15} = \frac{-300}{-15}$$

$x = 20$ liters

Number of liters of 20% solution: $x = 20$ liters

24. $(42{,}000)(30{,}000)$

$$= (4.2 \times 10^4)(3.0 \times 10^4)$$
$$= 12.6 \times 10^8$$
$$= 1.26 \times 10^9$$

25. $\frac{0.0008}{4{,}000} = \frac{8 \times 10^{-4}}{4 \times 10^3} = 2 \times 10^{-7}$

Cumulative Review Test

1. $16 \div (4 - 6)\cdot 5$

$$= 16 \div (-2)\cdot 5$$
$$= (-8)\cdot 5$$
$$= -40$$

2. $2x + 5 = 3(x - 5)$

$$2x + 5 = 3x - 15$$
$$-x + 5 = -15$$
$$-x = -20$$
$$x = 20$$

3. $3(x - 2) - (x + 4) = 2x - 10$

$$3x - 6 - x - 4 = 2x - 10$$
$$2x - 10 = 2x - 10$$
$$2x = 2x$$

Solution: all real numbers

4. $2x - 14 > 5x + 1$

$$-14 > 3x + 1$$
$$-15 > 3x$$
$$-5 > x \text{ or}$$
$$x < -5$$

-5

5. $V - lwh$

$$\frac{V}{lh} = w$$

6. $4x - 3y = 6$

$$-3y = -4x + 6$$
$$y = \frac{-4x + 6}{-3} = \frac{-4x}{-3} + \frac{6}{-3}$$
$$= \frac{4}{3}x - 2$$

7. $(3x^4)(2x^5) = 6x^{4+5} = 6x^9$

8. $(3x^2y^4)^3(5x^2y)$

$$= (3^3x^6y^{12})(5x^2y)$$
$$= 135x^{6+2}y^{12+1} = 135x^8y^{13}$$

9. $-2x + 3x^2 - 5$

$$= 3x^2 - 2x - 5, \text{ degree } 2$$

10. $(2x^2 - 9x - 7) - (6x^2 - 3x + 4)$

$$= 2x^2 - 9x - 7 - 6x^2 + 3x - 4$$
$$= -4x^2 - 6x - 11$$

11. $(2x^2 + 4x - 3) + (6x^2 - 7x + 12)$

$$= 2x^2 + 4x - 3 + 6x^2 - 7x + 12$$

$$= 8x^2 - 3x + 9$$

12. $(4x^2 - 5x - 2) - (3x^2 - 2x + 5)$

$$= 4x^2 - 5x - 2 - 3x^2 + 2x - 5$$

$$= x^2 - 3x - 7$$

13. $(2x - 3)(3x - 5)$

$$= 6x^2 - 10x - 9x + 15$$

$$= 6x^2 - 19x + 15$$

14.
$$\begin{array}{r} 2x^2 + 4x + 8 \\ x - 5 \\ \hline 2x^3 + 4x^2 + 8x \\ - 10x^2 - 20x - 40 \\ \hline 2x^3 - 6x^2 - 12x - 40 \end{array}$$

15. $\dfrac{9x^2 - 6x + 8}{3x}$

$$= \frac{9x^2}{3x} - \frac{6x}{3x} + \frac{8}{3x}$$

$$= 3x - 2 + \frac{8}{3x}$$

16.
$$\begin{array}{r} 2x - 3 \\ x+2 \,/\, 2x^2 + x - 6 \\ -2x^2 \mp 4x \\ \hline -3x - 6 \\ \pm 3x \pm 6 \\ \hline -0- \end{array}$$

17. $\dfrac{3}{1.25} = \dfrac{8}{x}$

$$3x = 10$$

$$x = \frac{10}{3} = 3.33$$

$$x = \$3.33$$

18.
$$11 + 2x = 19$$
$$2x = 8$$
$$x = 4$$

19. x = width

$2x + 4$ = length

$$2l + 2w = P$$
$$2(2x + 4) + 2(x) = 26$$
$$4x + 8 + 2x = 26$$
$$6x + 8 = 26$$
$$6x = 18$$
$$x = 3$$
$$2x + 4 = 2(3) + 4$$
$$= 6 + 4 = 10$$

width: 3 ft.

length: 10 ft.

20.

Runners	rate	time	distance
Fast	8	x	8x
Slow	6	x	6x

$$8x + 6x = 28$$
$$14x = 28$$
$$x = 2 \text{ hours}$$

CHAPTER FIVE

Exercise Set 5.1

1.

40

10 4

$2 \cdot 5$ $2 \cdot 2$

$40 = 2^3 \cdot 5$

3.

90

10 9

$2 \cdot 5$ $3 \cdot 3$

$90 = 2 \cdot 3^2\ 5$

5.

200

100 2

$10 \cdot 10$ 2

$2 \cdot 5$ $2 \cdot 5$ 2

$200 = 2^3 \cdot 5^2$

7. $36 = 2^2 \cdot 3^2$

$20 = 2^2 \cdot 5$

$GCF = 2^2 = 4$

9. $60 = 2^2 \cdot 3 \cdot 5$

$84 = 2^2 \cdot 3 \cdot 7$

$GCF = 2^2 \cdot 3 = 12$

11. $72 = 2^3 \cdot 3^2$

$90 = 2 \cdot 3^2 \cdot 5$

$GCF = 2 \cdot 3^2 = 18$

13. x^2, x, x^3
$GCF = x$

15. $3x = 3x$

$6x^2 = 2 \cdot 3x^2$

$9x^3 = 3^2 \cdot x^3$
$GCF = 3x$

17. x, y, z
$GCF = 1$

19. xy, xy^2, xy^3
$GCF = xy$

21. $x^3y^7, x^7y^{12}, x^5y^5$

$GCF = x^3y^5$

23. $-8 = -1 \cdot 2^3$

$24x = 2^3 \cdot 3x$

$48x^2 = 2^4 \cdot 3x^2$

$GCF = 2^3 = 8$

25. $12x^4y^7 = 2^2 \cdot 3x^4y^7$

$6x^3y^9 = 2 \cdot 3x^3y^9$

$9x^{12}y^7 = 3^2x^{12}y^7$

$GCF = 3x^3y^7$

27. $36x^2y = 2^2 \cdot 3^2x^2y^2$

$15x^3 = 3 \cdot 5x^3$

$14x^2y = 2 \cdot 7x^2y$

$GCF = x^2$

29. $2(x + 3), 3(x + 3)$
$GCF = (x + 3)$

31. $x^2(2x - 3), 5(2x - 3)$

GCF = $2x - 3$

33. $3x - 4 = 1(3x - 4)$
$y(3x - 4) = y(3x - 4)$
GCF = $3x - 4$

35. $2x + 4$
GCF = 2
$2x + 4 = 2 \cdot x + 2 \cdot 2$
$= 2(x + 2)$

37. $15x - 5$
GCF = 5
$15x - 5 = 5 \cdot 3x - 5 \cdot 1$
$= 5(3x - 1)$

39. $13x + 5$
GCF = 1
Cannot be factored.

41. $16x^2 - 12x$
GCF = $4x$

$16x^2 - 12x = 4x \cdot 4x$
$- 4x \cdot 3$
$= 4x(4x - 3)$

43. $20p - 18p^2$
GCF = $2p$

$20p - 18p^2 = 2p \cdot 10$
$- 2p \cdot 9p$
$= 2p(10 - 9p)$

45. $6x^3 - 8x$
GCF = $2x$
$6x^3 - 8x = 2x(3x^2) - 2x(4)$

$= 2x(3x^2 - 4)$

47. $36x^{12} - 24x^6$

GCF = $12x^8$

$36x^{12} - 24x^8$

$= 12x^8(3x^4) - 12x^8(2)$

$= 12x^8(3x^4 - 2)$

49. $24y^{15} - 9y^3$

GCF = $3y^3$

$24y^{15} - 9y^3$

$= 3y^3 \cdot 8y^{12} - 3y^3 \cdot 3$

$= 3y^3(8y^{12} - 3)$

51. $x + 3xy^2$
GCF = x

$x + 3xy^2$

$= x \cdot 1 + x \cdot 3y^2$

$= x(1 + 3y^2)$

53. $6x + 5y$
GCF = 1
Cannot be factored.

55. $16xy^2z + 4x^3y$
GCF = $4xy$

$16xy^2z$
$= 4xy \cdot 4yz + 4xy \cdot x^2$

$= 4xy(4yz + x^2)$

57. $34x^2y^2 + 16xy^4$

GCF = $2xy^2$

$34x^2y^2 + 16xy^4$

$= 2xy^2 \cdot 17x + 2xy^2$
$\cdot 8y^2$

$= 2xy^2(17x + 8y^2)$

59. $36xy^2z^3 + 36x^3y^2z$

GCF = $36xy^2z$

$36xy^2z^3 + 36x^3y^2z$

$= 36xy^2z \cdot z^2 + 36xy^2z \cdot x^2$

$= 36xy^2z(z^2 + x^2)$

61. $14y^3z^5 - 9xy^3z^5$

$GCF = y^3z^5$

$14y^3z^5 - 9xy^3z^5$

$= y^3z^5 \cdot 14 - y^3z^5 \cdot 9x$

$= y^3z^5(14 - 9x)$

63. $3x^2 + 6x + 9$

$GCF = 3$

$3x^2 + 6x + 9$

$= 3 \cdot x^2 + 3 \cdot 2x + 3 \cdot 3$

$= 3(x^2 + 2x + 3)$

65. $9x^2 + 18x + 3$

$GCF = 3$

$9x^2 + 18x + 3$

$= 3 \cdot 3x^2 + 3 \cdot 6x + 3 \cdot 1$

$= 3(3x^2 + 6x + 1)$

67. $3x^3 - 6x^2 + 12x$
$GCF = 3x$

$3x^3 - 6x^2 + 12x$

$= 3x \cdot x^2 - 3x \cdot 2x$
$+ 3x \cdot 4$

$= 3x(x^2 - 2x + 4)$

69. $45x^2 - 16x + 10$
$GCF = 1$
Cannot be factored.

71. $15p^2 - 6p + 9$
$GCF = 3$

$15p^2 - 6p + 9$

$= 3 \cdot 5p^2 - 3 \cdot 2p + 3 \cdot 3$

$= 3(5p^2 - 2p + 3)$

73. $24x^6 + 8x^4 - 4x^3$

$GCF = 4x^3$

$24x^6 + 8x^4 - 4x^3$

$= 4x^3 - 6x^3 + 4x^3 \cdot 2x$

$- 4x^3 \cdot 1$

$= 4x^3(6x^3 + 2x - 1)$

75. $48x^2y + 16xy^2 + 33xy$
$GCF = xy$

$48x^2y + 16xy^2 + 33xy$

$= xy \cdot 48x + xy \cdot 16y$
$+ xy \cdot 33$
$= xy(48x + 16y + 33)$

77. $x(x + 2) + 3(x + 2)$
$GCF = x + 2$
$x(x + 2) + 3(x + 2)$
$= (x + 2)(x + 3)$

79. $7x(4x - 3) - 4(4x - 3)$
$GCF = 4x - 3$
$7x(4x - 3) - 4(4x - 3)$
$= (7x - 4)(4x - 3)$

81. $4x(2x + 1) + 1(2x + 1)$
$GCF = 2x + 1$
$4x(2x + 1) + 1(2x + 1)$
$= (4x + 1)(2x + 1)$

83. $4x(2x + 1) + 2x + 1$
$GCF = 2x + 1$
$4x(2x + 1) + 2x + 1$
$= 4x(2x + 1) + 1(2x + 1)$
$= (4x + 1)(2x + 1)$

85. A factored expression is an expression written as a product of factors.

87. To factor a monomial from a polynomial first determine the greatest common factor of all terms in the polynomial. Second, write each term as the product of the GCF and its other factor. Third, use the distributive property to

factor out the GCF.

Cumulative Review Exercises

89. $3x - (x - 6) + 4(3 - x)$
$= 3x - x + 6 + 12 - 4x$
$= 3x - x - 4x + 6 + 12$
$= -2x + 18$

91. $A = P(1 + rt)$
$1000 = 500(1 + r \cdot 2)$
$1000 = 500 + 1000r$
$500 = 1000r$
$$\frac{500}{1000} = \frac{1000r}{1000}$$
$$\frac{1}{2} = r$$

Just for fun

1. $4x^2(x - 3)^3 - 6x(x - 3)^2 + 4(x - 3)$
GCF $= 2(x - 3)$
$4x^2(x - 3)^3 - 6x(x - 3)^2 + 4(x - 3)$
$= 2(x - 3) \cdot 2x^2(x - 3)^2 - 2(x - 3)(3x)(x - 3) + 2(x - 3) \cdot 2$
$= 2(x - 3)[2x^2(x - 3)^2 - 3x(x - 3) + 2]$

2. $6x^5(2x + 7) + 4x^3(2x + 7) - 2x^2(2x + 7)$
GCF $= 2x^2(2x + 7)$
$6x^5(2x + 7) + 4x^3(2x + 7) - 2x^2(2x + 7)$
$= 2x^2(2x + 7) \cdot 3x^3 + 2x^2(2x + 7) \cdot 2x - 2x^2(2x + 7) \cdot 1$
$= 2x^2(2x + 7) \cdot (3x^3 + 2x - 1)$

3. a) $1 + 2 - 3 + 4 + 5 - 6 + 7 + 8 - 9 + 10 + 11 - 12 + 13 + 14 - 15$
$= (1 + 2 - 3) + (4 + 5 - 6) + (7 + 8 - 9) + (10 + 11 - 12) + (13 + 14 - 15)$
$= (0) + (3) + (6) + (9) + (12)$
$= 3(0) + 3(1) + 3(2) + 3(3) + 3(4)$

b) $3(0 + 1 + 2 + 3 + 4)$

c) $3(10) = 30$

d) Since $31 + 32 - 33 = 30$ or $3 \cdot 10$ we need to extend $3(0) + 3(1) + 3(2) + 3(3) + 3(4)$ to $3(0) + 3(1) + 3(2) + 3(3) + 3(4) + 3(5) + 3(6) + 3(7) + 3(8) + 3(9) + 3(10)$
$3[0 + 1 + 2 + 3 + 4 + 5 + 6 + 7 + 8 + 9 + 10]$
$3[55] = 165$

Exercise Set 5.2

1. $x^2 + 4x + 3x + 12$
$= x(x + 4) + 3(x + 4)$
$= (x + 3)(x + 4)$

3. $x^2 + 2x + 5x + 10$
$= x(x + 2) + 5(x + 2)$
$= (x + 5)(x + 2)$

5. $x^2 + 3x + 2x + 6$
$= x(x + 3) + 2(x + 3)$
$= (x + 2)(x + 3)$

7. $x^2 + 3x - 5x - 15$
$= x(x + 3) - 5(x + 3)$
$= (x - 5)(x + 3)$

9. $4x^2 + 6x - 6x - 9$
$= 2x(2x + 3) - 3(2x + 3)$
$= (2x - 3)(2x + 3)$

11. $3x^2 + 9x + x + 3$
$= 3x(x + 3) + 1(x + 3)$
$= (3x + 1)(x + 3)$

13. $4x^2 - 2x - 2x + 1$
$= 2x(2x - 1) - 1(2x - 1)$
$= (2x - 1)(2x - 1)$

$= (2x - 1)^2$

15. $8x^2 + 32x + x + 4$
$= 8x(x + 4) + 1(x + 4)$
$= (8x + 1)(x + 4)$

17. $3x^2 - 2x + 3x - 2$
$= x(3x - 2) + 1(3x - 2)$
$= (x + 1)(3x - 2)$

19. $3x^2 - 2x - 3x + 2$
$= x(3x - 2) - 1(3x - 2)$
$= (x - 1)(3x - 2)$

21. $15x^2 - 18x - 20x + 24$
$= 3x(5x - 6) - 4(5x - 6)$
$= (3x - 4)(5x - 6)$

23. $x^2 + 2xy - 3xy - 6y^2$
$= x(x + 2y) - 3y(x + 2y)$
$= (x - 3y)(x + 2y)$

25. $6x^2 - 9xy + 2xy - 3y^2$
$= 3x(2x - 3y)$
$+ y(2x - 3y)$
$= (3x + y)(2x - 3y)$

27. $10x^2 - 12xy - 25xy + 30y^2$
$= 2x(5x - 6y)$
$- 5y(5x - 6y)$
$= (2x - 5y)(5x - 6y)$

29. $x^2 + bx + ax + ab$
$= x(x + b) + a(x + b)$
$= (x + a)(x + b)$

31. $xy + 4x - 2y - 8$
$= x(y + 4) - 2(y + 4)$
$= (x - 2)(y + 4)$

33. $a^2 + 2a + ab + 2b$
$= a(a + 2) + b(a + 2)$
$= (a + b)(a + 2)$

35. $xy - x + 5y - 5$
$= x(y - 1) + 5(y - 1)$
$= (x + 5)(y - 1)$

37. $6 + 4y - 3x - 2xy$
$= 2(3 + 2y) - x(3 + 2y)$
$= (2 - x)(3 + 2y)$

39. $a^3 + 2a^2 + a + 2$

$= a^2(a + 2) + 1(a + 2)$

$= (a^2 + 1)(a + 2)$

41. $x^3 + 4x^2 - 3x - 12$

$= x^2(x + 4) - 3(x + 4)$

$= (x^2 - 3)(x + 4)$

43. $2x^2 - 12x + 8x - 48$

$= 2(x^2 - 6x + 4x - 24)$
$= 2[x(x - 6) + 4(x - 6)]$
$= 2(x + 4)(x - 6)$

45. $4x^2 + 8x + 8x + 16$

$= 4(x^2 + 2x + 2x + 4)$
$= 4[x(x + 2) + 2(x + 2)]$
$= 4(x + 2)(x + 2)$

$= 4(x + 2)^2$

47. $6x^3 + 9x^2 - 2x^2 - 3x$

$= x(6x^2 + 9x - 2x - 3)$
$= x[3x(2x+3) - 1(2x+3)]$
$= x(3x - 1)(2x + 3)$

49. $2x^2 - 4xy + 8xy - 16y^2$

$= 2(x^2 - 2xy + 4xy - 8y^2)$
$= 2[x(x-2y) + 4y(x-2y)]$
$= 2(x + 4y)(x - 2y)$

51. Determine if all terms have a common factor; if so, factor out the GCF.

53. $x^2 + 4x - 2x - 8$
The expression was determined by multiplying the factors using the FOIL method.

Cumulative Review Exercises

55. Let x = age of willow tree.

$3.5x = 25$

$\frac{3.5x}{3.5} = \frac{25}{3.5}$

$x = 7.14$ years

57. $\frac{15x^3 - 6x^2 - 9x + 5}{3x}$

$= \frac{15x^3}{3x} - \frac{6x^2}{3x} - \frac{9x}{3x} + \frac{5}{3x}$

$= 5x^2 - 2x - 3 + \frac{5}{3x}$

Just for fun

1. $3x^5 - 15x^3 + 2x^3 - 10x$

$= x(3x^4 - 15x^2 + 2x^2 - 10)$

$= x[3x^2(x^2-5) + 2(x^2-5)]$

$= x(3x^2 + 2)(x^2 - 5)$

2. $x^3 + xy - x^2y - y^2$

$= x(x^2 + y) - y(x^2 + y)$

$= (x - y)(x^2 + y)$

3. $18a^2 + 3ax^2 - 6ax - x^3$

$= 3a(6a + x^2) - x(6a + x^2)$

$= (3a - x)(6a + x^2)$

Exercise Set 5.3

1. $x^2 + 7x + 10$

Factors of 10

1	10
-1	-10
5	2
-5	-2

Sum of Factors

$1 + 10 = 11$
$-1 + (-10) = -11$
$5 + 2 = 7$ ✓
$-5 + (-2) = -7$

$x^2 + 7x + 10$
$= (x + 5)(x + 2)$
or
$(x + 2)(x + 5)$

3. $x^2 + 5x + 6$

Factors of 6

1	6
-1	-6
2	3
-2	-3

Sum of Factors

$1 + 6 = 7$
$-1 + (-6) = -7$
$2 + 3 = 5$ ✓
$-2 + (-3) = -5$

$x^2 + 5x + 6$
$= (x + 2)(x + 3)$
or
$(x + 3)(x + 2)$

5. $x^2 + 7x + 12$

Factors of 12

1	12
-1	-12
2	6
-2	-6
3	4
-3	-4

Sum of Factors

$1 + 12 = 13$
$-1 + (-12) = -13$
$2 + 6 = 8$
$-2 + (-6) = -8$
$3 + 4 = 7$ ✓
$-3 + (-4) = -7$

$x^2 + 7x + 12$
$= (x + 3)(x + 4)$

7. $x^2 - 7x + 9$

Factors of 9

9	1
3	3
-9	-1
-3	-3

Sum of Factors

$9 + 1 = 10$
$3 + 3 = 6$
$-9 + (-1) = -10$
$-3 + (-3) = -6$

Since none of the sum of the factors of 9 is -7, the trinomial cannot be factored.

9. $y^2 - 16y + 15$

Factors of 15

1	15
-1	-15
3	5
-3	-5

Sum of Factors

$1 + 15 = 16$
$-1 + (-15) = -16$ ✓
$3 + 5 = 8$
$-3 + (-5) = -8$

$y^2 - 16y + 15$
$= (y - 1)(y - 15)$
or
$(y - 15)(y - 1)$

11. $x^2 + x - 6 = x^2 + 1x - 6$

Factors of -6

1	-6
-1	6
2	-3
-2	3

Sum of Factors

$1 + (-6) = -5$
$-1 + 6 = 5$
$2 + (-3) = -1$
$-2 + 3 = 1$ ✓

$x^2 + 1x - 6$
$= (x - 2)(x + 3)$
or
$(x + 3)(x - 2)$

13. $k^2 - 2k - 15$

Factors of -15

1	-15
-1	15
3	-5
-3	5

Sum of Factors

$1 + (-15) = -14$
$-1 + 15 = 14$
$3 + (-5) = -2$ ✓
$-3 + 5 = 2$

$k^2 - 2k - 15$
$= (k + 3)(k - 5)$
or
$(k - 5)(k + 3)$

15. $b^2 - 11b + 18$

Factors of 18

1	18
-1	-18
2	9
-2	-9
3	6
-3	-6

Sum of Factors

$1 + 18 = 19$
$-1 + (-18) = -19$
$2 + 9 = 11$
$-2 + (-9) = -11$ ✓
$3 + 6 = 9$
$-3 + (-6) = -9$

$b^2 - 11b + 18$
$= (b - 2)(b - 9)$
or
$(b - 9)(b - 2)$

17. $x^2 - 8x - 15$

Factors of -15

1	-15
-1	15
-3	5
3	-5

Sum of Factors

$1 + (-15) = -14$
$-1 + 15 = 14$
$-3 + 5 = 2$
$3 + (-5) = -2$

Since none of the sum of factors of -15 is -8, the trinomial cannot be factored.

19. $a^2 + 12a + 11$

Factors of 11

1	11
-1	-11

Sum of Factors

$1 + 11 = 12$ ✓
$-1 + (-11) = -12$

$a^2 + 12a + 11$
$= (a + 1)(a + 11)$
or
$(a + 11)(a + 1)$

21. $x^2 + 13x - 30$

Factors of -30

1	-30
-1	30
2	-15
-2	15
5	-6
-5	6

Sum of Factors

$1 + (-30) = -29$
$-1 + 30 = 29$
$2 + (-15) = -13$
$-2 + 15 = 13$ ✓
$5 + (-6) = -1$
$-5 + 6 = 1$

$x^2 + 13x - 30$
$= (x - 2)(x + 15)$
or
$(x + 15)(x - 2)$

23. $x^2 + 4x + 4$

Factors of 4

1	4
-1	-4
2	2
-2	-2

Sum of Factors

$1 + 4 = 5$
$-1 + (-4) = -5$
$2 + 2 = 4$ ✓
$-2 + (-2) = -4$

$x^2 + 4x + 4$
$= (x + 2)(x + 2)$

$= (x + 2)^2$

25. $x^2 + 6x + 9$

Factors of 9

1	9
-1	-9
3	3
-3	-3

Sum of Factors

$1 + 9 = 10$
$-1 + (-9) = -10$
$3 + 3 = 6$ ✓
$-3 + (-3) = -6$

$x^2 + 6x + 9$
$= (x + 3)(x + 3)$

$= (x + 3)^2$

27. $x^2 + 10x + 25$

Factors of 25

1	25
-1	-25
5	5
-5	-5

Sum of Factors
$1 + 25 = 26$
$-1 + (-25) = -26$
$5 + 5 = 10$ ✓
$-5 + (-5) = -10$

$x^2 + 10x + 25$
$= (x + 5)(x + 5)$

$= (x + 5)^2$

29. $w^2 - 18w + 45$

Factors of 45

1	45
-1	-45
3	15
-3	-15
5	9
-5	-9

Sum of Factors
$1 + 45 = 46$
$-1 + (-45) = -46$
$3 + 15 = 18$
$-3 + (-15) = -18$ ✓
$5 + 9 = 14$
$-5 + (-9) = -14$

$w^2 - 18w + 45$
$= (w - 3)(w - 15)$
or
$(w - 15)(w - 3)$

31. $x^2 + 22x - 48$

Factors of -48

1	-48
-1	48
2	-24
-2	24
3	-16
-3	16
4	-12
-4	12
6	-8
-6	8

Sum of Factors
$1 + (-48) = -47$
$-1 + 48 = 47$
$2 + (-24) = -22$
$-2 + 24 = 22$ ✓
$3 + (-16) = -13$
$-3 + 16 = 13$
$4 + (-12) = -8$
$-4 + 12 = 8$
$6 + (-8) = -2$
$-6 + 8 = 2$

$x^2 + 22x - 48$
$= (x - 2)(x + 24)$
or
$(x + 24)(x - 2)$

33. $x^2 - x - 20 = x^2 - 1x - 20$

Factors of -20

1	-20
-1	20
2	-10
-2	10
4	-5
-4	5

Sum of Factors
$1 + (-20) = -19$
$-1 + 20 = 19$
$2 + (-10) = -8$
$-2 + 10 = 8$
$4 + (-5) = -1$ ✓
$-4 + 5 = 1$

$x^2 - x - 20$
$= (x + 4)(x - 5)$
or
$(x - 5)(x + 4)$

35. $y^2 - 9y + 14$

Factors of 14

1	14
-1	-14
2	7
-2	-7

Sum of Factors

$1 + 14 = -15$
$-1 + (-14) = -15$
$2 + 7 = 9$
$-2 + (-7) = -9$ ✓

$y^2 - 9y + 14$
$= (y - 2)(y - 7)$
or
$(y - 7)(y - 2)$

37. $x^2 + 12x - 64$

Factors of -64

1	-64
-1	64
2	-32
-2	32
4	-16
-4	16
8	-8

Sum of Factors

$1 + (-64) = -63$
$-1 + 64 = 63$
$2 + (-32) = -30$
$-2 + 32 = 30$
$4 + (-16) = -12$
$-4 + 16 = 12$ ✓
$8 + (-8) = 0$

$x^2 + 12x - 64$
$= (x - 4)(x + 16)$
or
$(x + 16)(x - 4)$

39. $x^2 - 14x + 24$

Factors of 24

1	24
-1	-24
2	12
-2	-12
3	8
-3	-8
4	6
-4	-6

Sum of Factors

$1 + 24 = 25$
$-1 + (-24) = -25$
$2 + 12 = 14$
$-2 + (-12) = -14$ ✓
$3 + 8 = 11$
$-3 + (-8) = -11$
$4 + 6 = 10$
$-4 + (-6) = -10$

$x^2 - 14x + 24$
$(x - 2)(x - 12)$
or
$(x - 12)(x - 2)$

41. $x^2 - 2x - 80$

Factors of -80

1	-80
-1	80
2	-40
-2	40
4	-20
-4	20
5	-16
-5	16
8	-10
-8	10

Sum of Factors

$1 + (-80) = -79$
$-1 + 80 = 79$
$2 + (-40) = -38$
$-2 + 40 = 38$
$4 + (-20) = -16$
$-4 + 20 = 16$
$5 + (-16) = -11$
$-5 + 16 = 11$
$8 + (-10) = -2$ ✓
$-8 + 10 = 2$

$x^2 - 2x - 80$
$= (x + 8)(x - 10)$
or
$(x - 10)(x + 8)$

43. $x^2 - 17x + 60$

Factors of 60

1	60
-1	-60
2	30
-2	-30
3	20
-3	-20
4	15
-4	-15
5	12
-5	-12
6	10
-6	-10

Sum of Factors

$1 + 60 = 61$
$-1 + (-60) = -61$
$2 + 30 = 32$
$-2 + (-30) = -32$
$3 + 20 = 23$
$-3 + (-20) = -23$
$4 + 15 = 19$
$-4 + (-15) = -19$
$5 + 12 = 17$
$-5 + (-12) = -17$ ✓
$6 + 10 = 16$
$-6 + (-10) = -16$

$x^2 - 17x + 60$
$= (x - 5)(x - 12)$
or
$(x - 12)(x - 5)$

45. $x^2 + 30x + 56$

Factors of 56

1	56
-1	-56
2	28
-2	-28
4	14
-4	-14
7	8
-7	-8

Sum of Factors

$1 + 56 = 57$
$-1 + (-56) = -57$
$2 + 28 = 30$ ✓
$-2 + (-28) = -30$
$4 + 14 = 18$
$-4 + (-14) = -18$
$7 + 8 = 15$
$-7 + (-8) = -15$

$x^2 + 30x + 56$
$= (x + 2)(x + 28)$
or
$(x + 28)(x + 2)$

47. $x^2 - 4xy + 4y^2$

Factors of 4

1	4
-1	-4
2	2
-2	-2

Sum of Factors

$1 + 4 = 5$
$-1 + (-4) = -5$
$2 + 2 = 4$
$-2 + (-2) = -4$ ✓

$x^2 - 4xy + 4y^2$
$= (x - 2y)(x - 2y)$

$= (x - 2y)^2$

49. $x^2 + 8xy + 15x^2$

Factors of 15

1	15
-1	-15
3	5
-3	-5

Sum of Factors

$1 + 15 = 16$
$-1 + (-15) = -16$
$3 + 5 = 8$ ✓
$-3 + (-5) = -8$

$x^2 + 8xy + 15y^2$
$= (x + 3y)(x + 5y)$
or
$(x + 5y)(x + 3y)$

51. $2x^2 - 14x + 12$
GCF = 2
Factor out the GCF.

$2(x^2 - 7x + 6)$
Factor the trinomial.

Factors of 6

1	6
-1	-6
2	3
-2	-3

Sum of Factors
$1 + 6 = 7$
$-1 + (-6) = -7$ ✓
$2 + 3 = 5$
$-2 + (-3) = -5$

$2x^2 - 14x + 12$

$= 2(x^2 - 7x + 6)$
$= 2(x - 6)(x - 1)$
or
$2(x - 1)(x - 6)$

53. $5x^2 + 20x + 15$
GCF = 5
Factor out the GCF.

$5(x^2 + 4x + 3)$
Factor the trinomial.

Factors of 3

1	3
-1	-3

Sum of Factors
$1 + 3 = 4$ ✓
$-1 + (-3) = -4$

$5x^2 + 20x + 15$

$= 5(x^2 + 4x + 3)$
$= 5(x + 1)(x + 3)$
or
$5(x + 3)(x + 1)$

55. $2x^2 - 14x + 24$
GCF = 2
Factor out the GCF.

$2(x^2 - 7x + 12)$
Factor the trinomial.

Factors of 12

1	12
-1	-12
2	6
-2	-6
3	4
-3	-4

Sum of Factors
$1 + 12 = 13$
$-1 + (-12) = -13$
$2 + 6 = 8$
$-2 + (-6) = -8$
$3 + 4 = 7$
$-3 + (-4) = -7$ ✓

$2x^2 - 14x + 24$

$= 2(x^2 - 7x + 12)$
$= 2(x - 3)(x - 4)$
or
$2(x - 4)(x - 3)$

57. $x^3 - 3x^2 - 18x$
GCF = x
Factor out the GCF.

$x(x^2 - 3x - 18)$
Factor the trinomial.

Factors of -18

1	-18
-1	18
2	-9
-2	9
3	-6
-3	6

Sum of Factors
$1 + (-18) = -17$
$-1 + 18 = 17$
$2 + (-9) = -7$
$-2 + 9 = 7$
$3 + (-6) = -3$ ✓
$-3 + 6 = 3$

$x^3 - 3x^2 - 18x$

$= x(x^2 - 3x - 18)$
$= x(x + 3)(x - 6)$
or

$x(x - 6)(x + 3)$

59. $2x^3 + 6x^2 - 56x$
GCF = 2x
Factor out the GCF.

$2x(x^2 + 3x - 28)$
Factor the trinomial.

Factors of -28

1	-28
-1	28
2	-14
-2	14
4	-7
-4	7

Sum of Factors
$1 + (-28) = -27$
$-1 + 28 = 27$
$2 + (-14) = -12$
$-2 + 14 = 12$
$4 + (-7) = -3$
$-4 + 7 = 3$ ✓

$2x^3 + 6x^2 - 56x$

$= 2x(x^2 + 3x - 28)$
$= 2x(x - 4)(x + 7)$
or
$2x(x + 7)(x - 4)$

61. $x^3 + 4x^2 + 4x$
GCF = x
Factor out the GCF.

$x(x^2 + 4x + 4)$
Factor out the trinomial.

Factors of 4

1	4
-1	-4
2	2
-2	-2

Sum of Factors
$1 + 4 = 5$
$-1 + (-4) = -5$
$2 + 2 = 4$ ✓
$-2 + (-2) = -4$

$x^3 + 4x^2 + 4x$

$= x(x^2 + 4x + 4)$
$= x(x + 2)(x + 2)$
or
$x(x + 2)^2$

63. Multiply factors using the FOIL method.

65. $(x - 3)(x - 8)$

$= x^2 - 11x + 24$

The answer was determined by multiplying the factors and combining like terms.

67. $2(x - 5y)(x + y)$

$= 2(x^2 - 4xy - 5y^2)$

$= 2x^2 - 8xy - 10y^2$

The answer was determined by multiplying the factors and combining like terms.

Cumulative Review Exercises

69.
$$\begin{array}{r} 2x^2 + 5x - 6 \\ x - 2 \\ \hline 2x^3 + 5x^2 - 6x \\ - 4x^2 - 10x + 12 \\ \hline 2x^3 + x^2 - 16x + 12 \end{array}$$

71.

Solution	Strength	Liters	Amount of Pure Acid
18%	0.18	4	0.18(4)
26%	0.26	1	0.26(1)
mixture	x	5	5x

$$0.18(4) + 0.26(1) = 5x$$
$$0.72 + 0.26 = 5x$$
$$0.98 = 5x$$
$$\frac{0.98}{5} = \frac{5x}{5}$$
$$0.196 = x$$

Strength of mixture: 19.6%

Just for fun

1. $x^2 + 0.6x + 0.08$

Factors of 0.08

0.1	0.8
-0.1	-0.8
0.2	0.4
-0.2	-0.4

Sum of Factors

$0.1 + 0.8 = 0.9$
$-0.1 + -0.8 = -0.9$
$0.2 + 0.4 = 0.6$ ✓
$-0.2 + (-0.4) = -0.6$

$x^2 + 0.6x + 0.08$
$= (x + 0.2)(x + 0.4)$
or
$(x + 0.4)(x + 0.2)$

2. $x^2 - 0.5x - 0.06$

Factors of -0.06

0.1	-0.6
-0.1	0.6
0.2	-0.3
-0.2	0.3

Sum of factors

$0.1 + (-0.6) = -0.5$ ✓
$-0.1 + 0.6 = 0.5$
$0.2 + (-0.3) = -0.1$
$-0.2 + 0.3 = 0.1$

$x^2 - 0.5x - 0.06$
$= (x + 0.1)(x - 0.6)$
or
$(x - 0.6)(x + 0.1)$

3. $x^2 + \frac{2}{5}x + \frac{1}{25}$

Factors of $\frac{1}{25}$

$\frac{1}{1}$ $\frac{1}{25}$

$\frac{-1}{1}$ $-\frac{1}{25}$

$\frac{1}{5}$ $\frac{1}{5}$

$\frac{-1}{5}$ $\frac{-1}{5}$

Sum of Factors

$$\frac{1}{1} + \frac{1}{25} = \frac{25}{25} + \frac{1}{25} = \frac{26}{25}$$

$$-\frac{1}{1} + \left(\frac{1}{-25}\right) = \frac{-25}{25} + \frac{-1}{25} = \frac{-26}{25}$$

$$\frac{1}{5} + \frac{1}{5} = \frac{2}{5} \quad ✓$$

$$\frac{-1}{5} + \left(\frac{-1}{5}\right) = \frac{-2}{5}$$

$$x^2 + \frac{2}{5}x + \frac{1}{25}$$
$$= \left(x + \frac{1}{5}\right)\left(x + \frac{1}{5}\right)$$
$$= \left(x + \frac{1}{5}\right)^2$$

4. $x^2 - \frac{2}{3}x + \frac{1}{9}$

Factors of $\frac{1}{9}$

$\frac{1}{1}$ $\frac{1}{9}$

$\frac{-1}{1}$ $\frac{-1}{9}$

$\frac{1}{3}$ $\frac{1}{3}$

$\frac{-1}{3}$ $\frac{-1}{3}$

Sum of factors

$\frac{1}{1} + \frac{1}{9} = \frac{9}{9} + \frac{1}{9} = \frac{10}{9}$

$\frac{-1}{1} + (\frac{-1}{9}) = \frac{-9}{9} + \frac{-1}{9} = \frac{-10}{9}$

$\frac{1}{3} + \frac{1}{3} = \frac{2}{3}$

$\frac{-1}{3} + (\frac{-1}{3}) = \frac{-2}{3}$ ✓

$x^2 - \frac{-2}{3}x + \frac{1}{9}$

$= (x - \frac{1}{3})(x - \frac{1}{3})$

$(x - \frac{1}{3})^2$

Exercise Set 5.4

1. $3x^2 + 5x + 2$
The product of a(c)
$= 3(2) = 6$

Factors of 6

1	6
-1	-6
2	3
-2	-3

Sum of Factors

$1 + 6 = 7$
$-1 + (-6) = -7$
$2 + 3 = 5$ ✓
$-2 + (-3) = -5$

$3x^2 + 5x + 2$

$= 3x^2 + 2x + 3x + 2$
$= x(3x + 2) + 1(3x + 2)$
$= (x + 1)(3x + 2)$
or
$(3x + 2)(x + 1)$

3. $6x^2 + 13x + 6$
The product of a · c
$= 6(6) = 36$

Factors of 36

1	36
-1	-36
2	18
-2	-18
3	12
-3	-12
4	9
-4	-9
6	6
-6	-6

Sum of Factors

$1 + 36 = 37$
$-1 + (-37) = -37$
$2 + 18 = 20$
$-2 + (-18) = -20$
$3 + 12 = 15$
$-3 + (-12) = -15$
$4 + 9 = 13$ ✓
$-4 + (-9) = -13$
$6 + 6 = 12$
$-6 + (-6) = -12$

$6x^2 + 13x + 6$

$= 6x^2 + 4x + 9x + 6$
$= 2x(3x + 2) + 3(3x + 2)$
$= (2x + 3)(3x + 2)$
or
$(3x + 2)(2x + 3)$

5. $2x^2 + 5x + 3$
The product of 2 · c is
$2 \cdot 3 = 6$.

Factors of 6

1	6
-1	-6
2	3
-2	-3

Sum of Factors

$1 + 6 = 7$
$-1 + (-6) = -7$
$2 + 3 = 5$ ✓
$-2 + (-3) = -5$

$2x^2 + 5x + 3$

$= 2x^2 + 2x + 3x + 3$
$= 2x(x + 1) + 3(x + 1)$
$= (2x + 3)(x + 1)$
or
$(x + 1)(2x + 3)$

7. $2x^2 + 11x + 15$
The product of $a \cdot c$ is $2(15) = 30$.

Factors of 30

1	30
-1	-30
2	15
-2	-15
3	10
-3	-10
5	6
-5	-6

Sum of Factors

$1 + 30 = 31$
$-1 + (-30) = -31$
$2 + 15 = 17$
$-2 + (-15) = -17$
$3 + 10 = 13$
$-3 + (-10) = -13$
$5 + 6 = 11$ ✓
$-5 + (-6) = -11$

$2x^2 + 11x + 15$

$= 2x^2 + 5x + 6x + 15$
$= x(2x + 5) + 3(2x + 5)$
$= (x + 3)(2x + 5)$
or
$(2x + 5)(x + 3)$

9. $3x^2 - 10x - 8$
The product of $a(c) = 3(-8) = -24$.

Factors of -24

1	-24
-1	24
2	-12
-2	12
3	-8
-3	8
4	-6
-4	6

Sum of Factors

$1 + (-24) = -23$
$-1 + 24 = 23$
$2 + (-12) = -10$
$-2 + 12 = 10$ ✓
$3 + (-8) = -5$
$-3 + 8 = 5$
$4 + (-6) = -2$
$-4 + 6 = 2$

$3x^2 - 10x - 8$

$= 3x^2 + 2x - 12x - 8$
$= x(3x + 2) - 4(3x + 2)$
$= (x - 4)(3x + 2)$
or
$(3x + 2)(x - 4)$

11. $5x^2 - 8y + 3$
The product of $a \cdot c = 5(3) = 15$.

Factors of 15

1	15
-1	-15
3	5
-3	-5

Sum of Factors

$1 + 15 = 16$
$-1 + (-15) = -16$
$3 + 5 = 8$
$-3 + (-5) = -8$ ✓

$5y^2 - 8y + 3$

$= 5y^2 - 3y - 5y + 3$
$= y(5y - 3) - 1(5y - 3)$
$= (y - 1)(5y - 3)$
or

$(5y - 3)(y - 1)$

13. $5a^2 - 12a + 6$
The product of a ' c
= 5 ' 6 = 30.

Factors of 30

1	30
-1	-30
2	15
-2	-15
3	10
-3	-10
5	6
-5	-6

Sum of Factors

$1 + 30 = 31$
$-1 + (-30) = -31$
$2 + 15 = 17$
$-2 + (-15) = -17$
$3 + 10 = 13$
$-3 + (-10) = -13$
$5 + 6 = 11$
$-5 + (-6) = -11$

Since none of the sums of factors of 30 equal -12, this trinomial cannot be factored.

15. $4x^2 + 13x + 3$
The product of a ' c = 4(3) = 12.

Factors of 12

1	12
-1	-12
2	6
-2	-6
3	4
-3	-4

Sum of Factors

$1 + 12 = 13$ ✓
$-1 + (-12) = -13$
$2 + 6 = 8$
$-2 + (-6) = -8$
$3 + 4 = 7$
$-3 + (-4) = -7$

$4x^2 + 13x + 3$

$= 4x^2 + 1x + 12x + 3$
$= x(4x + 1) + 3(4x + 1)$
$= (x + 3)(4x + 1)$
or
$(4x + 1)(x + 3)$

17. $5x^2 + 11x + 4$
The product of a ' c is 5(4) = 20.

Factors of 20

1	20
-1	-20
2	10
-2	-10
4	5
-4	-5

Sum of Factors

$1 + 20 = 21$
$-1 + (-20) = -21$
$2 + 10 = 12$
$-2 + (-10) = -12$
$4 + 5 = 9$
$-4 + (-5) = -9$

Since none of the sums of the factors of 20 equal 11, this trinomial cannot be factored.

19. $5y^2 - 16y + 3$
The product of a ' c is 5(3) = 15.

Factors of 15

1	15
-1	-15
3	5
-3	-5

Sum of Factors

$1 + 15 = 16$
$-1 + (-15) = -16$ ✓
$3 + 5 = 8$
$-3 + (-5) = -8$

$5y^2 - 16y + 3$

$= 5y^2 - 1y - 15y + 3$
$= y(5y - 1) - 3(5y - 1)$
$= (y - 3)(5y - 1)$
or

$(5y - 1)(y - 3)$

21. $3x^2 + 14x - 5$
The product of $a \cdot c$ is
$3(-5) = -15$.

Factors of -15

1	-15
-1	15
3	-5
-3	5

Sum of Factors

$1 + (-15) = -14$
$-1 + 15 = 14$ ✓
$3 + (-5) = -2$
$-3 + 5 = 2$

$3x^2 + 14x - 5$

$= 3x^2 - 1x + 15x - 5$
$= x(3x - 1) + 5(3x - 1)$
$= (x + 5)(3x - 1)$
or
$(3x - 1)(x + 5)$

23. $7x^2 - 16x + 4$
The product of $a \cdot c$ is
$7(4) = 28$.

Factors of 28

1	28
-1	-28
2	14
-2	-14
4	7
-4	-7

Sum of Factors

$1 + 28 = 29$
$-1 + (-28) = -29$
$2 + 14 = 16$
$-2 + (-14) = -16$ ✓
$4 + 7 = 11$
$-4 + (-7) = -11$

$7x^2 - 16x + 4$

$= 7x^2 - 2x - 14x + 4$
$= x(7x - 2) - 2(7x - 2)$
$= (x - 2)(7x - 2)$
or
$(7x - 2)(x - 2)$

25. $3x^2 - 10x + 7$
The product of $a \cdot c$ is
$3(7) = 21$.

Factors of 21

1	21
-1	-21
3	7
-3	-7

Sum of Factors

$1 + 21 = 22$
$-1 + (-21) = -22$
$3 + 7 = 10$
$-3 + (-7) = -10$ ✓

$3x^2 - 10x + 7$

$= 3x^2 - 3x - 7x + 7$
$= 3x(x - 1) - 7(x - 1)$
$= (3x - 7)(x - 1)$
or
$(x - 1)(3x - 7)$

27. $5z^2 - 33z - 14$
The product of $a \cdot c$ is
$5(-14) = -70$

Factors of -70

1	-70
-1	70
2	-35
-2	35
5	-14
-5	14
7	-10
-7	10

Sum of Factors

$1 + (-70) = -69$
$-1 + 70 = 69$
$2 + (-35) = -33$ ✓
$-2 + 35 = 33$
$5 + (-14) = -9$
$-5 + 14 = 9$
$7 + (-10) = -3$
$-7 + 10 = 3$

$5z^2 - 33z - 14$

$= 5z^2 - 35z + 2z - 14$
$= 5z(z - 7) + 2(z - 7)$
$= (5z + 2)(z - 7)$

or
$(z - 7)(5z + 2)$

29. $8x^2 + 2x - 3$
The product of a ' c
$= 8(-3) = -24.$

Factors of -24

1 -24
-1 24
2 -12
-2 12
3 -8
-3 8
4 -6
-4 6

Sum of Factors

$1 + (-24) = -23$
$-1 + 24 = 23$
$2 + (-12) = -10$
$-2 + 12 = 10$
$3 + (-8) = -5$
$-3 + 8 = 5$
$4 + (-6) = -2$
$-4 + 6 = 2$ ✓

$8x^2 + 2x - 3$

$= 8x^2 - 4x + 6x - 3$
$= 4x(2x - 1) + 3(2x - 1)$
$= (4x + 3)(2x - 1)$

or
$(2x - 1)(4x + 3)$

31. $10x^2 - 27x + 5$
The product of a ' c
$= 10(5) = 50.$

Factors of 50

1 50
-1 -50
2 25
-2 -25
5 10
-5 -10

Sum of Factors

$1 + 50 = 51$
$-1 + (-50) = -51$
$2 + 25 = 27$
$-2 + (-25) = -27$ ✓
$5 + 10 = 15$
$-5 + (-10) = -15$

$10x^2 - 27x + 5$

$= 10x^2 - 2x - 25x + 5$
$= 2x(5x - 1) - 5(5x - 1)$
$= (2x - 5)(5x - 1)$

or
$(5x - 1)(2x - 5)$

33. $8x^2 - 2x - 15$
The product of a ' c
$= 8(-15) = -120.$

Factors of -120

1 -120
-1 120
2 -60
-2 60
3 -40
-3 40
4 -30
-4 30
5 -24
-5 24
6 -20
-6 20
8 -15
-8 15
10 -12
-10 12

Sum of Factors

$1 + (-120) = -119$
$-1 + 120 = 119$
$2 + (-60) = -58$
$-2 + 60 = 58$
$3 + (-40) = -37$
$-3 + 40 = 37$
$4 + (-30) = -26$
$-4 + 30 = 26$
$5 + (-24) = -19$
$-5 + 24 = 19$
$6 + (-20) = -14$
$-6 + 20 = 14$
$8 + (-15) = -7$
$-8 + 15 = 7$
$10 + (-12) = -2$ ✓
$-10 + 12 = 2$

$8x^2 - 2x - 15$

$= 8x^2 + 10x - 12x - 15$
$= 2x(4x + 5) - 3(4x + 5)$
$= (2x - 3)(4x + 5)$

or

$(4x + 5)(2x - 3)$

35. $6x^2 + 33x + 15$

Factor out the GCF of 3.

$3(2x^2 + 11x + 5)$

Factor the trinomial.

The product of a ' c is $2(5) = 10$.

Factors of 10

1	10
-1	-10
2	5
-2	-5

Sum of Factors

$1 + 10 = 11$ ✓
$-1 + (-10) = -11$
$2 + 5 = 7$
$-2 + (-5) = -7$

$6x^2 + 33x + 15$

$= 3(2x^2 + 11x + 5)$

$= 3(2x^2 + 1x + 10x + 5)$
$= 3[x(2x+1) + 5(2x+1)]$
$= 3(x + 5)(2x + 1)$

or

$3(2x + 1)(x + 5)$

37. $6x^2 + 4x - 10$

Factor out a GCF of 2.

$2(3x^2 + 2x - 5)$

Factor the trinomial.

The product of a ' c is $3(-5) = -15$.

Factors of -15

1	-15
-1	15
3	-5
-3	5

Sum of Factors

$1 + (-15) = -14$
$-1 + 15 = 14$
$3 + (-5) = -2$
$3 + 5 = 2$ ✓

$6x^2 + 4x - 10$

$= 2(3x^2 + 2x - 5)$

$= 2(3x^2 - 3x + 5x - 5)$
$= 2[3x(x-1) + 5(x-1)]$
$= 2(3x + 5)(x - 1)$

or

$2(x - 1)(3x + 5)$

39. $6x^3 + 5x^2 - 4x$

Factor out a GCF of x.

$x(6x^2 + 5x - 4)$

Factor the trinomial.

The product of a ' c is $6(-4) = -24$.

Factors of -24

1	-24
-1	24
2	-12
-2	12
3	-8
-3	8
4	-6
-4	6

Sum of Factors

$1 + (-24) = -23$
$-1 + 24 = 23$
$2 + (-12) = -10$
$-2 + 12 = 10$
$3 + (-8) = -5$
$-3 + 8 = 5$ ✓
$4 + (-6) = -2$
$-4 + 6 = 2$

$6x^3 + 5x - 4x$

$= x(6x^2 + 5x - 4)$

$= x(6x^2 - 3x + 8x - 4)$
$= x[3x(2x-1) + 4(2x-1)]$
$= x(3x + 4)(2x - 1)$
or
$x(2x - 1)(3x + 4)$

41. $4x^3 + 2x^2 - 6x$
Factor out a GCF of 2x.

$2x(2x^2 + x - 3)$
Factor the trinomial.
The product of a · c is
$2(-3) = -6$.

Factors of -6

1	-6
-1	6
2	-3
-2	3

Sum of Factors

$1 + (-6) = -5$
$-1 + 6 = 5$
$2 + (-3) = -1$
$-2 + 3 = 1$ ✓

$4x^3 + 2x^2 - 6x$

$= 2x(2x^2 + x - 3)$

$= 2x(2x^2 - 2x + 3x - 3)$
$= 2x[2x(x-1) + 3(x-1)]$
$= 2x(2x + 3)(x - 1)$
or
$2x(x - 1)(2x + 3)$

43. $6x^3 + 4x^2 - 10x$
Factor out a GCF of 2x.

$2x(3x^2 + 2x - 5)$
Factor the trinomial.
The product of a · c is
$3(-5) = -15$

Factors of -15

1	-15
-1	15
3	-5
-3	5

Sum of Factors

$1 + (-15) = -14$
$-1 + 15 = 14$
$3 + (-5) = -2$
$-3 + 5 = 2$ ✓

$6x^3 + 4x^2 - 10x$

$= 2x(3x^2 + 2x - 5)$

$= 2x(3x^2 - 3x + 5x - 5)$
$= 2x[3x(x-1) + 5(x-1)]$
$= 2x(3x + 5)(x - 1)$
or
$2x(x - 1)(3x + 5)$

45. $60x^2 + 40x + 5$
Factor out a GCF of 5.

$5(12x^2 + 8x + 1)$
Factor the trinomial.
The product of a · c is
$12(1) = 12$.

Factors of 12

1	12
-1	-12
2	6
-2	-6
3	4
-3	-4

Sum of Factors

$1 + 12 = 13$
$-1 + (-12) = -13$
$2 + 6 = 8$ ✓
$-2 + (-6) = -8$
$3 + 4 = 7$
$-3 + (-4) = -7$

$60x^2 + 40x + 5$

$= 5(12x^2 + 8x + 1)$

$= 5(12x^2 + 2x + 6x + 1)$
$= 5[2x(6x+1) + 1(6x+1)]$
$= 5(2x + 1)(6x + 1)$
or
$5(6x + 1)(2x + 1)$

47. $2x^2 + 5xy + 2y^2$
The product of a · c is 4.

Factors of 4

1	4
-1	-4
2	2
-2	-2

Sum of Factors
$1 + 4 = 5$ ✓
$-1 + (-4) = -5$
$2 + 2 = 4$
$-2 + (-2) = -4$

$2x^2 + 5xy + 2y^2$

$= 2x^2 + 1xy + 4xy + 2y^2$
$= x(2x+y) + y(2x+y)$
$= (x + 2y)(2x + y)$
or
$(2x + y)(x + 2y)$

49. $2x^2 - 7xy + 3y^2$
The product of a · c is $2(3) = 6$.

Factors of 6

1	6
-1	-6
2	3
-2	-3

Sum of Factors
$1 + 6 = 7$
$-1 + (-6) = -7$ ✓
$2 + 3 = 5$
$-2 + (-3) = -5$

$2x^2 - 7xy + 3y^2$

$= 2x^2 - 1xy - 6xy + 3y^2$
$= x(2x-y) - 3y(2x-y)$
$= (x - 3y)(2x - y)$
or
$(2x - y)(x - 3y)$

51. $18x^2 + 18xy - 8y^2$
Factor out a GCF of 2.

$2(9x^2 + 9xy - 4y^2)$
Factor the trinomial.
The product of a · c is $9(-4) = -36$.

Factors of -36

1	-36
-1	36
2	-18
-2	18
3	-12
-3	12
4	-9
-4	9
6	-6

Sum of Factors
$1 + (-36) = -35$
$-1 + 36 = 35$
$2 + (-18) = -16$
$-2 + 18 = 16$
$3 + (-12) = -9$
$-3 + 12 = 9$ ✓
$4 + (-9) = -5$
$-4 + 9 = 5$
$6 + (-6) = 0$

$18x^2 + 18xy - 8y^2$

$= 2(9x^2 + 9xy - 4y^2)$

$= 2(9x^2 - 3xy + 12xy - 4y^2)$
$= 2[3x(3x-y) + 4y(3x-y)]$
$= 2(3x + 4y)(3x - y)$
or
$2(3x - y)(3x + 4y)$

53. Factor out the GCF, if there is one.

55. 1. Factor out the GCF, if there is one.

2. Find the product of a · c.

3. Write out all the

factors of a·c and the sum of those factors.

4. Find the two numbers whose product is equal to the product of a·c and whose sum is equal to b.

5. Rewrite the bx term using the a factors from step 4.

6. Factor by grouping.

Sum of Factors

$1 + 54 = 55$
$(-1) + (-54) = -55$
$2 + 27 = 29$
$-2 + (-27) = -29$
$3 + 18 = 21$
$-3 + (-18) = -21$
$6 + 9 = 15$
$-6 + (-9) = -15$ ✓

$x^2 - 15x + 54$
$= (x - 6)(x - 9)$
or
$(x - 9)(x - 6)$

Cumulative Review Exercises

57. Let x = the width.
$2x + 2$ = the length.

$P = 2l + 2w$
$22 = 2(2x + 2) + 2x$
$22 = 4x + 4 + 2x$
$22 = 6x + 4$
$18 = 6x$
$3 = x$
width: 3 ft.
length: $2x + 2 = 2(3) + 2$
$= 6 + 2 = 8$ ft.

59. $x^2 - 15x + 54$

Factors of 54

1	54
-1	-54
2	27
-2	-27
3	18
-3	-18
6	9
-6	-9

Just for fun

1. $18x^2 + 9x - 20$
The product of a · c
$= 18(-20) = -360$.

Factors of -360

1	-360
-1	360
2	-180
-2	180
3	-120
-3	120
4	-90
-4	90
5	-72
-5	72
6	-60
-6	60
8	-45
-8	45
9	-40
-9	40
10	-36
-10	36
12	-30
-12	30
15	-24
-15	24
18	-20
-18	20

Sum of Factors

$1 + (-360) = -359$
$-1 + 360 = 359$
$2 + (-180) = -178$
$-2 + 180 = 178$
$3 + (-120) = -117$
$-3 + 120 = 117$
$4 + (-90) = -86$
$-4 + 90 = 86$
$5 + (-72) = -67$
$-5 + 72 = 67$
$6 + (-60) = -54$
$-6 + 60 = 54$
$8 + (-45) = -37$
$-8 + 45 = 37$
$9 + (-40) = -39$
$-9 + 40 = 39$
$10 + (-36) = -26$
$-10 + 36 = 26$
$12 + (-30) = -18$
$-12 + 30 = 18$
$15 + (-24) = -9$
$-15 + 24 = 9$ ✓
$18 + (-20) = -2$
$-18 + 20 = 2$

$18x^2 + 9x - 20$

$= 18x^2 - 15x + 24x - 20$
$= 3x(6x - 5) + 4(6x - 5)$
$= (3x + 4)(6x - 5)$

or

$(6x - 5)(3x + 4)$

2. $8x^2 - 99x + 36$

The product of a ' c
$= 8(36) = 288.$

Factors of 288

1	288
-1	-288
2	144
-2	-144
3	96
-3	-96
4	72
-4	-72
6	48
-6	-48
8	36
-8	-36
9	32
-9	-32
12	24
-12	-24
16	18
-16	-18

Sum of Factors

$1 + 288 = 289$
$-1 + (-288) = -289$
$2 + 144 = 146$
$-2 + (-144) = -146$
$3 + 96 = 99$
$-3 + (-96) = -99$ ✓
$4 + 72 = 76$
$-4 + (-72) = -76$
$6 + 48 = 54$
$-6 + (-48) = -54$
$8 + 36 = 44$
$-8 + (-36) = -44$
$9 + 32 = 41$
$-9 + (-32) = -41$
$12 + 24 = 36$
$-12 + (-24) = -36$
$16 + 18 = 34$
$-16 + (-18) = -34$

$8x^2 - 99x + 36$

$= 8x^2 - 3x - 96x + 36$
$= x(8x - 3) - 12(8x - 3)$
$= (x - 12)(8x - 3)$

or

$(8x - 3)(x - 12)$

Exercise Set 5.5

1. $x^2 - 4 = x^2 - (2)^2$
$= (x + 2)(x - 2)$

3. $y^2 - 25 = y^2 - (5)^2$
$= (y + 5)(y - 5)$

5. $x^2 - 49 = x^2 - (7)^2$
$= (x + 7)(x - 7)$

7. $x^2 - y^2 = (x + y)(x - y)$

9. $9y^2 - 16 = (3y)^2 - (4)^2$
$= (3y + 4)(3y - 4)$

11. $64a^2 - 36b^2 = 4(16a^2 - 9b^2)$

$= 4[(4a)^2 - (3b)^2]$

$= 4(4a + 3b)(4a - 3b)$

13. $25x^2 - 16 = (5x)^2 - (4)^2$
$= (5x + 4)(5x - 4)$

15. $z^4 - 81x^2 = (z^2)^2 - (9x)^2$

$= (z^2 + 9x)(z^2 - 9x)$

17. $9x^4 - 81y^2 = 9(x^4 - 9y^2)$

$= 9[(x^2)^2 - (3y)^2]$

$= 9(x^2 + 3y)(x^2 - 3y)$

19. $49m^4 - 16n^2 = (7m^2)^2 - (4n)^2$

$= (7m^2 + 4n)(7m^2 - 4n)$

21. $20x^2 - 180 = 20(x^2 - 9)$

$= 20(x^2 - (3)^2)$
$= 20(x + 3)(x - 3)$

23. $x^3 + y^3$

$= (x+y)[(x)^2-(x)(y)+(y)^2]$

$= (x + y)(x^2 - xy + y^2)$

25. $a^3 - b^3$

$= (a-b)[(a)^2+(a)(b)+(b)^2]$

$= (a - b)(a^2 + ab + b^2)$

27. $x^3 + 8 = x^3 + (2)^2$

$= (x+2)[(x)^2-(x)(2)+(2)^2]$

$= (x + 2)(x^2 - 2x + 4)$

29. $x^3 - 27 = x^3 - (3)^3$

$= (x-3)[(x)^2+(x)(3)+(3)^2]$

$= (x - 3)(x^2 + 3x + 9)$

31. $a^3 + 1 = a^3 + (1)^3$

$= (a+1)[(a)^2-a(1)+(1)^2]$

$= (a + 1)(a^2 - a + 1)$

33. $8x^3 + 27 = (2x)^3 + (3)^3$

$= (2x+3)[(2x)^2$

$- (2x)(3)+(3)^2]$

$= (2x + 3)(4x^2 - 6x + 9)$

35. $27a^3 - 64 = (3a)^3 - (4)^3$

$= (3a-4)[(3a)^2+(3a)(4)$

$+ (4)^2]$

$= (3a - 4)(9a^2 + 12a + 16)$

37. $27 - 8y^3 = (3)^3 - (2y)^3$

$= (3 - 2y)[3^2+3(2y)$

$+ (2y)^2]$

$= (3 - 2y)(9 + 6y + 4y^2)$

39. $8x^3 - 27y^3 = (2x)^3 - (3y)^3$

$= (2x-3y)[(2x)^2+2x)(3y)$

$+ (3y)^2]$

$= (2x - 3y)(4x^2 + 6xy$

$+ 9y^2)$

41. $2x^2 - 2x - 12$
Factor out the GCF of 2.

$2(x^2 - x - 6)$
Factor the trinomial
The product of a(c) - 1(-6) = -6.

Factors of -6

1	-6
-1	6
2	-3
-2	3

Sum of Factors
$1 + (-6) = -5$
$-1 + 6 = 5$
$2 + (-3) = -1$ ✓
$-2 + 3 = 1$

$2(x^2 - 2x - 12)$

$= 2(x^2 - x - 6)$

$= 2(x^2 - 3x + 2x - 6)$
$= 2[x(x-3) + 2(x-3)]$
$= 2(x + 2)(x - 3)$
or
$2(x - 3)(x + 2)$

43. Factor out GCF of y.
Factor using the difference of squares.

$x^2y - 16y = y(x^2 - 16)$

$= y[x^2 - (4)^2]$
$= y(x + 4)(x - 4)$

45. $3x^2 + 6x + 3$
Factor out a GCF of 3.

$3(x^2 + 2x + 1)$
Factor the trinomial.
The product of a ' c = 1 ' 1 = 1.

Factors of 1

1	1
-1	-1

Sum of Factors
$1 + 1 = 2$ ✓
$-1 + (-1) = -2$

$3x^2 + 6x + 3$

$= 3(x^2 + 2x + 1)$

$= 3(x^2 + 1x + 1x + 1)$
$= 3[x(x+1) + 1(x+1)]$
$= 3(x + 1)(x + 1)$

$= 3(x + 1)^2$

47. $5x^2 + 10x - 15$
Factor out a GCF of 5.

$5(x^2 + 2x - 3)$
Factor the trinomial.
The product of a ' c is 1(-3) = -3.

Factors of -3

1	-3
-1	3

Sum of Factors
$1 + (-3) = -2$
$-1 + 3 = 2$ ✓

$5x^2 + 10x - 15$

$= 5(x^2 + 2x - 3)$

$= 5(x^2 - 1x + 3x - 3)$
$= 5[x(x-1) + 3(x-1)]$
$= 5(x + 3)(x - 1)$
or
$5(x - 1)(x + 3)$

49. $3xy - 6x + 9y - 18$
Factor out a GCF of 3.
$= 3(xy - 2x + 3y - 6)$

Factor by grouping.
$= 3[x(y-2) + 3(y-2)]$
$= 3(x + 3)(y - 2)$

51. Factor out a GCF of 2, then

factor using difference of squares.

$2x^2 - 7x = 2(x^2 - 36)$

$= 2[x^2 - (6)^2]$
$= 2(x + 6)(x - 6)$

53. Factor out a GCF of 3y, then factor using difference of squares.

$3x^2y - 27y = 3y(x^2 - 9)$

$= 3y[x^2 - (3)^2]$
$= 3y(x + 3)(x - 3)$

55. Factor out a GCF of $3y^2$, then factor using sum of cubes.

$3x^3y^2 + 3y^2 = 3y^2(x^3 + 1)$

$= 3y^2(x^3 + 1^3)$

$= 3y^2(x + 1)[(x)^2$

$- (x)(1) + 1^2]$

$= 3y^2(x + 1)(x^2 - x + 1)$

57. Factor out a GCF of 2, then factor using difference of cubes.

$2x^3 - 16 = 2(x^3 - 8)$

$= 2[x^3 - (2)^3]$

$= 2(x - 2)[x^2 + x(2) + 2^2]$

$= 2(x - 2)(x^2 + 2x + 4)$

59. Factor out a GCF of 2, then factor by grouping.

$6x^2 - 4x + 24x - 16$

$= 2(3x^2 - 2x + 12x - 8)$
$= 2[x(3x-2) + 4(3x-2)]$
$= 2(x + 4)(3x - 2)$

61. $3x^3 - 10x^2 - 8x$

Factor out a GCF of x.

$x(3x^2 - 10x - 8)$
Factor the trinomial.
The product of a ' c is $3(-8) = -24$.

Factors of -24

1	-24
-1	24
2	-12
-2	12
3	-8
-3	8
4	-6
-4	6

Sum of Factors
$1 + (-24) = -23$
$-1 + 24 = 23$
$2 + (-12) = -10$ ✓
$-2 + 12 = 10$
$3 + (-8) = -5$
$-3 + 8 = 5$
$4 + (-6) = -2$
$-4 + 6 = 2$

$3x^3 - 10x^2 - 8x$

$= x(3x^2 - 10x - 8)$

$= x(3x^2 - 12x + 2x - 8)$
$= x[3x(x-4) + 2(x-4)]$
$= x(3x + 2)(x - 4)$

63. $4x^2 + 5x - 6$
The product of a ' c is $4(-6) = -24$.

Factors of -24

1	-24
-1	24
2	-12
-2	12
3	-8
-3	8
4	-6
-4	6

Sum of Factors

$1 + (-24) = -23$

$-1 + 24 = 23$

$2 + (-12) = -10$

$-2 + 12 = 10$

$3 + (-8) = -5$

$-3 + 8 = 5$ ✓

$4 + (-6) = -2$

$-4 + 6 = 2$

$4x^2 + 5x - 6$

$= 4x^2 - 3x + 8x - 6$
$= x(4x - 3) + 2(4x - 3)$
$= (x + 2)(4x - 3)$

65. Factor out a GCF of 25, then factor using difference of squares.

$25b^2 - 100 = 25(b^2 - 4)$

$= 25[b^2 - (2)^2]$
$= 25(b + 2)(b - 2)$

67. Factor out a GCF of a^3b^2, then factor using difference of squares.

$a^5b^2 - 4a^3b^4 = a^3b^2(a^2 - 4b^2)$

$= a^3b^2[a^2 - (2b)^2]$

$= a^3b^2(a + 2b)(a - 2b)$

69. $3x^4 - 18x^3 + 27x^2$
Factor out GCF of $3x^2$.

$3x^2(x^2 - 6x + 9)$
Factor the trinomial.
The product of a ' c is
1(9) = 9.

Factors of 9

1	9
-1	-9
3	3
-3	-3

Sum of Factors

$1 + 9 = 10$

$-1 + (-9) = -10$

$3 + 3 = 6$

$-3 + (-3) = -6$ ✓

$3x^4 - 18x^3 + 27x^2$

$= 3x^2(x^2 - 6x + 9)$

$= 3x^2(x^2 - 3x - 3x + 9)$

$= 3x^2[x(x-3) - 3(x-3)]$

$= 3x^2(x - 3)(x - 3)$

$= 3x^2(x - 3)^2$

71. $x^3 + 25x = x(x^2 + 25)$
Factor out a GCF of x.

Since $x^2 + 25$ is the sum of 2 squares, the binomial will not factor further.

73. Factor using difference of

squares, then factor $y^2 - 4$ using difference of squares.

$y^4 - 16 = (y^2)^2 - (4)^2$

$= (y^2 + 4)(y^2 - 4)$

$= (y^2 + 4)[y^2 - (2)^2]$

$= (y^2 + 4)(y + 2)(y - 2)$

75. $10a^2 + 25ab - 60b^2$
Factor out a GCF of 5.

$5(2a^2 + 5ab - 12b^2)$
Factor the trinomial.
The product of a ' c

is 2(-12) = -24.

Factors of -24

1	-24
-1	24
2	-12
-2	12
3	-8
-3	8
4	-6
-4	6

Sum of Factors

$1 + (-24) = -23$
$-1 + 24 = 23$
$2 + (-12) = -10$
$-2 + 12 = 10$
$3 + (-8) = -5$
$-3 + 8 = 5$ ✓
$4 + (-6) = -2$
$-4 + 6 = 2$

$10a^2 + 25ab - 60b^2$

$= 5(2a^2 + 5ab - 12b^2)$

$= 5(2a^2 - 3ab + 8ab$

$- 12b^2)$

$= 5[a(2a-3b) + 4b(2a-3b)]$

$= 5(2a - 3b)(a + 4b)$

77. $9x^2 + 12x - 5$
The product of $a \cdot c$ is $9(-5) = -45$.

Factors of -45

1	-45
-1	45
3	-15
-3	15
5	-9
-5	9

Sum of Factors

$1 + (-45) = -44$
$-1 + 45 = 44$
$3 + (-15) = -12$
$-3 + 15 = 12$ ✓
$5 + (-9) = -4$
$-5 + 9 = 4$

$9x^2 + 12x - 5 = 9x^2 - 3x$
$+ 15x - 5$
$= 3x(3x - 1) + 5(3x - 5)$
$= (3x + 5)(3x - 1)$

79. Factor out a GCF of x, then factor using difference of squares.

$x^3 - 25x = x(x^2 - 25)$

$= x[x^2 - (5)^2]$
$= x(x + 5)(x - 5)$

81. a) $a^2 - b^2 = (a + b)(a - b)$

b) The difference of two squares is factorable into the product of the sum and difference of the square roots of the two squares.

83. a) $a^3 - b^3$
$= (a - b)(a^2 + ab + b^2)$

b) The difference of two cubes is factorable into the product of a binomial and a trinomial. The binomial is the difference of the cube roots of the two cubes. The trinomial is the square of the first term of the binomial plus the product of the terms of the binomial plus the square of the second term of the binomial.

Cumulative Review Exercises

85. $2x - 5y = 6$

$-5y = -2x + 6$

$$\frac{-5y}{-5} = \frac{-2x + 6}{-5}$$

$$y = \frac{2x - 6}{5}$$

87. $x^{-2}x^{-3} = x^{-2-3} = x^{-5}$

$= \frac{1}{x^5}$

Just for fun

1. Factor using sum of cubes.

$x^6 + 1 = (x^2)^3 + (1)^3$

$= (x^2 + 1)[(x^2)^2 - x^2(1) + 1^2]$

$= (x^2 + 1)(x^4 - x^2 + 1)$

2. $x^6 - 27y^9$
Factor using difference of cubes.

$= (x^2)^3 - (3y^3)^3$

$= (x^2 - 3y^3)[(x^2)^2 + x^2(3y^3) + (3y^3)^2]$

$= (x^2 - 3y^3)(x^4 + 3x^2y^3 + 9y^6$

3. We are given that a = b. If b is subtracted from both sides, the following results:

$a = b$
$a - b = b - b$
$a - b = 0$

In line 6 of the proof, both sides of the equation are divided by a - b, but a - b is zero and since division by zero is undefined, this step is invalid.

Exercise Set 5.6

1. $x(x + 3) = 0$

$x = 0$	$x + 3 = 0$
	$x + 3 - 3 = 0 - 3$
	$x = -3$

$x = 0$ or $x = -3$

3. $5x(x - 9) = 0$

$5x = 0$	$x - 9 = 0$
$\frac{5x}{5} = \frac{0}{5}$	$x - 9 + 9 = 0 + 9$
$x = 0$	$x = 9$

$x = 0$ or $x = 9$

5. $(2x + 5)(x - 3) = 0$
Set each factor to zero and solve.

$2x + 5 = 0$	$x - 3 = 0$
$2x+5-5 = 0-5$	$x-3+3 = 0+3$
$2x = -5$	$x = 3$
$\frac{2x}{2} = \frac{-5}{2}$	

$x = \frac{-5}{2}$ or $x = 3$

7. $x^2 - 16 = 0$
$(x - 4)(x + 4) = 0$
Set each factor to zero and solve.

$x - 4 = 0$	$x + 4 = 0$
$x-4+4 = 0+4$	$x+4-4 = 0-4$
$x = 4$	$x = -4$

$x = 4$ or $x = -4$

9. $x^2 - 12x = 0$
Factor out the GCF x.
$x(x - 12) = 0$
Set each factor to zero and solve

$x = 0$ | $x - 12 = 0$, $x-12+12 = 0+12$, $x = 12$

$x = 0$ or $x = 12$

11. $9x^2 + 18x = 0$
Factor out the GCF 9x.
$9x(x + 2) = 0$
Set each factor to zero and solve.

$9x = 0$, $\frac{9x}{9} = \frac{0}{9}$, $x = 0$ | $x + 2 = 0$, $x+2-2 = 0-2$, $x = -2$

$x = 0$ or $x = -2$

13. $x^2 + x - 12 = 0$
Factor
$(x + 4)(x - 3) = 0$
Set each factor to zero and solve.

$x + 4 = 0$, $x+4-4 = 0-4$, $x = -4$ | $x - 3 = 0$, $x-3+3 = 0+3$, $x = 3$

$x = -4$ or $x = 3$

15. $x^2 - 12x = -20$
Write in standard form.

$x^2 - 12x + 20 = -20 + 20$

$x^2 - 12x + 20 = 0$
Factor
$(x - 10)(x - 2) = 0$
Set each factor to zero and solve.

$x - 10 = 0$, $x-10+10 = 0+10$, $x = 10$ | $x - 2 = 0$, $x-2+2 = 0+2$, $x = 2$

$x = 10$ or $x = 2$

17. $z^2 + 3z = 18$
Write in standard form.

$z^2 + 3z - 18 = 18 - 18$

Factor

$z^2 + 3z - 18 = 0$
$(z + 6)(z - 3) = 0$
Set each factor to zero and solve.

$z + 6 = 0$, $z+6-6 = 0-6$, $z = -6$ | $z - 3 = 0$, $z-3+3 = 0+3$, $z = 3$

$z = -6$ or $z = 3$

19. $3x^2 - 6x - 72 = 0$
Factor out the GCF.

$3(x^2 - 2x - 24) = 0$
Factor again
$3(x - 6)(x + 4) = 0$
Set each factor to zero and solve.

$x - 6 = 0$, $x-6+6 = 0+6$, $x = 6$ | $x + 4 = 0$, $x+4-4 = 0-4$, $x = -4$

$x = 6$ or $x = -4$

21. $x^2 + 19x = 42$
Write in standard form.

$x^2 + 19x - 42 = 42 - 42$

$x^2 + 19x - 42 = 0$
Factor
$(x + 21)(x - 2) = 0$
Set each factor to zero and solve.

$x + 21 = 0$, $x+21-21 = 0-21$, $x = -21$ | $x - 2 = 0$, $x-2+2 = 0+2$, $x = 2$

$x = -21$ or $x = 2$

23. $2y^2 + 22y + 60 = 0$
Factor out the GCF.

$2(y^2 + 11y + 30) = 0$
Factor
$2(y + 5)(y + 6) = 0$
Set each factor to zero and solve.

$$\begin{array}{r|r} y + 5 = 0 & y + 6 = 0 \\ y+5-5 = 0-5 & y+6-6 = 0-6 \\ y = -5 & y = -6 \end{array}$$

$y = -5$ or $y = -6$

25. $-2x - 8 = -x^2$
Write in standard form.

$x^2 - 2x - 8 = -x^2 + x^2$

$x^2 - 2x - 8 = 0$
Factor
$(x - 4)(x + 2) = 0$
Set each factor to zero and solve.

$$\begin{array}{r|r} x - 4 = 0 & x + 2 = 0 \\ x-4+4 = 0+4 & x+2-2 = 0-2 \\ x = 4 & x = -2 \end{array}$$

$x = 4$ or $x = -2$

27. $-x^2 + 30x + 64 = 0$
Write in standard form.

$-1(x^2 + 30x + 64) = -1(0)$

$x^2 - 30x - 64 = 0$
Factor
$(x - 32)(x + 2) = 0$
Set each factor to zero and solve.

$$\begin{array}{r|r} x - 32 = 0 & x + 2 = 0 \\ x-32+32 = 0+32 & x+2-2 = 0+2 \\ x = 32 & x = -2 \end{array}$$

$x = 32$ or $x = -2$

29. $x^2 - 3x - 18 = 0$
Factor
$(x - 6)(x + 3) = 0$
Set each factor to zero and solve.

$$\begin{array}{r|r} x - 6 = 0 & x + 3 = 0 \\ x-6+6 = 0+6 & x+3-3 = 0-3 \\ x = 6 & x = -3 \end{array}$$

$x = 6$ or $x = -3$

31. $3p^2 = 22p - 7$
Write in standard form.

$3p^2 - 22p = 22p - 22p - 7$

$3p^2 - 22p = -7$

$3p^2 - 22p + 7 = -7 + 7$

$3p^2 - 22p + 7 = 0$
Factor
$(3p - 1)(p - 7) = 0$
Set each factor to zero and solve.

$$\begin{array}{r|r} 3p - 1 = 0 & p - 7 = 0 \\ 3p-1+1 = 0+1 & p-7+7 = 0+7 \\ 3p = 1 & p = 7 \\ \frac{3p}{3} = \frac{1}{3} & \\ p = \frac{1}{3} & \end{array}$$

$p = \frac{1}{3}$ or $p = 7$

33. $3r^2 + r = 2$
Write in standard form.

$3r^2 + r - 2 = 2 - 2$

$3r^2 + r - 2 = 0$
Factor
$(3r - 2)(r + 1) = 0$
Set each factor to zero and solve.

$$\begin{array}{r|r} 3r - 2 = 0 & r + 1 = 0 \\ 3r-2+2 = 0+2 & r+1-1 = 0-1 \\ 3r = 2 & r = -1 \\ \frac{3r}{3} = \frac{2}{3} & \\ r = \frac{2}{3} & \end{array}$$

$r = \frac{2}{3}$ or $r = -1$

35. $4x^2 + 4x - 48 = 0$
Factor out the GCF.

$4(x^2 + x - 12) = 0$
Factor
$4(x + 4)(x - 3) = 0$
Set each factor to zero and solve.

$x + 4 = 0$	$x - 3 = 0$
$x+4-4 = 0-4$	$x-3+3 = 0+3$
$x = -4$	$x = 3$

$x = -4$ or $x = 3$

37. $6x^2 - 5x = 4$
Write in standard form.

$6x^2 - 5x - 4 = 4 - 4$

$6x^2 - 5x - 4 = 0$
Factor
$(2x + 1)(3x - 4) = 0$
Set each factor to zero and solve.

$2x + 1 = 0$	$3x - 4 = 0$
$2x+1 - 1= 0-1$	$3x-4+4 = 0+4$
$2x = -1$	$3x = 4$
$\frac{2x}{2} = \frac{-1}{2}$	$\frac{3x}{3} = \frac{4}{3}$
$x = \frac{-1}{2}$	$x = \frac{4}{3}$

$x = \frac{-1}{2}$ or $x = \frac{4}{3}$

39. $2x^2 - 10x = -12$
Write in standard form.

$2x^2 - 10x + 12 = -12 + 12$

$2x^2 - 10x + 12 = 0$
Factor out the GCF.

$2(x^2 - 5x + 6) = 0$
Factor
$2(x - 3)(x - 2) = 0$
Set each factor to zero and solve.

$x - 3 = 0$	$x - 2 = 0$
$x-3+3 = 0+3$	$x-2+2 = 0+2$
$x = 3$	$x = 2$

$x = 3$ or $x = 2$

41. $2x^2 = 32x$
Write in standard form.

$2x^2 - 32x = 32x - 32x$

$2x^2 - 32x = 0$
Factor
$2x(x - 16) = 0$
Set each factor to zero and solve.

$2x = 0$	$x - 16 = 0$
$\frac{2x}{2} = \frac{0}{2}$	$x-16+16 = 16$
$x = 0$	$x = 16$

$x = 0$ or $x = 16$

43. $x^2 = 36$
Write in standard form.

$x^2 - 36 = 36 - 36$

$x^2 - 36 = 0$
Factor
$(x - 6)(x + 6) = 0$
Set each factor to zero and solve.

$x - 6 = 0$	$x + 6 = 0$
$x-6+6 = 0+6$	$x+6-6 = 0-6$
$x = 6$	$x = -6$

$x = -6$ or $x = 6$

45. $x^2 = 9$
Write in standard form.

$x^2 - 9 = 9 - 9$

$x^2 - 9 = 0$
Factor
$(x - 3)(x + 3) = 0$
Set each factor to zero and solve.

$x - 3 = 0$	$x + 3 = 0$
$x-3+3 = 0+3$	$x+3-3 = 0-3$
$x = 3$	$x = -3$

$x = 3$ or $x = -3$

47. Let x = 1st consecutive even positive integer.
Let $x + 2$ = next even positive integer.
$x(x + 2) = 80$
Write in standard form.

$x^2 + 2x = 80$

$x^2 + 2x - 80 = 80 - 80$
$x^2 + 2x - 80 = 0$
Factor
$(x + 10)(x - 8) = 0$
Set each factor to zero and solve.

$x + 10 = 0$	$x - 8 = 0$
$x+10-10 = 0-10$	$x-8+8 = 0+8$
$x = -10$	$x = 8$

-10 is not positive and not part of the answer.	1st integer = 8. 2nd integer is $x+2 = 8+2 = 10$. Two consecutive even positive integers are: 8 and 10.

49. Let x = 1st positive odd integer.
Let $x + 2$ = 2nd positive odd integer.
$x(x + 2) = 63$
Write in standard form.

$x^2 + 2x = 63$

$x^2 + 2x - 63 = 63 - 63$
$x^2 + 2x - 63 = 0$
Factor
$(x + 9)(x - 7) = 0$
Set each factor to zero and solve.

$x + 9 = 0$	$x - 7 = 0$
$x+9-9 = 0-9$	$x-7+7 = 0+7$
$x = -9$	$x = 7$

-9 is not positive and not part of the answer.	1st integer = 7, 2nd integer is: $x+2 = 7+2 = 9$. Two consecutive positive odd integers are: 7 and 9.

51. Let x = smaller positive integer.
Let $2x - 3$ = larger positive integer.
$x(2x - 3) = 35$

$2x^2 - 3x = 35$

$2x^2 - 3x - 35 = 35 - 35$

$2x^2 - 3x - 35 = 0$
Factor
$(2x + 7)(x - 5) = 0$
Set each factor to zero and solve.

$2x + 7 = 0$	$x - 5 = 0$
$2x+7-7 = 0-7$	$x-5+5 = 0+5$
$2x = -7$	$x = 5$
$\frac{2x}{2} = \frac{-7}{2}$	
$x = \frac{-7}{2}$	

Since the question asked for a positive integer, $\frac{-7}{2}$ is not part of the solution.	The smaller integer is 5, and the larger integer is: $2x - 3 = 2(5) - 3 = 10 - 3 = 7$

53. Let w = width and $4w$ = length.
Area = length · width.

$36 = w(4w)$

$36 = 4w^2$

$0 = 4w^2 - 36$

$0 = 4(w^2 - 9)$
$0 = 4(w - 3)(w + 3)$

$w - 3 = 0$ $w-3+3 = 0+3$ $w = 3$	$w + 3 = 0$ $w+3-3 = 0-3$ $w = -3$
The width is 3. The length is $4w = 4(3) = 12$.	Since dimensions are always positive, -3 is not part of the solution.

width = 3 feet
length = 12 feet

55. Let x = length of side for the smaller square.
x + 6 = length of side for the larger square.
side · side = Area of a square.
$(x + 6)(x + 6) = 64$

$x^2 + 12x + 36 = 64$

$x^2 + 12x + 36 - 64 = 64 - 64$

$x^2 + 12x - 28 = 0$
$(x + 14)(x - 2) = 0$

$x + 14 = 0$ $x+14-14 = 0-14$ $x = -14$	$x - 2 = 0$ $x-2+2 = 0+2$ $x = 2$
Since dimensions are always positive -14 is not part of the solution.	The smaller square has a side of 2 meters.

57. $s = 12$

$s = n^2 + n$

$12 = n^2 + n$

$0 = n^2 + n - 12$
$0 = (n + 4)(n - 3)$

$n + 4 = 0$ $n+4-4 = 0-4$ $n = -4$	$n - 3 = 0$ $n-3+3 = 0+3$ $n = 3$
Since n equals the number of even integers in the sum, n must be a counting number. Therefore, n must be positive and -4 cannot be the solution.	The sum of the first 3 even numbers is $2 + 4 + 6 = 12$.

59. a) The zero factor property may only be used when one side of the equation is equal to 0.

b) $(x + 1)(x - 2) = 4$
Write in standard form.

$x^2 - x - 2 = 4$

$x^2 - x - 2 - 4 = 4 - 4$

$x^2 - x - 6 = 0$
Factor
$(x - 3)(x + 2) = 0$
Set each factor to zero and solve.

$x - 3 = 0$ $x-3+3 = 3$ $x = 3$	$x + 2 = 0$ $x+2-2 = -2$ $x = -2$

$x = 3$ or $x = -2$

Cumulative Review Exercises

61. $(3x + 2) - (x^2 - 4x + 6)$

$= 3x + 2 - x^2 + 4x - 6$
$= -x + 7x - 4$

63.

$$\begin{array}{r} 2x - 3 \\ 3x-5\,\overline{)6x^2 - 19x + 15} \\ -6x^2 \pm 10x \\ \hline -9x + 15 \\ +9x \mp 15 \\ \hline \end{array}$$

Just for fun

1. a) $h = -16t^2 + 128t$

$h = -16(2)^2 + 128(2)$
$h = -16(4) + 256$
$h = -64 + 256$
$h = 192$ feet

b) $h = -16t^2 + 128t$

$0 = -16t^2 + 128t$
$0 = -16t(t - 8)$
Factor
Set each factor to zero and solve.

$-16t = 0$	$t - 8 = 0$
$\frac{-16t}{-16} = \frac{0}{-16}$	$t - 8 + 8 = 8$
$t = 0$	$t = 8$ sec

$t = 8$ seconds

2. $x^3 + 3x^2 - 10x = 0$
Factor

$x(x^2 + 3x - 10) = 0$
$x(x + 5)(x - 2) = 0$
Set each factor to 0 and solve.

$x = 0$	$x + 5 = 0$	$x - 2 = 0$
	$x+5-5 = -5$	$x-2+2 = 2$
	$x = -5$	$x = 2$

$x = 0$ or $x = -5$ or $x = 2$

Review Exercises

1. $x^3, x^5, 2x^2$

GCF is x^2

3. $18x, 24, 36y^2$

$18x = 2 \cdot 3^2x$

$24 = 2^3 \cdot 3$

$36y^2 = 2^2 \cdot 3^2 \cdot y^2$
GCF $= 2 \cdot 3 = 6$

5. $9xyz, 12xz, 36, x^2y$

$9xyz = 3^2xyz$

$12xz = 2^2 \cdot 3xz$

$36 = 2^2 \cdot 3^2$

$x^2y = x^2y$
GCF $= 1$

7. $x(2x - 5), 3(2x - 5)$
GCF $= 2x - 5$

9. $5x - 20 = 5x - 5(4)$
$= 5(x - 4)$

11. $16y^2 - 12y = 4y \cdot 4y - 4y \cdot 3$
$= 4y(4y - 3)$

13. $24x^2y + 18x^3y^2$

$= 6x^2y \cdot 4 + 6x^2y \cdot 3xy$

$= 6x^2y(4 + 3xy)$

15. $2x^2 + 4x - 8$

$= 2 \cdot x^2 + 2 \cdot 2x - 2 \cdot 4$

$= 2(x^2 + 2x - 4)$

17. $24x^2 - 13y^2 + 6xy$
Cannot be factored further.
GCF is 1.

19. $3x(x - 1) - 2(x - 1)$
$= (3x - 2)(x - 1)$

21. $x^2 + 3x + 2x + 6$
$= x(x + 3) + 2(x + 3)$
$= (x + 2)(x + 3)$

23. $x^2 - 7x + 7x - 49$
$= x(x - 7) + 7(x - 7)$
$= (x + 7)(x - 7)$

25. $3xy + 3x + 2y + 2$
$= 3x(y + 1) + 2(y + 1)$
$= (3x + 2)(y + 1)$

27. $5x^2 + 20x - x - 4$
$= 5x(x + 4) - 1(x + 4)$
$= (5x - 1)(x + 4)$

29. $12x^2 - 8xy + 15xy - 10y^2$
$= 4x(3x - 2y)$
$+ 5y(3x - 2y)$
$= (4x + 5y)(3x - 2y)$

31. $ab - a + b - 1$
$= a(b - 1) + 1(b - 1)$
$= (a + 1)(b - 1)$

33. $20x^2 - 12x + 15x - 9$
$= 4x(5x - 3) + 3(5x - 3)$
$= (4x + 3)(5x - 3)$

35. $x^2 + 6x + 8$
The product of $a \cdot c$ is
$1 \cdot 8 = 8$.

Factors of 8

1	8
-1	-8
2	4
-2	-4

Sum of Factors
$1 + 8 = 9$
$-1 + (-8) = -9$
$2 + 4 = 6$ ✓
$-2 + (-4) = -6$

$x^2 + 6x + 8$
$= (x + 2)(x + 4)$

37. $x^2 - x - 20 = x^2 - 1x - 20$
The product of $a \cdot c$ is
$1(-20) = -20$.

Factor of -20

1	-20
-1	20
2	-10
-2	10
4	-5
-4	5

Sum of Factors
$1 + (-20) = -19$
$-1 + 20 = 19$
$2 + (-10) = -8$
$-2 + 10 = 8$
$4 + (-5) = -1$ ✓
$-4 + 5 = 1$

$x^2 - x - 20 = (x - 5)(x + 4)$

39. $x^2 - 3x - 18$
The product of $a \cdot c$ is
$1(-18) = -18$.

Factors of -18

1	-18
-1	18
2	-9
-2	9
3	-6
-3	6

Sum of Factors
$1 + (-18) = -17$
$-1 + 18 = 17$
$2 + (-9) = -7$
$-2 + 9 = 7$
$3 + (-6) = -3$ ✓
$-3 + 6 = 3$

$x^2 - 3x - 18$
$= (x - 6)(x + 3)$

41. $x^2 - 12x - 45$
The product of a ' c is
$1(-45) = -45$.

Factors of -45

1	-45
-1	45
3	-15
-3	15
5	-9
-5	9

Sum of Factors
$1 + (-45) = -44$
$-1 + 45 = 45$
$3 + (-15) = -12$ ✓
$-3 + 15 = 12$
$5 + (-9) = -4$
$-5 + 9 = 4$

$x^2 - 12x - 45$
$= (x - 15)(x + 3)$

43. $x^3 + 5x^2 + 4x$
Factor out a GCF of x.

$x(x^2 + 5x + 4)$
Factor the trinomial.
The product of a ' c is
$1(4) = 4$.

Factors of 4

1	4
-1	-4
2	2
-2	-2

Sum of Factors
$1 + 4 = 5$ ✓
$-1 + (-4) = -5$
$2 + 2 = 4$
$-2 + (-2) = -4$

$x^3 + 5x^2 + 4x$

$x(x^2 + 5x + 4)$
$= x(x + 4)(x + 1)$

45. $x^2 - 2xy - 15y^2$
The product of a(c) is
$1(-15) = -15$.

Factors of -15

1	-15
-1	15
3	-5
-3	5

Sum of Factors
$1 + (-15) = -14$
$-1 + 15 = 14$
$3 + (-5) = -2$ ✓
$-3 + 5 = 2$

$x^2 - 2xy - 15y^2$
$= (x + 3y)(x - 5y)$

47. $2x^2 + 7x - 4$
The product of a ' c is
$2(-4) = -8$.

Factors of -8

1	-8
-1	8
2	-4
-2	4

Sum of Factors
$1 + (-8) = -7$
$-1 + 8 = 7$ ✓
$2 + (-4) = -2$
$-2 + 4 = 2$

$2x^2 + 7x - 4$

$= 2x^2 - 1x + 8x - 4$
$= x(2x - 1) + 4(2x - 1)$
$= (2x - 1)(x + 4)$

49. $4x^2 - 9x + 5$
The product of a(c) is
$4(5) = 20$.

Factors of 20

1	20
-1	-20
2	10
-2	-10
4	5
-4	-5

Sum of Factors

$1 + 20 = 21$
$-1 + (-20) = -21$
$2 + 10 = 12$
$-2 + (-10) = -12$
$4 + 5 = 9$
$-4 + (-5) = -9$ ✓

$4x^2 - 9x + 5$

$= 4x^2 - 4x - 5x + 5$
$= 4x(x - 1) - 5(x - 1)$
$= (4x - 5)(x - 1)$

51. $4x^2 + 4x - 15$
The product of a(c) is
$4(-15) = 60$.

Factors of 60

1	-60
-1	60
2	-30
-2	30
3	-20
-3	20
4	-15
-4	15
5	-12
-5	12
6	-10
-6	10

Sum of Factors

$1 + (-60) = -59$
$-1 + 60 = 59$
$2 + (-30) = -28$
$-2 + 30 = 28$
$3 + (-20) = -17$
$-3 + 20 = 17$
$4 + (-15) = -11$
$-4 + 15 = 11$
$5 + (-12) = -7$
$-5 + 12 = 7$
$6 + (-10) = -4$
$-6 + 10 = 4$ ✓

$4x^2 + 4x - 15$

$= 4x^2 - 6x + 10x - 15$
$= 2x(2x - 3) + 5(2x - 3)$
$= (2x + 3)(2x - 5)$

53. $3x^2 + 13x + 12$
The product of a(c) is
$3(12) = 36$.

Factors of 36

1	36
-1	-36
2	18
-2	-18
3	12
-3	-12
4	9
-4	-9
6	6
-6	-6

Sum of Factors

$1 + 36 = 37$
$-1 + (-36) = -37$
$2 + 18 = 20$
$-2 + (-18) = -20$
$3 + 12 = 15$
$-3 + (-12) = -15$
$4 + 9 = 13$ ✓
$-4 + (-9) = -13$
$6 + 6 = 12$
$-6 + (-6) = -12$

$3x^2 + 13x + 12$

$= 3x^2 + 9x + 4x + 12$
$= 3x(x + 3) + 4(x + 3)$
$= (3x + 4)(x + 3)$

55. $2x^2 + 9x - 35$
The product of a ' c is
$2(-35) = -70$.

Factors of -70

-1	70
1	-70
-2	35
2	-35
-5	14
5	-14
-7	10
7	-10

Sum of Factors

$-1 + 70 = 69$
$1 + (-70) = -69$
$-2 + 35 = 33$
$2 + (-35) = -33$
$-5 + 14 = 9$ ✓
$5 + (-14) = -9$
$-7 + 10 = 3$
$7 + (-10) = -3$

$2x^2 + 9x - 35$

$= 2x^2 - 5x + 14x - 35$
$= x(2x - 5) + 7(2x - 5)$
$= (2x - 5)(x + 7)$

57. $8x^2 - 18x - 35$
The product of a ' c is $8(-35) = -280$.

Factors of -280

1	-280
-1	280
2	-140
-2	140
4	-70
-4	70
5	-56
-5	56
7	-40
-7	40
10	-28
-10	28
14	-20
-14	20

Sum of Factors

$1 + (-280) = -279$
$-1 + 280 = 279$
$2 + (-140) = -138$
$-2 + 140 = 138$
$4 + (-70) = -66$
$-4 + 70 = 66$
$5 + (-56) = -51$
$-5 + 56 = 51$
$7 + (-40) = -33$
$-7 + 40 = 33$
$10 + (-28) = -18$ ✓
$-10 + 28 = 18$
$14 + (-20) = -6$
$-14 + 20 = 6$

$8x^2 - 18x - 35$

$= 8x^2 - 28x + 10x - 35$
$= 4x(2x - 7) + 5(2x - 7)$
$= (4x + 5)(2x - 7)$

59. $9x^3 - 12x^2 = 4x$
Factor out a GCF of x.

$x(9x^2 - 12x + 4)$
Factor the trinomial.
The product of a ' c is $9(4) = 36$.

Factors of 36

1	36
-1	-36
2	18
-2	-18
3	12
-3	-12
4	9
-4	-9
6	6
-6	-6

Sum of Factors

$1 + 36 = 37$
$-1 + (-36) = -37$
$2 + 18 = 20$
$-2 + (-18) = -20$
$3 + 12 = 15$
$-3 + (-12) = -15$
$4 + 9 = 13$
$-4 + (-9) = -13$
$6 + 6 = 12$
$-6 + (-6) = -12$ ✓

$9x^3 - 12x^2 + 4x$

$= x(9x^2 - 12x + 4)$

$= x(9x^2 - 6x - 6x + 4)$
$= x[3x(3x-2) - 2(3x-2)]$
$= x(3x - 2)(3x - 2)$

$= x(3x - 2)^2$

61. $4x^2 - 16xy + 15y^2$
The product of a(c) is $4(15) = 60$.

Factors of 60

1	60
-1	-60
2	30
-2	-30
3	20
-3	-20
4	15
-4	-15
5	12
-5	-12
6	10
-6	-10

Sum of Factors

$1 + 60 = 61$
$-1 + (-60) = -61$
$2 + 30 = 32$
$-2 + (-30) = -32$
$3 + 20 = 23$
$-3 + (-20) = -23$
$4 + 15 = 19$
$-4 + (-15) = -19$
$5 + 12 = 17$
$-5 + (-12) = -17$
$6 + 10 = 16$
$-6 + (-10) = -16$ ✓

$4x^2 - 16xy + 15y^2$

$= 4x^2 - 6xy - 10xy + 15y^2$
$= 2x(2x-3y) - 5y(2x-3y)$
$= (2x - 3y)(2x - 5y)$

63. $x^2 - 25 = x^2 - (5)^2$
$= (x + 5)(x - 5)$

65. $4x^2 - 16 = 4(x^2 - 4)$

$= 4[x^2 - (2)^2]$
$= 4(x + 2)(x - 2)$

67. $64x^4 - 81y^4$

$= (8x^2)^2 - (9y^2)^2$

$= (8x^2 + 9y^2)(8x^2 - 9y^2)$

69. $4x^4 - 9y^4$

$= (2x^2)^2 - (3y^2)^2$

$= (2x^2 + 3y^2)(2x^2 - 3y^2)$

71. $x^3 - y^3$

$= (x - y)[x^2 + x(y) + y^2]$

$= (x - y)(x^2 + xy + y^2)$

73. $a^3 + 8 = a^3 + 2^3$

$= (a + 2)[a^2 - a(2) + 2^2]$

$= (a + 2)(a^2 - 2a + 4)$

75. $a^3 + 27 = a^3 + 3^3$

$= (a + 3)[a^2 - a(3) + 3^2]$

$= (a + 3)(a^2 - 3a + 9)$

77. $8x^3 = y^3 = (2x)^3 - y^3$

$= (2x-y)[(2x)^2+2x(y)+y^2]$

$= (2x - y)(4x^2 + 2xy + y^2)$

79. $8x^2 + 16x - 24$
Factor out a GCF of 8.

$8(x^2 + 2x - 3)$
Factor the trinomial.
The product of a ' c is
$1(-3) = -3$.

Factors of -3

1	-3
-1	3

Sum of Factors
$1 + (-3) = -2$
$-1 + 3 = 2$ ✓

$8x^2 + 16x - 24$

$= 8(x^2 + 2x - 3)$
$= 8(x + 3)(x - 1)$

81. Factor out a GCF of 4, then factor using difference of squares.

$4x^2 - 36 = 4(x^2 - 9)$

$= 4[x^2 - (3)^2]$
$= 4(x + 3)(x - 3)$

83. $x^2 - 10x + 24$

The product of $a \cdot c$ is $1(24) = 24$.

Factors of 24

1	24
-1	-24
2	12
-2	-12
3	8
-3	-8
4	6
-4	-6

Sum of Factors

$1 + 24 = 25$

$-1 + (-24) = -25$

$2 + 12 = 14$

$-2 + (-12) = -14$

$3 + 8 = 11$

$-3 + (-8) = -11$

$4 + 6 = 10$

$-4 + (-6) = -10$ ✓

$x^2 - 10x + 24$

$= (x - 6)(x - 4)$

85. $4x^2 - 4x - 15$

The product of $a \cdot c$ is $4(-15) = -60$.

Factors of -60

1	-60
-1	60
2	-30
-2	30
3	-20
-3	20
4	-15
-4	15
5	-12
-5	12
6	-10
-6	10

Sum of Factors

$1 + (-60) = -59$

$-1 + 60 = 59$

$2 + (-30) = -28$

$-2 + 30 = 28$

$3 + (-20) = -17$

$-3 + 20 = 17$

$4 + (-15) = -11$

$-4 + 15 = 11$

$5 + (-12) = -7$

$-5 + 12 = 7$

$6 + (-10) = -4$ ✓

$-6 + 10 = 4$

$4x^2 - 4x - 15$

$= 4x^2 - 10x + 6x - 15$

$= 2x(2x-5)(+ 3(2x-5)$

$= (2x + 3)(2x - 5)$

87. Factor out a GCF of 8, then factor using difference of cubes.

$8x^3 - 8 = 8(x^3 - 1)$

$= 8[x^3 - (1)^3]$

$= 8(x - 1)[(x^2+x(1)+1^2]$

$= 8(x - 1)(x^2 + x + 1)$

89. Factor out a GCF of y, then factor by grouping.

$x^2y - xy + 4xy - 4y$

$= y(x^2 - x + 4x - 4)$

$= y[x(x-1) + 4(x-1)]$

$= y(x + 4)(x - 1)$

91. $x^2 + 5xy + 6y^2$

The product of $a \cdot c$ is $1(6) = 6$.

Factors of 6

1	6
-1	-6
2	3
-2	-3

Sum of Factors

$1 + 6 = 7$
$-1 + (-6) = -7$
$2 + 3 = 5$ ✓
$-2 + (-3) = -5$

$x^2 + 5xy + 6y^2$
$= (x + 2y)(x + 3y)$

93. $4x^2 - 20xy + 25y^2$
The product of a ' c is
$4(25) = 100$.

Factors of 100

1	100
-1	-100
2	50
-2	-50
4	25
-4	-25
5	20
-5	-20
10	10
-10	-10

Sum of Factors

$1 + 100 = 101$
$-1 + (-100) = -101$
$2 + 50 = 52$
$-2 + (-50) = -52$
$4 + 25 = 29$
$-4 + (-25) = -29$
$5 + 20 = 25$
$-5 + (-20) = -25$
$10 + 10 = 20$
$-10 + (-10) = -20$ ✓

$4x^2 - 20xy + 25y^2$

$= 4x^2 - 10xy - 10xy + 25y^2$
$= 2x(2x - 5y)$
$\quad - 5y(2x - 5y)$
$= (2x - 5y)(2x - 5y)$

$= (2x - 5y)^2$

95. Factor by grouping.

$ab + 7a + 6b + 42$
$= a(b + 7) + 6(b + 7)$
$= (a + 6)(b + 7)$

97. $2x^3 + 12x^2y + 16xy^2$

Factor out GCF of 2x.

$2x(x^2 + 6xy + 8y^2)$
Factor the trinomial.
The product of a ' c
$= 1(8) = 8$.

Factors of 8

1	8
-1	-8
2	4
-2	-4

Sum of Factors

$1 + 8 = 9$
$-1 + (-8) = -9$
$2 + 4 = 6$ ✓
$-2 + (-4) = -6$

$2x^3 + 12x^2y = 16xy^2$

$= 2x(x^2 + 6xy + 8y^2)$
$= 2x(x + 2y)(x + 4y)$

99. $32x^3 + 32x^2 + 6x$
Factor out a GCF of 2x.

$2x(16x^2 + 16x + 3)$
Factor the trinomial.
The product of a ' c is
$16(3) = 48$.

Factors of 48

1	48
-1	-48
2	24
-2	-24
3	16
-3	-16
4	12
-4	-12
6	8
-6	-8

Sum of Factors

$1 + 48 = 49$
$-1 + (-48) = -49$
$2 + 24 = 26$
$-2 + (-24) = -26$
$3 + 16 = 19$
$-3 + (-16) = -19$
$4 + 12 = 16$ ✓
$-4 + (-12) = -16$
$6 + 8 = 14$
$-6 + (-8) = -14$

$32x^3 + 32x^2 + 6x$

$= 2x(16x^2 + 16x + 3)$

$= 2x(16x^2 + 4x + 12x + 3)$
$= 2x[4x(4x+1) + 3(4x+1)]$
$= 2x(4x + 1)(4x + 3)$

101. $x(x - 4) = 0$
Set each factor to zero and solve.

$x = 0$	$x - 4 = 0$
	$x - 4 + 4 = 0 + 4$
	$x = 4$

$x + 0$ or $x = 4$

103. $(x - 5)(3x + 2) = 0$
Set each factor to zero and solve.

$x - 5 = 0$	$3x + 2 = 0$
$x-5+5 = 0+5$	$3x+2-2 = 0-2$
$x = 5$	$3x = -2$
	$\frac{3x}{3} = \frac{-2}{3}$
	$x = -\frac{2}{3}$

$x = 5$ or $x = -\frac{2}{3}$

105. $5x^2 + 20x = 0$
Factor out a GCF of 5x.
$5x(x + 4) = 0$
Set each factor to zero and solve.

$5x = 0$	$x + 4 = 0$
$\frac{5x}{5} = \frac{0}{5}$	$x + 4 - 4 = 0 - 4$
$x = 0$	$x = -4$

$x = 0$ or $x = -4$

107. $x^2 + 8x + 15 = 0$
Factor the trinomial.
$(x + 5)(x + 3) = 0$
Set each factor to zero and solve.

$x + 5 = 0$	$x + 3 = 0$
$x+5-5 = 0-5$	$x+3-3 = 0-3$
$x = -5$	$x = -3$

$x = -5$ or $x = -3$

109. $x^2 - 12 = -x$
Write in standard form

$x^2 + x - 12 = -x + x$

$x^2 + x - 12 = 0$
Factor the trinomial.
$(x + 4)(x - 3) = 0$
Set each factor to zero and solve.

$x + 4 = 0$	$x - 3 = 0$
$x+4-4 = 0-4$	$x-3+3 = 0+3$
$x = -4$	$x = 3$

$x = -4$ or $x = 3$

111. $x^2 - 6x + 8 = 0$
Factor the trinomial.
$(x - 4)(x - 2) = 0$
Set each factor to zero and solve.

$x - 4 = 0$	$x - 2 = 0$
$x-4+4 = 0+4$	$x-2+2 = 0+2$
$x = 4$	$x = 2$

$x = 4$ or $x = 2$

113. $8x^2 - 3 = -10x$
Write in standard form.

$8x^2 + 10x - 3 = -10x + 10x$

$8x^2 + 10x - 3 = 0$
Factor the trinomial.
$(4x - 1)(2x + 3) = 0$
Set each factor to zero and solve.

$4x - 1 = 0$	$2x + 3 = 0$
$4x-1+1 = 0+1$	$2x+3-3 = 0-3$
$4x = 1$	$2x = -3$
$\frac{4x}{4} = \frac{1}{4}$	$\frac{2x}{2} = \frac{-3}{2}$
$x = \frac{1}{4}$	$x = \frac{-3}{2}$

$x = \frac{1}{4}$ or $x = -\frac{3}{2}$

115. $4x^2 - 16 = 0$
Factor out a GCF of 4.

$4(x^2 - 4) = 0$
Factor the difference of squares.
$4(x + 2)(x - 2)$
Set each factor to zero and solve.

$x + 2 = 0$	$x - 2 = 0$
$x+2-2 = 0-2$	$x-2+2 = 0+2$
$x = -2$	$x = 2$

$x = -2$ or $x = 2$

117. Let x = the first consecutive positive integer and x + 1 = the second consecutive positive integer.
$x(x + 1) = 110$
Write in standard form.

$x^2 + 1x = 110$

$x^2 + 1x - 110 = 0$
Factor the trinomial
$(x + 11)(x - 10) = 0$
Set each factor to zero and solve.

$x + 11 = 0$	$x - 10 = 0$
$x+11-11 = 0-11$	$x-10+10 = 0+10$
$x = -11$	$x = 10$
Since the problem asks for positive integers, -11 is not part of the solution.	The first integer is 10 and the second is: $x + 1 = (10)+1 = 11$.

10 and 11 are the positive consecutive integers.

119. Let x = smaller positive integer.
Let 2x - 2 = larger positive integer.

$x(2x - 2) = 40$
Write in standard form.

$2x^2 - 2x = 40$
$2x^2 - 2x - 40 = 40 - 40$
$2x^2 - 2x - 40 = 0$
Factor out a GCF of 2.

$2(x^2 - x - 20) = 0$
Factor the trinomial.
$2(x - 5)(x + 4) = 0$
Set each factor to zero and solve.

$x - 5 = 0$	$x + 4 = 0$
$x-5+5 = 0+5$	$x+4-4 = 0-4$
$x = 5$	$x = -4$
Since the smaller number is 5, the larger number is: $2x - 2 = 2(5) - 2 = 8$	Since the problem asks for positive integers, -4 is not part of the solution.

The two positive integers are 5 and 8.

121. Let x = side of the smaller square.

$x + 4$ = side of the larger square.

Length = area of a square.
$(x + 4)(x + 4) = 81$

$x^2 + 8x + 16 = 81$
Write in standard form.

$x^2 + 8x + 16 - 81 = 81 - 81$
$x^2 + 8x - 65 = 0$
Factor
$(x + 13)(x - 5) = 0$
Set each factor to zero and solve.

$x + 13 = 0$	$x - 5 = 0$
$x+13-13 = 0-13$	$x-5+5 = 0+5$
$x = -13$	$x = 5$

Since dimensions are always positive -13 is not part of the solution.	The side of the smaller square is 5 inches and the side of the larger square is $x+4 = 5+4 = 9$ inches.

Practice Test

1. $4x^4,\ 12x^5,\ 10x^2$

$4x^4 = 2^2x^4$

$12x^5 = 2^2 \cdot 3x^5$

$10x^7 = 2 \cdot 5x^7$

GCF $= 2x^2$

2. $6x^2y^3,\ 9xy^2,\ 12xy^5$

$6x^2y^3 = 2 \cdot 3x^2y^3$

$9xy^2 = 3^2xy^2$

$12xy^5 = 2^2 \cdot 3xy^5$

GCF $= 3xy^2$

3. $4x^2y - 8xy = 4xy(x - 2)$
Factor out a GCF of $4xy$.

4. $24x^2y - 6xy + 9x$
$= 3x(8xy - 2y + 3)$
Factor out a GCF of $3x$.

5. $x^2 - 3x + 2x - 6$
$= x(x - 3) + 2(x - 3)$
$= (x + 2)(x - 3)$
Factor by grouping.

6. $3x^2 - 12x + x - 4$
$= 3x(x - 4) + 1(x - 4)$
$= (3x + 1)(x - 4)$
Factor by grouping.

7. $5x^2 - 15xy - 3xy + 9y^2$
$= 5x(x - 3y) - 3y(x - 3y)$
$= (5x - 3y)(x - 3y)$
Factor by grouping.

8. $x^2 + 12x + 32$
The product of $a \cdot c$ is $1(32) = 32$.

Factors of 32

1	32
-1	-32
2	16
-2	-16
4	8
-4	-8

Sum of Factors
$1 + 32 = 33$
$-1 + (-32) = -33$
$2 + 16 = 18$
$-2 + (-16) = -18$
$4 + 8 = 12$ ✓
$-4 + (-8) = -12$

$x^2 + 12x + 32$
$= (x + 4)(x + 8)$

9. $x^2 + 5x - 24$
The product of $a \cdot c$ is $1(-24) = -24$.

Factors of -24

1	-24
-1	24
2	-12
-2	12
3	-8
-3	8
4	-6
-4	6

Sum of Factors

$1 + (-24) = -23$
$-1 + 24 = 23$
$2 + (-12) = -10$
$-2 + 12 = 10$
$3 + (-8) = -5$
$-3 + 8 = 5$ ✓
$4 + (-6) = -2$
$-4 + 6 = 2$

$x^2 + 5x - 24$
$= (x + 8)(x - 3)$

10. $x^2 - 9xy + 20y^2$
The product of a · c is
$1(20) = 20$.

Factors of 20

1	20
-1	-20
2	10
-2	-10
4	5
-4	-5

Sum of Factors

$1 + 20 = 21$
$-1 + (-20) = -21$
$2 + 10 = 12$
$-2 + (-10) = -12$
$4 + 5 = 9$
$-4 + (-5) = -9$ ✓

$x^2 - 9xy + 20y^2$
$= (x - 5y)(x - 4y)$

11. $2x^2 - 22x + 60$
Factor out a GCF of 2.

$2(x^2 - 11x + 30)$
Factor the trinomial.
The product of a · c is
$1(30) = 30$.

Factors of 30

1	30
-1	-30
2	15
-2	-15
3	10
-3	-10
5	6
-5	-6

Sum of Factors

$1 + 30 = 31$
$-1 + (-30) = -31$
$2 + 15 = 17$
$-2 + (-15) = -17$
$3 + 10 = 13$
$-3 + (-10) = -13$
$5 + 6 = 11$
$-5 + (-6) = -11$ ✓

$2x^2 - 22x + 60$

$= 2(x^2 - 11x + 30)$
$= 2(x - 5)(x - 6)$

12. $2x^3 - 3x^2 + x$
Factor out a GCF of x.

$x(2x^2 - 3x + 1)$
Factor the trinomial.
The product of a · c is
$2(1) = 2$.

Factors of 2

1	2
-1	-2

Sum of Factors

$1 + 2 = 3$
$-1 + (-2) = -3$ ✓

$2x^3 - 3x^2 + x$

$= x(2x^2 - 3x + 1)$

$= x[2x^2 - 1x + (-2x) + 1]$
$= x[x(2x-1) - 1(2x-1)]$
$= x(2x - 1)(x - 1)$

13. $12x^2 - xy - 6y^2$

$= 12x^2 - 1xy - 6y^2$
The product of a · c is

$12(-6) = -72$

Factors of -72

1	−72
−1	72
2	−36
−2	36
3	−24
−3	24
4	−18
−4	18
6	−12
−6	12
8	−9
−8	9

Sum of Factors

$1 + (-72) = -71$
$-1 + 72 = 71$
$2 + (-36) = -34$
$-2 + 36 = 34$
$3 + (-24) = -21$
$-3 + 24 = 21$
$4 + (-18) = -14$
$-4 + 18 = 14$
$6 + (-12) = -6$
$-6 + 12 = 6$
$8 + (-9) = -1$ ✓
$-8 + 9 = 1$

$12x^2 - xy - 6y^2$

$= 12x^2 - 9xy + 8xy - 6y^2$
$= 3x(4x-3y) + 2y(4x-3y)$
$= (3x + 2y)(4x - 3y)$

14. $x^2 - 9y^2 = x^2 - (3y)^2$
$= (x + 3y)(x - 3y)$
Factor using difference of squares.

15. $x^3 + 27 = x^3 + 3^3$

$= (x + 3)[x^2-x(3)+3^2]$

$= (x + 3)(x^2 - 3x + 9)$
Factor using sum of cubes.

16. $(x - 2)(2x - 5) = 0$
Set each factor to zero and solve.

$x - 2 = 0$	$2x - 5 = 0$
$x-2+2 = 0+2$	$2x-5+5 = 0+5$
$x = 2$	$2x = 5$
	$\frac{2x}{2} = \frac{5}{2}$
	$x = \frac{5}{2}$

$x = 2$ or $x = \frac{5}{2}$

17. $x^2 + 6 = -5x$
Write in standard form.

$x^2 + 5x + 6 = -5x + 5x$

$x^2 + 5x + 6 = 0$
Factor
$(x + 3)(x + 2) = 0$
Set each factor to zero and solve.

$x + 3 = 0$	$x + 2 = 0$
$x+3-3 = 0-3$	$x+2-2 = 0-2$
$x = -3$	$x = -2$

$x = -3$ or $x = -2$

18. $x^2 + 4x - 5 = 0$
Factor
$(x + 5)(x - 1) = 0$
Set each factor to zero and solve.

$x + 5 = 0$	$x - 1 = 0$
$x+5-5 = 0-5$	$x-1+1 = 0+1$
$x = -5$	$x = 1$

$x = -5$ or $x = 1$

19. Let x - the smaller positive integer.
$2x + 1$ = the larger positive integer.
$x(2x + 1) = 36$
Write in standard form.

$2x^2 + x = 36$

$2x^2 + x - 36 = 36 - 36$

$2x^2 + x - 36 = 0$
Factor
$(x - 4)(2x + 9) = 0$
Set each factor to zero and solve.

$x - 4 = 0$ $x-4+4 = 0+4$ $x = 4$	$2x + 9 = 0$ $2x+9-9 = 0-9$ $2x = -9$ $\frac{2x}{2} = \frac{-9}{2}$ $x = -\frac{9}{2}$
The smaller number is 4. The larger number is $2x+1$ $= 2(4)+1 = 9$	Since $-\frac{9}{2}$ is not positive, $-\frac{9}{2}$ is not part of the solution.

20. Let w = width.
$w + 2$ = length.

Length ' width = area
$(w + 2)(w) = 24$

$w^2 + 2w = 24$
Write in standard form.

$w^2 + 2w - 24 = 24 - 24$

$w^2 + 2w - 24 = 0$
Factor
$(w + 6)(w - 4) = 0$
Set each factor to zero and solve.

$w + 6 = 0$ $w+6-6 = 0-6$ $w = -6$	$w - 4 = 0$ $w-4+4 = 0+4$ $w = 4$
Since dimensions are always positive, −6 is not part of the solution.	The width is 4 and the length is $w+2 = 4+2 = 6$. Width = 4 meters. Length = 6 meters.

CHAPTER SIX

Exercise Set 6.1

1. $\frac{x + 4}{x}$ is defined for all real numbers except $x = 0$.

3. $\frac{4}{x - 6}$ is defined for all real numbers except when $x - 6 = 0$ or when $x = 6$.

5. $\frac{x + 4}{x^2 - 4} = \frac{x + 4}{(x - 2)(x + 2)}$

   The expression is defined for all real numbers except when $x - 2 = 0$ and $x + 2 = 0$ or when $x = 2$ and $x = -2$.

7. $\frac{x - 3}{x^2 + 6x - 16} = \frac{x - 3}{(x + 8)(x - 2)}$

   The expression is defined for all real numbers except when $x + 8 = 0$ and $x - 2 = 0$ or when $x = -8$ and $x = 2$.

9. $\frac{x}{x + xy} = \frac{\cancel{x}}{\cancel{x}(1 + y)}$

   $= \frac{1}{1 + y}$

11. $\frac{4x + 12}{x + 3} = \frac{4(\cancel{x + 3})}{\cancel{x + 3}}$

    $= 4$

13. $\frac{x^3 + 6x^2 + 3x}{2x}$

    $= \frac{\cancel{x}[x^2 + 6x + 3]}{2\cancel{x}}$

    $= \frac{x^2 + 6x + 3}{2}$

15. $\frac{x^2 + 2x + 1}{x + 1}$

    $= \frac{(x + 1)(\cancel{x + 1})}{\cancel{x + 1}}$

    $= x + 1$

17. $\frac{x^2 - 2x}{x^2 - 4x + 4}$

    $= \frac{x(\cancel{x - 2})}{(x - 2)(\cancel{x - 2})}$

    $= \frac{x}{x - 2}$

19. $\frac{x^2 - x - 6}{x^2 - 4}$

    $= \frac{(x - 3)(\cancel{x + 2})}{(x - 2)(\cancel{x + 2})}$

    $= \frac{x - 3}{x - 2}$

21. $\frac{2x^2 - 4x - 6}{x - 3}$

    $= \frac{2[x^2 - 2x - 3]}{x - 3}$

    $= \frac{2(\cancel{x - 3})(x + 1)}{(\cancel{x - 3})}$

    $= 2(x + 1)$

23. $\frac{2x - 3}{3 - 2x} = \frac{\overset{-1}{-(\cancel{3 - 2x})}}{\underset{1}{\cancel{3 - 2x}}}$

    $= -1$

25. $\frac{x^2 - 2x - 8}{4 - x}$

    $= \frac{(x - 4)(x + 2)}{4 - x}$

$= \dfrac{-(4 - x)(x + 2)}{4 - x}$

$= -(x + 2)$

27. $\dfrac{x^2 + 3x - 18}{-2x^2 + 6x}$

$= \dfrac{(x + 6)(x - 3)}{-2x(x - 3)}$

$= \dfrac{-1(x + 6)}{2x}$

29. $\dfrac{2x^2 + 5x - 3}{1 - 2x}$

$= \dfrac{(2x - 1)(x + 3)}{1 - 2x}$

$= \dfrac{-1(1 - 2x)(x + 3)}{1 - 2x}$

$= -(x + 3)$

31. $\dfrac{6x^2 + x - 2}{2x - 1}$

$= \dfrac{(2x - 1)(3x + 2)}{(2x - 1)}$

$= 3x + 2$

33. $\dfrac{6x^2 + 7x - 20}{2x + 5}$

$= \dfrac{(2x + 5)(3x - 4)}{(2x + 5)}$

$= 3x - 4$

35. $\dfrac{6x^2 - 13x + 6}{3x - 2}$

$= \dfrac{(3x - 2)(2x - 3)}{(3x - 2)}$

$= 2x - 3$

37. $\dfrac{x^2 - 3x + 4x - 12}{x - 3}$

$= \dfrac{x^2 + x - 12}{x - 3}$

$= \dfrac{(x + 4)(x - 3)}{x - 3}$

$= x + 4$

39. $\dfrac{2x^2 - 8x + 3x - 12}{2x^2 + 8x + 3x + 12}$

$= \dfrac{2x^2 - 5x - 12}{2x^2 + 11x + 12}$

$= \dfrac{(2x + 3)(x - 4)}{(2x + 3)(x + 4)}$

$= \dfrac{x - 4}{x + 4}$

41. $\dfrac{x^3 - 8}{x - 2}$

$= \dfrac{(x - 2)(x^2 + 2x + 4)}{x - 2}$

$= x^2 + 2x + 4$

43. $\dfrac{x + 3}{x^2 + 4}$ is defined for all real numbers, because the denominator, x^2+4, is never equal to 0.

45. $\dfrac{x}{(x - 4)^2}$ is undefined when $x - 4 = 0$ or when $x = 4$.

47. $-\dfrac{3x + 2}{-3x - 2} = 1$, because

$-3x - 2 = -(3x + 2)$.

$-\dfrac{3x + 2}{-3x - 2} = -\dfrac{3x + 2}{-(3x + 2)}$

$= \dfrac{3x + 2}{3x + 2} = 1$

49. $\dfrac{x^2 - x - 6}{(?)} = x - 3$

$\dfrac{(x - 3)(x + 2)}{(x + 2)} = x - 3$

The factor $(x + 2)$ makes the statement true.

Cumulative Review Exercises

51. $z = \frac{x - y}{2}$

$2z = \frac{x - y}{2} \cdot \frac{2}{1}$

$2z = x - y$

$2z + y = x$

$y = x - 2z$

53. $\left(\frac{3x^6y^2}{9x^4y^3}\right)^2 = \left(\frac{x^2}{3y}\right)^2$

$= \frac{x^4}{9y^2}$

Exercise Set 6.2

1. $\frac{3x}{2y} \cdot \frac{y^2}{6} = \frac{3xy^2}{2y6} = \frac{xy}{4}$

3. $\frac{16x^2}{y^4} \cdot \frac{5x^2}{y^2} = \frac{80x^4}{y^6}$

5. $\frac{6x^5y^3}{5z^3} \cdot \frac{6x^4}{5yz^4} = \frac{6x^5y^3}{5z^3} \cdot \frac{6x^4}{5yz^4}$

$= \frac{36x^9y^2}{25z^7}$

7. $\frac{3x - 2}{3x + 2} \cdot \frac{4x - 1}{1 - 4x}$

$= \frac{3x - 2}{3x + 2} \cdot \frac{(-1)(1 - 4x)}{1 - 4x}$

$= \frac{(-1)(3x - 2)}{3x + 2}$

$= \frac{-3x + 2}{3x + 2}$

9. $\frac{x^2 + 7x + 12}{x + 4} \cdot \frac{1}{x + 3}$

$= \frac{(x + 4)(x + 3)}{x + 4} \cdot \frac{1}{x + 3}$

$= 1$

11. $\frac{a^2 - b^2}{a} \cdot \frac{a^2 + ab}{a + b}$

$= \frac{(a + b)(a - b)}{a} \cdot \frac{a(a + b)}{a + b}$

$= (a - b)(a + b)$ or $a^2 - b^2$

13. $\frac{6x^2 - 14x - 12}{(6x + 4)} \cdot \frac{x + 3}{2x^2 - 2x - 12}$

$= \frac{2(3x + 2)(x - 3)}{2(3x + 2)} \cdot \frac{x + 3}{2(x - 3)(x + 2)}$

$= \frac{x + 3}{2(x + 2)}$

15. $\frac{x + 3}{x - 3} \cdot \frac{x^3 - 27}{x^2 + 3x + 9}$

$= \frac{(x + 3)}{(x - 3)} \cdot \frac{(x - 3)(x^2 + 3x + 9)}{(x^2 + 3x + 9)}$

$= x + 3$

17. $\frac{6x^3}{y} \div \frac{2x}{y^2} = \frac{6x^3}{y} \cdot \frac{y^2}{2x}$

$= 3x^2y$

19. $\frac{25xy^2}{7z} \div \frac{5x^2y^2}{14z^2}$

$= \frac{25xy^2}{7z} \cdot \frac{14z^2}{5x^2y^2}$

$= \frac{10z}{x}$

21. $\frac{7a^2b}{xy} \div \frac{7}{6xy} = \frac{7a^2b}{xy} \cdot \frac{6xy}{7}$

$= 6a^2b$

23. $\frac{3x^2 + 6x}{x} \div \frac{2x + 4}{x^2}$

$= \frac{3x^2 + 6x}{x} \cdot \frac{x^2}{2x + 4}$

$= \frac{3x(x + 2)}{x} \cdot \frac{x^2}{2(x + 2)}$

$= \frac{3x^2}{2}$

25. $(x - 3) \div \frac{x^2 + 3x - 18}{x}$

$= (x - 3) \cdot \frac{x}{x^2 + 3x - 18}$

$= \frac{x - 3}{1} \cdot \frac{x}{(x + 6)(x - 3)}$

$= \frac{x}{x + 6}$

27. $\frac{x^2 - 12x + 32}{x^2 - 6x - 16}$

$\div \frac{x^2 - x - 12}{x^2 - 5x - 24}$

$= \frac{x^2 - 12x + 32}{x^2 - 6x - 16}$

$\cdot \frac{x^2 - 5x - 24}{x^2 - x - 12}$

$= \frac{(x - 8)(x - 4)}{(x - 8)(x + 2)}$

$\cdot \frac{(x - 8)(x + 3)}{(x - 4)(x + 3)}$

$= \frac{x - 8}{x + 2}$

29. $\frac{2x^2 + 9x + 4}{x^2 + 7x + 12}$

$\div \frac{2x^2 - x - 1}{(x + 3)^2}$

$= \frac{2x^2 + 9x + 4}{x^2 + 7x + 12}$

$\cdot \frac{(x + 3)(x + 3)}{2x^2 - x - 1}$

$= \frac{(2x + 1)(x + 4)}{(x + 4)(x + 3)}$

$\cdot \frac{(x + 3)(x + 3)}{(2x + 1)(x - 1)}$

$= \frac{x + 3}{x - 1}$

31. $\frac{x^2 - y^2}{x^2 - 2xy + y^2} \div \frac{x + y}{x - y}$

$= \frac{x^2 - y^2}{x^2 - 2xy + y^2} \cdot \frac{x - y}{x + y}$

$= \frac{(x - y)(x + y)(x - y)}{(x - y)(x - y)(x + y)}$

33. $\frac{12x^2}{6y^2} \cdot \frac{36xy^5}{12}$

$= \frac{12x^2}{6y^2} \cdot \frac{36xy^{5}}{12}$ (cancelled: $36 \to 6$, $12 \to 1$, $y^5 \to y^3$)

$= 6x^3y^3$

35. $\frac{45a^2b^3}{12c^3} \cdot \frac{4c}{9a^3b^5}$

$= \frac{45a^2b^3}{12c^3} \cdot \frac{4c}{9a^3b^5}$ (cancelled: $\frac{5 \cdot 1 \cdot 1}{3 \cdot c^2} \cdot \frac{1 \cdot 1}{1 \cdot a \cdot b^2}$)

$= \frac{5}{3ab^2c^2}$

37. $\frac{-xy}{a} \div \frac{-2ax}{6y}$

$$= \frac{\overset{-1}{\cancel{-x}}y}{a} \cdot \frac{\overset{-3}{\cancel{6}}y}{-\cancel{2}a\cancel{x}} = \frac{3y^2}{a^2}$$

39. $$\frac{80m^4}{49x^5y^7} \cdot \frac{14x^{12}y^3}{25m^5}$$

$$= \frac{\overset{16}{\cancel{80}}\cancel{m^4}}{\underset{7}{\cancel{49}}x^5 y^{\overset{2}{\cancel{7}}}} \cdot \frac{\overset{2}{\cancel{14}}x^{\overset{7}{\cancel{12}}}y^5}{\underset{5}{\cancel{25}}m^{\cancel{5}}}$$

$$= \frac{32x^7}{35my^2}$$

41. $$(2x + 5) \cdot \frac{1}{4x + 10}$$

$$= \frac{\cancel{2x + 5}}{1} \cdot \frac{1}{2(\cancel{2x + 5})}$$

$$= \frac{1}{2}$$

43. $$\frac{1}{7x^2y} \div \frac{1}{21x^3y} = \frac{1}{\cancel{7}\cancel{x^2}\cancel{y}} \cdot \frac{\cancel{21}x^{\cancel{3}}\cancel{y}}{1}$$

$$= 3x$$

45. $$\frac{12a^2}{4bc} \div \frac{3a^2}{bc} = \frac{\overset{1}{\cancel{12}}\cancel{a^2}}{\cancel{4}\cancel{bc}} \cdot \frac{\cancel{bc}}{\cancel{3}\cancel{a^2}} = 1$$

47. $$\frac{5 - 2x}{x + 8} \cdot \frac{-x - 8}{2x - 5}$$

$$= \frac{(-1)(\cancel{2x - 5})}{\cancel{x + 8}} \cdot \frac{(-1)(\cancel{x+8})}{\cancel{2x - 5}}$$

$$= 1$$

49. $$\frac{6x + 6y}{a} \div \frac{12x + 12y}{a^2}$$

$$= \frac{6x + 6y}{a} \cdot \frac{a^2}{12x + 12y}$$

$$= \frac{\cancel{6}(\cancel{x + y})}{\cancel{a}} \cdot \frac{a^{\cancel{2}}}{\underset{2}{\cancel{12}}(\cancel{x + y})}$$

$$= \frac{a}{2}$$

51. $$\frac{a^2b^2}{6x + 6y} \div \frac{ab}{x^2 - y^2}$$

$$= \frac{a^2b^2}{6x + 6y} \cdot \frac{x^2 - y^2}{ab}$$

$$= \frac{a^{\cancel{2}}b^{\cancel{2}}}{6(\cancel{x + y})} \cdot \frac{(x - y)(\cancel{x + y})}{\cancel{ab}}$$

$$= \frac{ab(x - y)}{6}$$

53. $$\frac{x^2 - 5x - 24}{x^2 - x - 12} \cdot \frac{x^2 + x - 6}{x^2 - 10x + 16}$$

$$= \frac{(\cancel{x - 8})(\cancel{x + 3})}{(x - 4)(\cancel{x + 3})} \cdot \frac{(x + 3)(\cancel{x - 2})}{(\cancel{x - 8})(\cancel{x - 2})}$$

$$= \frac{x + 3}{x - 4}$$

55. $$\frac{a^2 + 6a + 9}{a^2 - 4} \cdot \frac{a - 2}{(a + 3)}$$

$$= \frac{(a + 3)(\cancel{a + 3})}{(a + 2)(\cancel{a - 2})} \cdot \frac{(\cancel{a - 2})}{\cancel{a + 3}}$$

$$= \frac{a + 3}{a + 2}$$

57. $$\frac{x^2 + 10x + 21}{x + 7} \div (x + 3)$$

$$= \frac{x^2 + 10x + 21}{x + 7} \cdot \frac{1}{x + 3}$$

$$= \frac{(\cancel{x + 7})(\cancel{x + 3})}{\cancel{x + 7}} \cdot \frac{1}{\cancel{x + 3}}$$

$$= 1$$

59. $$\frac{3x^2 - x - 2}{x + 7} \div \frac{x - 1}{4x^2 + 25x - 21}$$

$$= \frac{(3x + 2)(x - 1)}{(x + 7)} \div \frac{(x - 1)}{(4x - 3)(x + 7)}$$

$$= \frac{(3x + 2)(\cancel{x - 1})}{(\cancel{x + 7})} \cdot \frac{(4x - 3)(\cancel{x + 7})}{(\cancel{x - 1})}$$

$$= (3x + 2)(4x - 3)$$

61. $$\frac{9x^2 + 6x - 8}{x - 3} \cdot \frac{(x - 3)^2}{(3x + 4)}$$

$$= \frac{(3x - 2)(\cancel{3x + 4})}{(\cancel{x - 3})} \cdot \frac{(\cancel{x - 3})(x - 3)}{(\cancel{3x + 4})}$$

$$= (3x - 2)(x - 3)$$

63. $$\frac{2x + 4y}{x^2 + 4xy + 4y^2} \cdot \frac{x + 2y}{2}$$

$$= \frac{\overset{1}{\cancel{2}}(\cancel{x + 2y})}{(\cancel{x + 2y})(\cancel{x + 2y})} \cdot \frac{\cancel{x + 2y}}{\underset{1}{\cancel{2}}}$$

$$= 1$$

65. $$\frac{x^2 - 4}{2y} \div \frac{2 - x}{6xy}$$

$$= \frac{x^2 - 4}{2y} \cdot \frac{6xy}{2 - x}$$

$$= \frac{(\cancel{x - 2})(x + 2)}{\cancel{2y}} \cdot \frac{\overset{3}{\cancel{6xy}}}{(-1)(\cancel{x - 2})}$$

$$= \frac{3x(x + 2)}{-1}$$

$$= -3x(x + 2)$$

67. $$\frac{x^2 - y^2}{8x^2 - 16xy + 8y^2} \cdot \frac{4x - 4y}{(x + y)}$$

$$= \frac{(\cancel{x + y})(\cancel{x - y})}{\underset{2}{\cancel{8}}(\cancel{x - y})(\cancel{x - y})} \cdot \frac{\overset{1}{\cancel{4}}(\cancel{x - y})}{\cancel{x + y}}$$

$$= \frac{1}{2}$$

69. $$\frac{x^3 - 64}{x + 4} \div \frac{x^2 + 4x + 16}{x^2 + 8x + 16}$$

$$= \frac{x^3 - 64}{x + 4} \cdot \frac{x^2 + 8x + 16}{x^2 + 4x + 16}$$

$$= \frac{(x - 4)(\cancel{x^2 + 4x + 16})}{(\cancel{x + 4})} \cdot \frac{(\cancel{x + 4})(x + 4)}{(\cancel{x^2 + 4x + 16})}$$

$$= (x - 4)(x + 4)$$

71. $$\frac{x - 5}{x + 2} \cdot \frac{(x + 2)(2x - 3)}{x - 5}$$

$$= 2x - 3$$

73. To divide the rational expressions, invert the divisor and then multiply.

Cumulative Review Exercises

75. $2x^2 + x - 2 - 2/2x-1$

$$\begin{array}{r} 2x-1\,/\ 4x^3 + 0x^2 - 5x + 0 \\ -4x^3 \pm 2x^2 \\ \hline 2x^2 - 5x \\ -2x^2 \pm x \\ \hline -4x + 0 \\ \pm 4x \mp 2 \\ \hline -2 \end{array}$$

77. $3x^2 - 9x - 30 = 0$

$3(x - 5)(x + 2) = 0$

$x - 5 = 0 \qquad x + 2 = 0$

$x = 5 \qquad x = -2$

Just for fun

1. $\left(\dfrac{x + 2}{x^2 - 4x - 12} \cdot \dfrac{x^2 - 9x + 18}{x - 2}\right) \div \dfrac{x^2 + 5x + 6}{x^2 - 4}$

$= \dfrac{(x + 2)}{(x + 6)(x + 2)} \cdot \dfrac{(x - 3)(x + 6)}{(x + 2)} \cdot \dfrac{(x + 2)(x + 2)}{(x + 3)(x + 2)}$

$= \dfrac{x - 3}{x + 3}$

2. $\left(\dfrac{x^2 - x - 6}{2x^2 - 9x + 9} \div \dfrac{x^2 + x - 12}{x^2 + 3x - 4}\right) \cdot \dfrac{2x^2 - 5x + 3}{x^2 + x - 2}$

$= \dfrac{x^2 - x - 6}{2x^2 - 9x + 9} \cdot \dfrac{x^2 + 3x - 4}{x^2 + x - 12} \cdot \dfrac{2x^2 - 5x + 3}{x^2 + x - 2}$

$= \dfrac{(x + 3)(x + 2)}{(2x + 3)(x - 3)} \cdot \dfrac{(x + 4)(x + 1)}{(x + 4)(x + 3)} \cdot \dfrac{(2x + 3)(x - 1)}{(x + 2)(x + 1)}$

$= \dfrac{x - 1}{x - 3}$

Exercise Set 6.3

1. $\dfrac{x - 1}{6} + \dfrac{x}{6} = \dfrac{x - 1 + x}{6}$

$= \dfrac{2x - 1}{6}$

3. $\dfrac{x - 7}{3} - \dfrac{4}{3} = \dfrac{x - 7 - 4}{3}$

$= \dfrac{x - 11}{3}$

5. $\dfrac{x + 2}{x} - \dfrac{5}{x} = \dfrac{x + 2 - 5}{x}$

$= \dfrac{x - 3}{x}$

7. $\dfrac{1}{x} + \dfrac{x + 2}{x} = \dfrac{1 + (x + 2)}{x}$

$= \dfrac{x + 3}{x}$

9. $$\frac{4}{x + 2} + \frac{x + 3}{x + 2} = \frac{4 + (x + 3)}{x + 2}$$

$$= \frac{x + 7}{x + 2}$$

11. $$\frac{x - 4}{x} - \frac{x + 4}{x}$$

$$= \frac{x - 4 - (x + 4)}{x}$$

$$= \frac{x - 4 - x - 4}{x}$$

$$= \frac{-8}{x}$$

13. $$\frac{4x - 3}{x - 7} - \frac{2x + 8}{x - 7}$$

$$= \frac{4x - 3 - (2x + 8)}{x - 7}$$

$$= \frac{4x - 3 - 2x - 8}{x - 7}$$

$$= \frac{2x - 11}{x - 7}$$

15. $$\frac{9x + 7}{6x^2} - \frac{3x + 4}{6x^2}$$

$$= \frac{9x + 7 - (3x + 4)}{6x^2}$$

$$= \frac{9x + 7 - 3x - 4}{6x^2}$$

$$= \frac{6x + 3}{6x^2}$$

$$= \frac{\cancel{3}(2x + 1)}{\cancel{3}(2x^2)}$$

$$= \frac{2x + 1}{2x^2}$$

17. $$\frac{-2x - 4}{x^2 + 2x + 1} + \frac{3x + 5}{x^2 + 2x + 1}$$

$$= \frac{-2x - 4 + 3x + 5}{x^2 + 2x + 1}$$

$$= \frac{\cancel{x + 1}}{(x + 1)(\cancel{x + 1})}$$

$$= \frac{1}{x + 1}$$

19. $$\frac{4}{x^2 - 2x - 3} + \frac{x - 3}{x^2 - 2x - 3}$$

$$= \frac{4 + x - 3}{x^2 - 2x - 3}$$

$$= \frac{\cancel{x + 1}}{(x - 3)(\cancel{x + 1})}$$

$$= \frac{1}{x - 3}$$

21. $$\frac{x + 4}{3x + 2} - \frac{x + 4}{3x + 2}$$

$$= \frac{x + 4 - (x + 4)}{3x + 2}$$

$$= \frac{x + 4 - x - 4}{3x + 2}$$

$$= \frac{0}{3x + 2}$$

$$= 0$$

23. $$\frac{2x + 4}{x - 7} - \frac{6x + 5}{x - 7}$$

$$= \frac{2x + 4 - (6x + 5)}{x - 7}$$

$$= \frac{2x + 4 - 6x - 5}{x - 7}$$

$$= \frac{-4x - 1}{x - 7}$$

$$= \frac{-(4x + 1)}{x - 7}$$

25. $$\frac{x^2 + 4x + 3}{x + 2} - \frac{5x + 9}{x + 2}$$

$$= \frac{x^2 + 4x + 3 - (5x + 9)}{x + 2}$$

$$= \frac{x^2 + 4x + 3 - 5x - 9}{x + 2}$$

$$= \frac{x^2 - x - 6}{x + 2}$$

$$= \frac{(x - 3)\cancel{(x + 2)}}{\cancel{x + 2}}$$

$$= x - 3$$

27. $$\frac{4}{2x + 3} + \frac{6x + 5}{2x + 3} = \frac{4 + 6x + 5}{2x + 3}$$

$$= \frac{6x + 9}{2x + 3}$$

$$= \frac{3\cancel{(2x + 3)}}{\cancel{2x + 3}}$$

$$= 3$$

29. $$\frac{x^2}{x + 3} + \frac{9}{x + 3} = \frac{x^2 + 9}{x + 3}$$

31. $$\frac{4x + 12}{3 - x} - \frac{3x + 15}{3 - x}$$

$$= \frac{4x + 12 - (3x + 15)}{3 - x}$$

$$= \frac{4x + 12 - 3x - 15}{3 - x}$$

$$= \frac{x - 3}{3 - x}$$

$$= \frac{-1\cancel{(3 - x)}}{\cancel{3 - x}}$$

$$= -1$$

33. $$\frac{x^2 - 2}{x^2 + 6x - 7} - \frac{-4x + 19}{x^2 + 6x - 7}$$

$$= \frac{x^2 - x - (-4x + 19)}{x^2 + 6x - 7}$$

$$= \frac{x^2 - x - 4x - 19}{x^2 + 6x - 7}$$

$$= \frac{\cancel{(x + 7)}(x - 3)}{\cancel{(x + 7)}(x - 1)}$$

$$= \frac{x - 3}{x - 1}$$

35. $$\frac{x^2 - 13}{x + 4} - \frac{3}{x + 4}$$

$$= \frac{x^2 - 13 - 3}{x + 4}$$

$$= \frac{x^2 - 16}{x + 4}$$

$$= \frac{\cancel{(x + 4)}(x - 4)}{\cancel{x + 4}}$$

$$= x - 4$$

37. $$\frac{3x^2 - 7x}{4x^2 - 8x} + \frac{x}{4x^2 - 8x}$$

$$= \frac{3x^2 - 7x + x}{4x^2 - 8x}$$

$$= \frac{3x^2 - 6x}{4x^2 - 8x}$$

$$= \frac{3\cancel{x}\cancel{(x - 2)}}{4\cancel{x}\cancel{(x - 2)}}$$

$$= \frac{3}{4}$$

39. $$\frac{2x^2 - 6x + 5}{2x^2 + 18x + 16} - \frac{8x + 21}{2x^2 + 18x + 16}$$

$$= \frac{2x^2 - 6x + 5 - 8x - 21}{2x^2 + 18x + 16}$$

$$= \frac{2x^2 - 14x - 16}{2x^2 + 18x + 16}$$

$$= \frac{2x^2 - 7x - 8}{2x^2 + 9x + 8}$$

$$= \frac{\cancel{2}(x - 8)\cancel{(x + 1)}}{\cancel{2}(x + 8)\cancel{(x + 1)}}$$

$$= \frac{x - 8}{x + 8}$$

41. $\dfrac{x^2 + 3x - 6}{x^2 - 5x + 4} - \dfrac{-2x^2 + 4x - 4}{x^2 - 5x + 4}$

$= \dfrac{x^2 + 3x - 6 - [-2x^2 + 4x - 4]}{[x^2 - 5x + 4]}$

$= \dfrac{x^2 + 3x - 6 + 2x^2 - 4x + 4}{[x^2 - 5x + 4]}$

$= \dfrac{3x^2 - x - 2}{[x^2 - 5x + 4]}$

$= \dfrac{(3x + 2)(x - 1)}{(x - 4)(x - 1)}$

$= \dfrac{3x + 2}{x - 4}$

43. $\dfrac{5x^2 + 40x + 8}{x^2 - 64} + \dfrac{x^2 + 9x}{x^2 - 64}$

$= \dfrac{5x^2 + 40x + 8 + x^2 + 9x}{x^2 - 64}$

$= \dfrac{6x^2 + 49x + 8}{x^2 - 64}$

$= \dfrac{(6x + 1)(x + 8)}{(x - 8)(x + 8)}$

$= \dfrac{6x + 1}{x - 8}$

45. The signs change.

47. Should be:

$\dfrac{6x - 2 - (3x^2 - 4x + 5)}{(x^2 - 4x + 3)}$

49. $\dfrac{x^2 - 6x + 3}{x + 3} + \dfrac{x^2 + x - 9}{x + 3}$

$= \dfrac{2x^2 - 5x - 6}{x + 3}$

The sum of the numerators is $2x^2 - 5x - 6$.

51. $\dfrac{4x^2 - 6x - 7}{x^2 - 4} - \dfrac{2x^2 - 7x - 4}{x^2 - 4}$

$= \dfrac{2x^2 + x - 3}{x^2 - 4}$

The difference of the numerators is $2x^2 + x - 3$.

53. To add or subtract rational expressions with the same denominator, you add or subtract the numerators over the given denominator. If possible, you then factor and reduce.

Cumulative Review Exercises

55. Let x = number of ounces of concentrate.

$\dfrac{6 \text{ ounces}}{128 \text{ ounces}} = \dfrac{x \text{ ounces}}{24 \text{ ounces}}$

$\dfrac{6}{128} = \dfrac{x}{24}$

$128x = 6(24)$
$128x = 144$

$x = \dfrac{144}{128} = 1.125$ ounces

Just for fun

1. $\dfrac{3x - 2}{x^2 - 9} - \dfrac{4x^2 - 6}{x^2 - 9} + \dfrac{5x - 1}{x^2 - 9}$

$= \dfrac{3x - 2 - (4x^2 - 6) + 5x - 1}{x^2 - 9}$

$= \dfrac{3x - 2 - 4x^2 + 6 + 5x - 1}{x^2 - 9}$

$= \dfrac{-4x^2 + 8x + 3}{x^2 - 9}$

2. $\dfrac{x^2 - 6x + 3}{x + 2} + \dfrac{x^2 - 2x}{x + 2}$

$- \dfrac{2x^2 - 3x + 5}{x + 2}$

$= \dfrac{x^2-6x+3+x^2-2x-(2x^2-3x+5)}{x + 2}$

$= \dfrac{x^2-6x+3+x^2-2x-2x^2+3x-5}{x + 2}$

$= \dfrac{-5x - 2}{x + 2}$

Exercise Set 6.4

1. $\dfrac{x}{3} + \dfrac{x - 1}{3}$

LCD = 3

3. $\dfrac{1}{2x} + \dfrac{1}{3}$

LCD $= 2 \cdot 3 \cdot x = 6x$

5. $\dfrac{3}{5x} + \dfrac{7}{2}$

LCD $= 2 \cdot 5 \cdot x = 10x$

7. $\dfrac{2}{x^2} + \dfrac{3}{x}$

LCD $= x^2$

9. $\dfrac{x + 4}{2x + 3} + x = \dfrac{x + 4}{2x + 3} + \dfrac{x}{1}$

LCD $= 1(2x + 3) = 2x + 3$

11. $\dfrac{x}{(x + 1)} + \dfrac{4}{x^2}$

LCD $= x^2 (x + 1)$

13. $\dfrac{x + 3}{16x^2 \cdot y} - \dfrac{5}{9x^3}$

LCD $= 9 \cdot 16x^3 \cdot y = 144x^3 \cdot y$

15. $\dfrac{x^2 + 3}{18x} - \dfrac{x - 7}{12(x + 5)}$

$= \dfrac{x^2 + 3}{3 \cdot 6x} - \dfrac{x - 7}{2 \cdot 6 \cdot (x + 5)}$

LCD $= 2 \cdot 3 \cdot 6 \cdot x(x + 5)$
$= 36x(x + 5)$

17. $\dfrac{2x - 7}{x^2 + x} - \dfrac{x^2}{(x + 1)}$

$= \dfrac{2x - 7}{x(x + 1)} - \dfrac{x^2}{x + 1}$

LCD $= x(x + 1)$

19. $\dfrac{15}{36x^2 \cdot y} + \dfrac{x + 3}{15xy^3}$

$= \dfrac{15}{2^2 \cdot 3^2 \cdot x^2 \cdot y} + \dfrac{x + 3}{3 \cdot 5 \cdot x \cdot y^3}$

LCD $= 2^2 \cdot 3^2 \cdot x^2 \cdot y^3 = 180x^2 \cdot y^3$

21. $\dfrac{6}{2x + 8} + \dfrac{6x + 3}{3x - 9}$

$= \dfrac{6}{2(x + 4)} + \dfrac{6x + 3}{3(x - 3)}$

LCD $= 2 \cdot 3(x + 4)(x - 3)$
$= 6(x + 4)(x - 3)$

23. $\dfrac{9x + 4}{x + 6} - \dfrac{3x - 6}{x + 5}$

LCD $= (x + 6)(x + 5)$

25. $\dfrac{x - 2}{x^2 - 5x - 24} + \dfrac{3}{x^2 + 11x + 24}$

$= \dfrac{x - 2}{(x - 8)(x + 3)}$

$+ \dfrac{3}{(x + 8)(x + 3)}$

LCD $= (x - 8)(x + 8)(x + 3)$

27. $\dfrac{6}{x + 3} - \dfrac{x + 5}{x^2 - 4x + 3}$

$= \dfrac{6}{x + 3} - \dfrac{x + 5}{(x - 3)(x - 1)}$

LCD $= (x + 3)(x - 1)(x - 3)$

29. $\dfrac{2x}{x^2 - x - 2} - \dfrac{3}{x^2 + 4x + 3}$

$$= \frac{2x}{(x-2)(x+1)}$$

$$- \frac{3}{(x+3)(x+1)}$$

LCD = $(x-2)(x+1)(x+3)$

31. $$\frac{3x-5}{x^2+4x+4} + \frac{3}{(x+2)}$$

$$= \frac{3x-5}{(x+2)^2} + \frac{3}{(x+2)}$$

LCD = $(x+2)^2$

33. $$\frac{x}{3x^2+16x-12}$$

$$+ \frac{6}{3x^2+17x-6}$$

$$= \frac{x}{(3x-2)(x+6)}$$

$$+ \frac{6}{(3x-1)(x+6)}$$

LCD = $(3x-2)(x+6)$ $(3x-1)$

35. $$\frac{2x-3}{4x^2+4x+1}$$

$$+ \frac{x^2-4}{8x^2+10x+3}$$

$$= \frac{2x-3}{(2x+1)^2}$$

$$+ \frac{x^2-4}{(4x+3)(2x+1)}$$

LCD = $(2x+1)^2(4x+3)$

Cumulative Review Exercises

37. $$4\frac{3}{5} - 2\frac{5}{9} = \frac{23}{5} - \frac{23}{9}$$

$$= \frac{207-115}{45} = \frac{92}{45} \text{ or } 2\frac{2}{45}$$

39. Distributive property.

41. a) To solve $3x + 5 = 0$, subtract 5 from each side, then divide each side by 3.

b)
$$3x + 5 = 0$$
$$3x + 5 - 5 = 0 - 5$$
$$3x = -5$$
$$\frac{3x}{3} = \frac{-5}{3}$$
$$x = -\frac{5}{3}$$

Just for fun

1. $$\frac{3}{2x^3y^6} - \frac{5}{6x^5y^9} + \frac{1}{5x^{12}y^2}$$

LCD = $30x^{12}y^9$

2. $$\frac{x}{x-2} - \frac{4}{x^2-4} + \frac{3}{x+2}$$

$$= \frac{x}{x-2} - \frac{4}{(x-2)(x+2)}$$

$$+ \frac{3}{x+2}$$

LCD = $(x-2)(x+2)$

3. $$\frac{4}{x^2-x-12} + \frac{3}{x^2-6x+8}$$

$$+ \frac{5}{x^2+x-6}$$

$$\frac{4}{(x-4)(x+3)}$$

$$+ \frac{3}{(x-4)(x-2)}$$

$$+ \frac{5}{(x+3)(x-2)}$$

LCD = $(x-4)(x+3)(x-2)$

Exercise Set 6.5

1. $\frac{4}{x} + \frac{3}{2x}$ LCD = 2x

$\frac{4}{x} + \frac{3}{2x} = [\frac{2}{2}] \cdot \frac{4}{x} + \frac{3}{2x}$

$= \frac{8 + 3}{2x}$

$= \frac{11}{2x}$

3. $\frac{6}{x^2} + \frac{3}{2x}$ LCD = $2x^2$

$\frac{6}{x^2} + \frac{3}{2x} = \frac{2}{2} \cdot \frac{6}{x^2} + \frac{3}{2x} \cdot \frac{x}{x}$

$= \frac{12 + 3x}{2x^2}$ or $\frac{3x + 12}{2x^2}$

5. $2 - \frac{1}{x^2}$ LCD = x^2

$2 - \frac{1}{x^2} = [\frac{x^2}{x^2}] \cdot [\frac{2}{1}] - \frac{1}{x^2}$

$= \frac{2x^2 - 1}{x^2}$

7. $\frac{1}{x^2} + \frac{3}{5x}$ LCD = $5x^2$

$\frac{1}{x^2} + \frac{3}{5x} = [\frac{5}{5}] \cdot \frac{1}{x^2} + \frac{3}{5x} \cdot \frac{x}{x}$

$= \frac{3x + 5}{5x^2}$

9. $\frac{3}{4x^2y} + \frac{7}{5xy^2}$ LCD = $20x^2y^2$

$\frac{3}{4x^2y} + \frac{7}{5xy^2} = [\frac{5y}{5y}] \cdot \frac{3}{4x^2y}$

$+ \frac{7}{5xy^2} \cdot [\frac{4x}{4x}]$

$= \frac{15y + 28x}{20x^2y^2}$

11. $x + \frac{x}{y}$ LCD = y

$x + \frac{x}{y} = \frac{y}{y} \cdot \frac{x}{1} + \frac{x}{y}$

$= \frac{xy + x}{y}$ or $\frac{x(y + 1)}{y}$

13. $\frac{3x - 1}{x} + \frac{2}{3x}$ LCD = 3x

$\frac{3x - 1}{x} + \frac{2}{3x}$

$= \frac{3}{3} \cdot \frac{(3x - 1)}{x} + \frac{2}{3x}$

$= \frac{9x - 3 + 2}{3x}$

$= \frac{9x - 1}{3x}$

15. $\frac{5x}{y} + \frac{y}{x}$ LCD = xy

$\frac{5x}{y} + \frac{y}{x} = \frac{x}{x} \cdot \frac{5x}{y} + \frac{y}{x} \cdot \frac{y}{y}$

$= \frac{5x^2 + y^2}{xy}$

17. $\frac{4}{5x^2} - \frac{6}{y}$ LCD = $5x^2y$

$\frac{4}{5x^2} - \frac{6}{y} = \frac{y}{y} \cdot \frac{4}{5x^2} - \frac{6}{y} \cdot \frac{5x^2}{5x^2}$

$= \frac{4y - 30x^2}{5x^2y}$

19. $\frac{5}{x} + \frac{3}{x - 2}$ LCD = x(x - 2)

$\frac{5}{x} + \frac{3}{x - 2} = \frac{(x - 2)}{(x - 2)} \cdot \frac{5}{x}$

$$+ \frac{3}{(x - 2)} \cdot \frac{x}{x}$$

$$= \frac{5 \cdot (x - 2) + 3x}{x(x - 2)}$$

$$= \frac{5x - 10 + 3x}{x(x - 2)}$$

$$= \frac{8x - 10}{x(x - 2)}$$

21. $\frac{9}{a + 3} + \frac{2}{a}$ LCD = $a(a + 3)$

$$\frac{9}{a + 3} + \frac{2}{a}$$

$$= \frac{a}{a} \cdot \frac{9}{(a + 3)} + \frac{2}{a} \cdot \frac{(a + 3)}{(a + 3)}$$

$$= \frac{9a + 2(a + 3)}{a(a + 3)}$$

$$= \frac{9a + 2a + 6}{a(a + 3)}$$

$$= \frac{11a + 6}{a(a + 3)}$$

23. $\frac{4}{3x} - \frac{2x}{3x + 6}$

LCD = $3x(x + 2)$

$$\frac{4}{3x} - \frac{2x}{3x + 6}$$

$$= \frac{(x + 2)}{(x + 2)} \cdot \frac{4}{3x} - \frac{2x}{(3x + 6)}$$

$$\cdot \frac{x}{x}$$

$$= \frac{4(x + 2) - 2x^2}{3x(x + 2)}$$

$$= \frac{4x + 8 - 2x^2}{3x(x + 2)}$$

$$= \frac{-2x^2 + 4x + 8}{3x(x + 2)}$$

25. $\frac{3}{x - 2} + \frac{1}{2 - x}$

LCD = $x - 2$

$$\frac{3}{x - 2} + \frac{1}{2 - x}$$

$$= \frac{3}{x - 2} + \frac{1}{(2 - x)} \cdot \frac{-1}{-1}$$

$$= \frac{3 - 1}{x - 2}$$

$$= \frac{2}{x - 2}$$

27. $\frac{5}{x + 3} - \frac{4}{-x - 3}$

LCD = $x + 3$

$$\frac{5}{x + 3} - \frac{4}{-x - 3}$$

$$= \frac{5}{x + 3} - [\frac{4}{-x - 3}] \cdot [\frac{-1}{-1}]$$

$$= \frac{5}{x + 3} + \frac{4}{x + 3}$$

$$= \frac{9}{x + 3}$$

29. $\frac{3}{x + 1} + \frac{4}{x - 1}$

LCD = $(x + 1)(x - 1)$

$$\frac{3}{x + 1} + \frac{4}{x - 1}$$

$$= \frac{(x - 1)}{(x - 1)} \cdot \frac{3}{(x + 1)}$$

$$+ \frac{4}{(x - 1)} \cdot \frac{(x + 1)}{(x + 1)}$$

$$= \frac{3x - 3 + 4x + 4}{(x - 1)(x + 1)}$$

$$= \frac{7x + 1}{(x - 1)(x + 1)}$$

31. $\frac{x + 5}{x - 5} - \frac{x - 5}{x + 5}$

LCD = $(x - 5)(x + 5)$

$$\frac{x + 5}{x - 5} - \frac{x - 5}{x + 5}$$

$$= \frac{(x + 5)}{(x + 5)} \cdot \frac{(x + 5)}{(x - 5)}$$

$$- \frac{(x - 5)}{(x + 5)} \cdot \frac{(x - 5)}{(x - 5)}$$

$$= \frac{x^2 + 10x + 25 - [x^2 - 10x + 25]}{(x - 5)(x + 5)}$$

$$= \frac{x^2 + 10x + 25 - x^2 + 10x - 25}{(x - 5)(x + 5)}$$

$$= \frac{20x}{(x - 5)(x + 5)}$$

33. $\frac{x}{x^2 - 9} + \frac{4}{x + 3}$

LCD = $x^2 - 9$

$$\frac{x}{x^2 - 9} + \frac{4}{x + 3}$$

$$= \frac{x}{(x - 3)(x + 3)}$$

$$+ \frac{4}{(x + 3)} \cdot \frac{(x - 3)}{(x - 3)}$$

$$= \frac{x + 4(x - 3)}{(x - 3)(x + 3)}$$

$$= \frac{x + 4x - 12}{(x - 3)(x + 3)}$$

$$= \frac{5x - 12}{(x - 3)(x + 3)}$$

35. $\frac{x + 2}{x^2 - 4} - \frac{2}{x + 2}$

LCD = $(x + 2)(x - 2)$

$$\frac{x + 2}{x^2 - 4} - \frac{2}{x + 2}$$

$$= \frac{x + 2}{(x + 2)(x - 2)}$$

$$- \frac{2}{(x + 2)} \cdot \frac{(x - 2)}{(x - 2)}$$

$$= \frac{x + 2 - 2(x - 2)}{(x + 2)(x - 2)}$$

$$= \frac{x + 2 - 2x + 4}{(x + 2)(x - 2)}$$

$$= \frac{-x + 6}{(x + 2)(x - 2)}$$

37. $\frac{2x + 3}{x^2 - 7x + 12} - \frac{2}{x - 3}$

$$= \frac{2x + 3}{(x - 3)(x - 4)} - \frac{2}{x - 3}$$

LCD = $(x - 3)(x - 4)$

$$\frac{2x + 3}{x^2 - 7x + 12} - \frac{2}{x - 3}$$

$$= \frac{2x + 3}{(x - 3)(x - 4)} - \frac{2}{x - 3}$$

$$= \frac{2x + 3}{(x - 3)(x - 4)}$$

$$- \frac{2}{(x - 3)} \cdot \frac{(x - 4)}{(x - 4)}$$

$$= \frac{2x + 3 - 2x + 8}{(x - 3)(x - 4)}$$

$$= \frac{11}{(x - 3)(x - 4)}$$

39. $\frac{x^2}{x^2 + 2x - 8} - \frac{x - 4}{x + 4}$

$$= \frac{x^2}{(x + 4)(x - 2)}$$

$$- \frac{x - 4}{x + 4}$$

$$= \frac{x^2}{(x + 4)(x - 2)}$$

$$- \frac{(x - 2)}{(x - 2)} \cdot \frac{(x - 4)}{(x + 4)}$$

$$= \frac{x^2}{(x + 4)(x - 2)}$$

$$- \frac{x^2 - 6x + 8}{(x + 4)(x - 2)}$$

$$= \frac{x^2 - (x^2 - 6x + 8)}{(x + 4)(x - 2)}$$

$$= \frac{x^2 - x^2 + 6x - 8}{(x + 4)(x - 2)}$$

$$= \frac{6x - 8}{(x + 4)(x - 2)}$$

$$= \frac{2(3x - 4)}{(x + 4)(x - 2)}$$

41. $\frac{x - 1}{x^2 + 4x + 4} + \frac{x - 1}{x + 2}$

$$= \frac{x - 1}{(x + 2)(x + 2)} + \frac{x - 1}{x + 2}$$

LCD = $(x + 2)(x + 2)$

$$\frac{x - 1}{x^2 + 4x + 4} + \frac{x - 1}{x + 2}$$

$$= \frac{x - 1}{(x + 2)(x + 2)} + \frac{x - 1}{x + 2}$$

$$= \frac{x - 1}{(x + 2)(x + 2)}$$

$$+ \frac{(x - 1)}{(x + 2)} \cdot \frac{(x + 2)}{(x + 2)}$$

$$= \frac{x - 1 + (x^2 + x - 2)}{(x + 2)(x + 2)}$$

$$= \frac{x - 1 + x^2 + x - 2}{(x + 2)(x + 2)}$$

$$= \frac{x^2 + 2x - 3}{(x + 2)(x + 2)}$$

$$= \frac{(x + 3)(x - 1)}{(x + 2)^2}$$

43. $\frac{3}{x^2 + 2x - 8} + \frac{2}{x^2 - 3x + 2}$

$$= \frac{3}{(x + 4)(x - 2)}$$

$$+ \frac{2}{(x - 1)(x - 2)}$$

LCD = $(x - 1)(x - 2)(x + 4)$

$$\frac{3}{x^2 + 2x - 8} + \frac{2}{x^2 - 3x + 2}$$

$$= \frac{3}{(x + 4)(x - 2)}$$

$$+ \frac{2}{(x - 1)(x - 2)}$$

$$= \frac{(x - 1)}{(x - 1)}$$

$$\cdot \frac{3}{(x + 4)(x - 2)}$$

$$+ \frac{2}{(x - 1)(x - 2)}$$

$$\cdot \frac{(x + 4)}{(x + 4)}$$

$$= \frac{3x - 3 + 2x + 8}{(x - 1)(x + 4)(x - 2)}$$

$$= \frac{5(x + 1)}{(x - 1)(x + 4)(x - 2)}$$

45. $\frac{1}{x^2 - 4} + \frac{3}{x^2 + 5x + 6}$

$$= \frac{1}{(x - 2)(x + 2)}$$

$$+ \frac{3}{(x + 2)(x + 3)}$$

LCD = $(x - 2)(x + 2)(x + 3)$

$$\frac{1}{x^2 - 4} + \frac{3}{x^2 + 5x + 6}$$

$$= \frac{1}{(x - 2)(x + 2)}$$

$$+ \frac{3}{(x + 2)(x + 3)}$$

$$= \frac{(x + 3)}{(x + 3)} \cdot \frac{1}{(x - 2)(x + 2)}$$

$$+ \frac{3}{(x + 2)(x + 3)}$$

$$\cdot \frac{(x - 2)}{(x - 2)}$$

$$= \frac{x + 3 + 3x - 6}{(x + 3)(x - 2)(x + 2)}$$

$$= \frac{4x - 3}{(x + 3)(x - 2)(x + 2)}$$

47. $$\frac{x}{3x^2 + 5x - 2}$$

$$- \frac{4}{2x^2 + 7x + 6}$$

$$= \frac{x}{(3x - 1)(x + 2)}$$

$$- \frac{4}{(2x + 3)(x + 2)}$$

$$= \frac{(2x + 3)}{(2x + 3)}$$

$$\cdot \frac{x}{(3x - 1)(x + 2)}$$

$$- \frac{(3x - 1)}{(3x - 1)}$$

$$\cdot \frac{4}{(2x + 3)(x + 2)}$$

$$= \frac{x(2x + 3)}{(2x + 3)(3x - 1)(x + 2)}$$

$$- \frac{4(3x - 1)}{(3x - 1)(2x + 3)(x + 2)}$$

$$= \frac{2x^2 + 3x - (12x - 4)}{(2x + 3)(3x - 1)(x + 2)}$$

$$= \frac{2x^2 + 3x - 12x + 4}{(2x + 3)(3x - 1)(x + 2)}$$

$$= \frac{2x^2 - 9x + 4}{(2x + 3)(3x - 1)(x + 2)}$$

$$= \frac{(2x - 1)(x - 4)}{(2x + 3)(3x - 1)(x + 2)}$$

49. $$\frac{x}{3x^2 + 5x - 2}$$

$$- \frac{3}{2x^2 + 7x + 6}$$

$$= \frac{x}{(3x - 1)(x + 2)}$$

$$- \frac{3}{(2x + 3)(x + 2)}$$

$$= \frac{(2x + 3)}{(2x + 3)}$$

$$\cdot \frac{x}{(3x - 1)(x + 2)}$$

$$- \frac{(3x - 1)}{(3x - 1)}$$

$$\cdot \frac{3}{(2x + 3)(x + 2)}$$

$$= \frac{x(2x + 3)}{(3x - 1)(2x + 3)(x + 2)}$$

$$- \frac{3(3x - 1)}{(3x - 1)(2x + 3)(x + 2)}$$

$$= \frac{2x^2 + 3x - (9x - 3)}{(3x - 1)(2x + 3)(x + 2)}$$

$$= \frac{2x^2 + 3x - 9x + 3}{(3x - 1)(2x + 3)(x + 2)}$$

$$= \frac{2x^2 - 6x + 3}{(3x - 1)(2x + 3)(x + 2)}$$

Cumulative Review Exercises

51. $$\frac{18}{2 \text{ min.}} = \frac{x}{90 \text{ min.}}$$

$$\frac{18}{2} = \frac{x}{90}$$

$$2x = (18)(90)$$
$$2x = 1620$$
$$x = 810$$

53. $$4x - 3 - 4/2x+3$$

$$2x+3/\ 8x^2 + 6x - 13$$
$$-8x^2 \mp 12x$$
$$-6x - 13$$
$$\pm 6x \pm 9$$
$$-4$$

Just for fun

1. $$\frac{x}{x - 2} + \frac{3}{x + 2} + \frac{4}{x^2 - 4}$$

$$= \frac{x}{x - 2} + \frac{3}{x + 2}$$

$$+ \frac{4}{(x - 2)(x + 2)}$$

$$= \frac{(x + 2)}{(x + 2)} \cdot \frac{x}{(x - 2)}$$

$$+ \frac{(x - 2)}{(x - 2)} \cdot \frac{3}{(x + 2)}$$

$$+ \frac{4}{(x - 2)(x + 2)}$$

$$= \frac{x(x + 2)}{(x + 2)(x - 2)}$$

$$+ \frac{3(x - 2)}{(x - 2)(x + 2)}$$

$$+ \frac{4}{(x - 2)(x + 2)}$$

$$= \frac{x^2 + 2x + 3x - 6 + 4}{(x + 2)(x - 2)}$$

$$= \frac{x^2 + 5x - 2}{(x + 2)(x - 2)}$$

2. $$\frac{4}{x^2 + x - 6} + \frac{x}{x + 3} - \frac{5}{x - 2}$$

$$= \frac{4}{(x + 3)(x - 2)} + \frac{x}{x + 3}$$

$$- \frac{5}{x - 2}$$

$$= \frac{4}{(x + 3)(x - 2)}$$

$$+ \frac{(x - 2)}{(x - 2)} \cdot \frac{x}{(x + 3)}$$

$$- \frac{(x + 3)}{(x + 3)} \cdot \frac{5}{(x - 2)}$$

$$= \frac{4}{(x + 3)(x - 2)}$$

$$+ \frac{x(x - 2)}{(x - 2)(x + 3)}$$

$$- \frac{5(x + 3)}{(x + 3)(x - 2)}$$

$$= \frac{4 + x^2 - 2x - (5x + 15)}{(x + 3)(x - 2)}$$

$$= \frac{4 + x^2 - 2x - 5x - 15}{(x + 3)(x - 2)}$$

$$= \frac{x^2 - 7x - 11}{(x + 3)(x - 2)}$$

3. $$\frac{x + 6}{4 - x^2} - \frac{x + 3}{x + 2} + \frac{x - 3}{2 - x}$$

$$= \frac{x + 6}{(2 - x)(2 + x)}$$

$$- \frac{x + 3}{2 + x} + \frac{x - 3}{2 - x}$$

LCD = $(2 - x)(2 + x)$

$$\frac{x + 6}{4 - x^2} - \frac{x + 3}{x + 2} + \frac{x - 3}{2 - x}$$

$$= \frac{x + 6}{(2 - x)(2 + x)}$$

$$- \frac{x + 3}{2 + x} + \frac{x - 3}{2 - x}$$

$$= \frac{x + 6}{(2 - x)(2 + x)}$$

$$- \frac{(x + 3)(2 - x)}{(2 + x)(2 - x)}$$

$$+ \frac{(x - 3)(2 + x)}{(2 - x)(2 + x)}$$

$$= \frac{x + 6 - [6 - x - x^2] + [x^2 - x - 6]}{(2 - x)(2 + x)}$$

$$= \frac{x + 6 - 6 + x + x^2 + x^2 - x - 6}{(2 - x)(2 + x)}$$

$$= \frac{2x^2 + x - 6}{(2 - x)(2 + x)}$$

$$= \frac{(2x - 3)\cancel{(x + 2)}}{(2 - x)\cancel{(2 + x)}}$$

$$= \frac{2x - 3}{2 - x}$$

4. $\frac{3x - 1}{x + 2} + \frac{x}{x - 3} - \frac{4}{2x + 3}$

LCD = $(x + 2)(x - 3)$
$(2x + 3)$

$$\frac{(3x - 1)(x - 3)(2x + 3)}{(x + 2)(x - 3)(2x + 3)}$$

$$+ \frac{x(x + 2)(2x + 3)}{(x + 2)(x - 3)(2x + 3)}$$

$$- \frac{4(x + 2)(x - 3)}{(x + 2)(x - 3)(2x + 3)}$$

$$= \frac{[6x^3 - 11x^2 - 24x + 9] + [2x^2 + 7x^2 + 6x] + [-4x^2 + 4x + 24]}{(x + 2)(x - 3)(2x + 3)}$$

$$= \frac{8x^3 - 8x^2 - 14x + 33}{(x + 2)(x - 3)(2x + 3)}$$

Exercise Set 6.6

1. $\dfrac{1 + \frac{3}{5}}{2 + \frac{1}{5}} = \dfrac{\frac{5}{5} + \frac{3}{5}}{\frac{10}{5} + \frac{1}{5}} = \dfrac{\frac{8}{5}}{\frac{11}{5}}$

$$= \frac{8}{5} \cdot \frac{5}{11}$$

$$= \frac{8}{11}$$

3. $\dfrac{2 + \frac{3}{8}}{1 + \frac{1}{3}} = \dfrac{\frac{16}{8} + \frac{3}{8}}{\frac{3}{3} + \frac{1}{3}} = \dfrac{\frac{19}{8}}{\frac{4}{3}}$

$$= \frac{19}{8} \cdot \frac{3}{4}$$

$$= \frac{57}{32}$$

5. $$\frac{\frac{4}{9} - \frac{3}{8}}{4 - \frac{3}{5}} = \frac{\frac{8}{8} \cdot \frac{4}{9} - \frac{3}{8} \cdot \frac{9}{9}}{\frac{20}{5} - \frac{3}{5}}$$

$$= \frac{\frac{32}{72} - \frac{27}{72}}{\frac{17}{5}}$$

$$= \frac{\frac{5}{72}}{\frac{7}{5}}$$

$$= \frac{5}{72} \cdot \frac{5}{17}$$

$$= \frac{25}{1224}$$

7. $$\frac{\frac{x^2y}{4}}{\frac{2}{x}} = \left[\frac{x^2y}{4}\right] \cdot \frac{x}{2} = \frac{x^3y}{8}$$

9. $$\frac{\frac{8x^2y}{3z^3}}{\frac{4xy}{9z^5}} = \frac{\overset{2}{\cancel{8}}x^{\cancel{2}}\cancel{y}}{\cancel{3}\cancel{z^3}} \cdot \frac{\overset{2}{\cancel{9}}z^{\cancel{5}\,2}}{\cancel{4}\cancel{x}\cancel{y}}$$

$$= 6xz^2$$

11. $$\frac{x + \frac{1}{y}}{\frac{x}{y}} = \frac{\frac{x}{1} \cdot \frac{y}{y} + \frac{1}{y}}{\frac{x}{y}} = \frac{\frac{xy}{y} + \frac{1}{y}}{\frac{x}{y}}$$

$$= \frac{\frac{xy + 1}{y}}{\frac{x}{y}}$$

$$= \frac{xy + 1}{\cancel{y}} \cdot \frac{\cancel{y}}{x}$$

$$= \frac{xy + 1}{x}$$

13. $$\frac{\frac{9}{x} + \frac{3}{x^2}}{3 + \frac{1}{x}} = \frac{\frac{x}{x} \cdot \frac{9}{x} + \frac{3}{x^2}}{\frac{x}{x} \cdot \frac{3}{1} + \frac{1}{x}}$$

$$= \frac{\frac{9x}{x^2} + \frac{3}{x^2}}{\frac{3x}{x} + \frac{1}{x}}$$

$$= \frac{\frac{9x + 3}{x^2}}{\frac{3x + 1}{x}}$$

$$= \frac{9x + 3}{x^2} \cdot \frac{x}{(3x + 1)}$$

$$= \frac{3\cancel{(3x + 1)}}{x^2} \cdot \frac{x}{\cancel{(3x + 1)}}$$

$$= \frac{3}{x}$$

15. $$\frac{3 - \frac{1}{y}}{2 - \frac{1}{y}} = \frac{\frac{y}{y} \cdot \frac{3}{1} - \frac{1}{y}}{\frac{y}{y} \cdot \frac{2}{1} - \frac{1}{y}}$$

$$= \frac{\frac{3y}{y} - \frac{1}{y}}{\frac{2y}{y} - \frac{1}{y}}$$

$$= \frac{\frac{3y - 1}{y}}{\frac{2y - 1}{y}}$$

$$= \frac{3y - 1}{\cancel{y}} \cdot \frac{\cancel{y}}{2y - 1}$$

$$= \frac{3y - 1}{2y - 1}$$

17. $$\frac{\frac{x}{y} - \frac{y}{x}}{\frac{x + y}{x}} = \frac{\frac{x}{x} \cdot \frac{x}{y} - \frac{y}{x} \cdot \frac{y}{y}}{\frac{x + y}{x}}$$

$$= \frac{\frac{x^2}{xy} - \frac{y^2}{xy}}{\frac{x + y}{x}}$$

$$= \frac{\frac{x^2 - y^2}{xy}}{\frac{x + y}{x}}$$

$$= \frac{(\cancel{x + y})(x - y)}{\cancel{x}y} \cdot \frac{\cancel{x}}{\cancel{x + y}}$$

$$= \frac{x - y}{y}$$

19. $$\frac{\frac{a^2}{b} - b}{\frac{b^2}{a} - a} = \frac{\frac{a^2}{b} - \frac{b}{1} \cdot \frac{b}{b}}{\frac{b^2}{a} - \frac{a}{1} \cdot \frac{a}{a}}$$

$$= \frac{\frac{a^2}{b} - \frac{b^2}{b}}{\frac{b^2}{a} - \frac{a^2}{a}}$$

$$= \frac{\frac{a^2 - b^2}{b}}{\frac{b^2 - a^2}{a}}$$

$$= \left[\frac{a^2 - b^2}{b}\right] \cdot \frac{a}{b^2 - a^2}$$

$$= \frac{(\cancel{a + b})\overset{-1}{(\cancel{a - b})}}{b} \cdot \frac{a}{(\cancel{b + a})(\cancel{b - a})}$$

$$= \frac{-a}{b}$$

21. $$\frac{\frac{a}{b} - 2}{\frac{-a}{b} + 2} = \frac{\frac{a}{b} - \frac{2}{1} \cdot \frac{b}{b}}{\frac{-a}{b} + \frac{2b}{b}}$$

$$= \frac{\frac{a}{b} - \frac{2b}{b}}{\frac{-a}{b} + \frac{2b}{b}}$$

$$= \frac{\dfrac{a - 2b}{b}}{\dfrac{2b - a}{b}}$$

$$= \frac{a - 2b}{\cancel{b}} \cdot \frac{\cancel{b}}{2b - a}$$

$$= \frac{a - 2b}{2b - a}$$

$$= \frac{-1(\cancel{2b - a})}{\cancel{2b - a}}$$

$$= -1$$

23. $$\frac{\dfrac{4x + 8}{3x^2}}{\dfrac{4x}{6}} = \frac{4x + 8}{3x^2} \cdot \frac{6}{4x}$$

$$= \frac{\cancel{4}(x + 2)}{\cancel{3}x^2} \cdot \frac{\cancel{6}^{2}}{\cancel{4}x}$$

$$= \frac{2(x + 2)}{x^3}$$

25. $$\frac{\dfrac{1}{a} + \dfrac{1}{b}}{\dfrac{1}{ab}} = \frac{\dfrac{b}{b} \cdot \dfrac{1}{a} + \dfrac{1}{b} \cdot \dfrac{a}{a}}{\dfrac{1}{ab}}$$

$$= \frac{\dfrac{b}{ab} + \dfrac{a}{ab}}{\dfrac{1}{ab}}$$

$$= \frac{\dfrac{a + b}{ab}}{\dfrac{1}{ab}}$$

$$= \frac{a + b}{\cancel{ab}} \cdot \frac{\cancel{ab}}{1}$$

$$= a + b$$

27. $$\frac{\dfrac{a}{b} + \dfrac{1}{a}}{\dfrac{b}{a} + \dfrac{1}{a}} = \frac{\dfrac{a}{a} \cdot \dfrac{a}{b} + \dfrac{1}{a} \cdot \dfrac{b}{b}}{\dfrac{b}{a} + \dfrac{1}{a}}$$

$$= \frac{\dfrac{a^2}{ab} + \dfrac{b}{ab}}{\dfrac{b + 1}{a}}$$

$$= \frac{a^2 + b}{\cancel{a}b} \cdot \frac{\cancel{a}}{b + 1}$$

$$= \frac{a^2 + b}{b(b + 1)}$$

29. $$\frac{\dfrac{1}{x} - \dfrac{1}{y}}{\dfrac{1}{x} + \dfrac{1}{y}} = \frac{\dfrac{y}{y} \cdot \dfrac{1}{x} - \dfrac{1}{y} \cdot \dfrac{x}{x}}{\dfrac{y}{y} \cdot \dfrac{1}{x} + \dfrac{1}{y} \cdot \dfrac{x}{x}}$$

$$= \frac{\dfrac{y}{xy} = \dfrac{x}{xy}}{\dfrac{y}{xy} + \dfrac{x}{xy}}$$

$$= \frac{\dfrac{y - x}{xy}}{\dfrac{x + y}{xy}}$$

$$= \frac{y - x}{\cancel{xy}} \cdot \frac{\cancel{xy}}{x + y}$$

$$= \frac{y - x}{x + y}$$

31. $$\frac{\frac{1}{x^2} + \frac{1}{x}}{\frac{1}{y} + \frac{1}{y^2}}$$

$$= \frac{x^2y^2[\frac{1}{x^2} + \frac{1}{x}]}{x^2y^2[\frac{1}{y} + \frac{1}{y^2}]}$$

$$= \frac{y^2 + xy^2}{x^2y + x^2}$$

$$= \frac{y^2(1 + x)}{x^2(y + 1)} = \frac{y^2(x + 1)}{x^2(y + 1)}$$

33. a) To simplify a complex fraction using Method 1:

1. Add or subtract the fractions in the numerator and denominator to obtain a single fraction in each.
2. Invert the denominator and multiply by the numerator.
3. Simplify when possible.

To simplify a complex fraction using Method 2:

1. Multiply the numerator and denominator by the least common denominator of all the denominators in the complex fraction.
2. Simplify when possible.

b) Method 1:

$$\frac{\frac{2}{x} - \frac{3}{y}}{x + \frac{1}{y}} = \frac{\frac{2y - 3x}{xy}}{\frac{xy + 1}{y}}$$

$$= \frac{2y - 3x}{xy} \cdot \frac{y}{xy + 1}$$

$$= \frac{2y - 3x}{x^2y + x}$$

Method 2:

$$\frac{\frac{2}{x} - \frac{3}{y}}{x + \frac{1}{y}} = \frac{xy[\frac{2}{x} - \frac{3}{y}]}{xy[x + \frac{1}{y}]}$$

$$= \frac{2y - 3x}{x^2y + x}$$

Cumulative Review Exercises

35. A polynomial is an expression containing a finite number of terms of the form ax^n, for any real number a and any whole number n.

37. $$\frac{x}{3x^2 + 17x - 6} - \frac{2}{x^2 + 3x - 18}$$

$$= \frac{x}{(3x - 1)(x + 6)} - \frac{2}{(x + 6)(x - 3)}$$

$$= \frac{x(x - 3)}{(3x - 1)(x + 6)(x - 3)} - \frac{2(3x - 1)}{(3x - 1)(x + 6)(x - 3)}$$

$$= \frac{x^2 - 3x - 6x + 2}{(3x - 1)(x + 6)(x - 3)}$$

$$= \frac{x^2 - 9x + 2}{(3x - 1)(x + 6)(x - 3)}$$

Just for fun

1. a) $$\frac{\frac{1}{2} \cdot h}{h + \frac{1}{2}} = \frac{\frac{1}{2} \cdot \frac{2}{3}}{\frac{2}{3} + \frac{1}{2}}$$

$$= \frac{\frac{2}{6}}{\frac{2}{2} \cdot \frac{2}{3} + \frac{1}{2} \cdot \frac{3}{3}}$$

$$= \frac{\frac{1}{3}}{\frac{4}{6} + \frac{3}{6}}$$

$$= \frac{\frac{1}{3}}{\frac{4}{6} + \frac{3}{6}}$$

$$= \frac{\frac{1}{3}}{\frac{7}{6}}$$

$$= \frac{1}{\cancel{3}} \cdot \frac{\cancel{6}^{\,2}}{7}$$

$$= \frac{2}{7}$$

b) $$\frac{\frac{1}{2} \cdot h}{h + \frac{1}{2}} = \frac{\frac{1}{2} \cdot \frac{4}{5}}{\frac{4}{5} + \frac{1}{2}}$$

$$= \frac{\frac{4}{10}}{\frac{2}{2} \cdot \frac{4}{5} + \frac{1}{2} \cdot \frac{5}{5}}$$

$$= \frac{\frac{2}{5}}{\frac{8}{10} + \frac{5}{10}}$$

$$= \frac{\frac{2}{5}}{\frac{13}{10}}$$

$$= \frac{2}{\cancel{5}} \cdot \frac{\cancel{10}^{\,2}}{13}$$

$$= \frac{4}{13}$$

Exercise Set 6.7

1. $$\frac{2}{5} = \frac{x}{10}$$

$$\frac{\overset{2}{\cancel{10}}}{1} \cdot \left[\frac{2}{\cancel{5}}\right] = \left[\frac{x}{\cancel{10}}\right] \cdot \frac{\cancel{10}}{1}$$

$$4 = x$$

3. $$\frac{5}{12} = \frac{20}{x}$$

$$\frac{\cancel{12}x}{1} \cdot \left[\frac{5}{\cancel{12}}\right] = \left[\frac{20}{\cancel{x}}\right] \cdot \frac{12\cancel{x}}{1}$$

$$5x = 240$$
$$x = 48$$

5. $$\frac{a}{25} = \frac{12}{10}$$

$$\frac{\overset{2}{\cancel{50}}}{1} \cdot \left[\frac{a}{\cancel{25}}\right] = \left[\frac{12}{\cancel{10}}\right] \cdot \frac{\overset{5}{\cancel{50}}}{1}$$

$$2a = 5(12)$$
$$2a = 60$$
$$a = 30$$

7. $$\frac{9}{3b} = \frac{-6}{2}$$

$$\frac{9}{3b} = \frac{-3}{1}$$

$$\frac{\cancel{3b}}{1} \cdot \left[\frac{9}{\cancel{3b}}\right] = \left[\frac{-3}{1}\right] \cdot \frac{3b}{1}$$

$$9 = -9b$$
$$-1 = b$$

9. $$\frac{x + 4}{9} = \frac{5}{9}$$

$$\frac{\cancel{9}}{1} \cdot \left[\frac{x + 4}{\cancel{9}}\right] = \frac{5}{\cancel{9}} \cdot \frac{\cancel{9}}{1}$$

$$x + 4 = 5$$
$$x = 1$$

11. $$\frac{4x + 5}{6} = \frac{7}{2}$$

$$\frac{\cancel{6}}{1} \cdot \left[\frac{4x + 5}{\cancel{6}}\right] = \frac{7}{\cancel{2}} \cdot \frac{\overset{3}{\cancel{6}}}{1}$$

$$4x + 5 = 7(3)$$

$$4x + 5 = 21$$
$$4x = 16$$
$$x = 4$$

13. $$\frac{6x + 7}{10} = \frac{2x + 9}{6}$$

$$\frac{\overset{3}{\cancel{30}}}{1} \cdot \left[\frac{6x + 7}{\cancel{10}}\right]$$

$$= \frac{2x + 9}{\cancel{6}} \cdot \frac{\overset{5}{\cancel{30}}}{1}$$

$$3(6x + 7) = 5(2x + 9)$$
$$18x + 21 = 10x + 45$$
$$8x + 21 = 45$$
$$8x = 24$$
$$x = 3$$

15. $$\frac{x}{3} - \frac{3x}{4} = \frac{1}{12}$$

$$\frac{12}{1} \cdot \left[\frac{x}{3} - \frac{3x}{4}\right] = \frac{1}{\cancel{12}} \cdot \frac{\cancel{12}}{1}$$

$$\frac{12x}{3} - \frac{36x}{4} = 1$$

$$4x - 9x = 1$$
$$-5x = 1$$

$$x = \frac{-1}{5}$$

17. $$\frac{3}{4} - x = 2x$$

$$\frac{3}{4} - x + x = 2x + x$$

$$\frac{3}{4} = 3x$$

$$\frac{1}{3} \cdot \frac{3}{4} = 3x \cdot \frac{1}{3}$$

$$\frac{1}{4} = x$$

19. $$\frac{5}{3x} + \frac{3}{x} = 1$$

$$\frac{3x}{1} \cdot \left[\frac{5}{3x} + \frac{3}{x}\right] = 1(3x)$$

$$\frac{15x}{3x} + \frac{9x}{x} = 3x$$

$$5 + 9 = 3x$$
$$14 = 3x$$

$$\frac{14}{3} = x$$

21. $$\frac{x - 1}{x - 5} = \frac{4}{x - 5}$$

$$\frac{\cancel{x - 5}}{1} \cdot \left[\frac{x - 1}{\cancel{x - 5}}\right]$$

$$= \left[\frac{4}{\cancel{x - 5}}\right] \cdot \frac{\cancel{x - 5}}{1}$$

$$x - 1 = 4$$
$$x = 5$$

Checking:

$$\frac{(5) - 1}{(5) - 5} = \frac{4}{(5) - 5}$$

$$\frac{4}{0} = \frac{4}{0}$$

Since $\frac{4}{0}$ is undefined, there is no solution.

23. $$\frac{5y - 3}{7} = \frac{15y - 2}{28}$$

$$\frac{\overset{4}{\cancel{28}}}{1} \cdot \left[\frac{5y - 3}{\cancel{7}}\right]$$

$$= \left[\frac{15y - 2}{\cancel{28}}\right] \cdot \frac{\cancel{28}}{1}$$

$$4(5y - 3) = 15y - 2$$
$$20y - 12 = 15y - 2$$
$$5y - 12 = -2$$
$$5y = 10$$
$$y = 2$$

25. $$\frac{5}{-x - 6} = \frac{2}{x}$$

$$x(\cancel{-x - 6}) \cdot \frac{5}{(\cancel{-x - 6})}$$

$$= \left[\frac{2}{\cancel{x}}\right] \cdot \cancel{x} \cdot (-x - 6)$$

$$5x = 2(-x - 6)$$
$$5x = -2x - 12$$
$$7x = -12$$
$$x = \frac{-12}{7}$$

27. $$\frac{2x - 3}{x - 4} = \frac{5}{x - 4}$$

$$(\cancel{x - 4}) \cdot \frac{2x - 3}{(\cancel{x - 4})}$$

$$= \frac{5}{(\cancel{x - 4})} \cdot (\cancel{x - 4})$$

$$2x - 3 = 5$$
$$2x = 8$$
$$x = 4$$

Since $x = 4$ will make the denominator equal zero, there is no solution.

29. $$\frac{x - 2}{x + 4} = \frac{x + 1}{x + 10}$$

$$(x + 4)(x + 10) \cdot \frac{x - 2}{x + 4}$$

$$= \frac{x + 1}{x + 10} \cdot (x + 4)(x + 10)$$

$$(x + 10)(x - 2)$$
$$= (x + 4)(x + 1)$$

$$x^2 + 8x - 20 = x^2 + 5x + 4$$
$$8x - 20 = 5x + 4$$
$$3x - 20 = 4$$
$$3x = 24$$
$$x = 8$$

31. $\frac{2x - 1}{3} - \frac{3x}{4} = \frac{5}{6}$

$\frac{12}{1} \cdot [\frac{2x - 1}{3} - \frac{3x}{4}]$

$= [\frac{5}{6}] \cdot \frac{\overset{2}{12}}{1}$

$$\begin{aligned} 4(2x - 1) - 3(3x) &= 5(2) \\ 8x - 4 - 9x &= 10 \\ -4 - x &= 10 \\ -x &= 14 \\ x &= -14 \end{aligned}$$

33. $x + \frac{6}{x} = -5$

$x[x + \frac{6}{x}] = -5 \cdot (x)$

$$\begin{aligned} x^2 + 6 &= -5x \\ x^2 + 5x + 6 &= 0 \\ x^2 + 5x + 6 &= 0 \\ (x + 3)(x + 2) &= 0 \end{aligned}$$

$x + 3 = 0 \quad x + 2 = 0$

$x = -3 \quad x = -2$

35. $\frac{3y - 2}{y + 1} = 4 \cdot \frac{y + 2}{y - 1}$

$(y + 1)(y - 1) \cdot \frac{3y - 2}{y + 1}$

$= [4 - \frac{y + 2}{y - 1}]$

$\cdot (y + 1)(y - 1)$

$(3y - 2)(y - 1)$

$= 4(y + 1)(y - 1)$

$- (y + 2)(y + 1)$

$3y^2 - 5y + 2$

$= 4y^2 - 4 - [y^2 + 3y + 2]$

$3y^2 - 5y + 2$

$= 4y^2 - 4 - y^2 - 3y - 2$

$$\begin{aligned} 3y^2 - 5y + 2 &= 3y^2 - 3y - 6 \\ -5y + 2 &= -3y - 6 \\ 2 &= 2y - 6 \\ 8 &= 2y \\ 4 &= y \end{aligned}$$

37. $\frac{1}{x + 3} + \frac{1}{x - 3} = \frac{-5}{x^2 - 9}$

$(x + 3)(x - 3)$

$\cdot [\frac{1}{x + 3} + \frac{1}{x - 3}]$

$= [\frac{-5}{(x + 3)(x + 3)}]$

$\cdot (x + 3)(x + 3)$

$$\begin{aligned} (x - 3) + (x + 3) &= -5 \\ 2x &= -5 \\ x &= \frac{-5}{2} \end{aligned}$$

39. $\frac{a}{a - 3} + \frac{3}{2} = \frac{3}{a - 3}$

$2(a - 3)[\frac{a}{a - 3} + \frac{3}{2}]$

$= \frac{3}{(a - 3)} \cdot 2(a - 3)$

$2(a + 3) \cdot \frac{a}{(a + 3)}$

$+ 2(a - 3)(\frac{3}{2})$

$= \frac{3}{(a + 3)} \cdot 2(a + 3)$

$$\begin{aligned} 2a + 3(a - 3) &= 6 \\ 2a + 3a - 9 &= 6 \\ 5a - 9 &= 6 \\ 5a &= 15 \\ a &= 3 \end{aligned}$$

Since a = 3 makes the denominator equal 0, there is no solution.

41. $\frac{2}{x - 3} - \frac{4}{x + 3} = \frac{8}{x^2 - 9}$

$(x - 3)(x + 3)$

$$\cdot \left[\frac{2}{x - 3} - \frac{4}{x + 3}\right]$$

$$= \left[\frac{8}{\cancel{(x - 3)}\cancel{(x + 3)}}\right] \cdot \cancel{(x - 3)}\cancel{(x + 3)}$$

$$\begin{aligned} 2(x + 3) - 4(x - 3) &= 8 \\ 2x + 6 - 4x + 12 &= 8 \\ -2x + 18 &= 8 \\ -2x &= -10 \\ x &= 5 \end{aligned}$$

43. $\dfrac{y}{2y + 2} + \dfrac{2y - 16}{4y + 4} = \dfrac{y - 3}{y + 1}$

$4(y + 1)$

$$\cdot \left[\frac{y}{2(y + 1)} + \frac{2(y - 8)}{4(y + 1)}\right]$$

$$= \left[\frac{y - 3}{\cancel{y + 1}}\right] \cdot \cancel{(y + 1)}4$$

$$\begin{aligned} 2y + 2(y - 8) &= 4(y - 3) \\ 2y + 2y - 16 &= 4y - 12 \\ 4y - 16 &= 4y - 12 \\ -16 &= -12 \end{aligned}$$

False

Since $-16 = -12$ is false, no solution.

45. $\dfrac{1}{2} + \dfrac{1}{x - 1} = \dfrac{2}{x^2 - 1}$

$2(x - 1)(x + 1)$

$$\left[\frac{1}{2} + \frac{1}{x - 1}\right]$$

$$= \frac{2}{(x - 1)(x + 1)}$$

$$\cdot 2(x - 1)(x + 1)$$

$$\cancel{2}(x - 1)(x + 1) \cdot \frac{1}{\cancel{2}}$$

$$+ 2\cancel{(x - 1)}(x + 1) \cdot \frac{1}{\cancel{(x - 1)}}$$

$$= \frac{2}{\cancel{(x - 1)}\cancel{(x + 1)}}$$

$$\cdot 2\cancel{(x - 1)}\cancel{(x + 1)}$$

$$\begin{aligned} x^2 - 1 + 2x + 2 &= 4 \\ x^2 + 2x + 1 &= 4 \\ x^2 + 2x - 3 &= 0 \\ (x + 3)(x - 1) &= 0 \end{aligned}$$

$x + 3 = 0 \quad x - 1 = 0$

$x = -3 \quad\quad x = 1$

Since $x = 1$ makes the denominator equal 0, the only solution is $x = -3$.

47. $\dfrac{x + 2}{x^2 - x} = \dfrac{6}{x^2 - 1}$

$$\frac{x + 2}{x(x - 1)} = \frac{6}{(x - 1)(x + 1)}$$

$\cancel{x}\cancel{(x - 1)}(x + 1)$

$$\cdot \frac{x + 2}{\cancel{x}\cancel{(x - 1)}}$$

$$= \frac{6}{\cancel{(x - 1)}\cancel{(x + 1)}}$$

$$\cdot x\cancel{(x - 1)}\cancel{(x + 1)}$$

$$\begin{aligned} (x + 1)(x + 2) &= 6x \\ x^2 + 3x + 2 &= 6x \\ x^2 - 3x + 2 &= 0 \\ (x - 2)(x - 1) &= 0 \end{aligned}$$

$x - 2 = 0 \quad x - 1 = 0$

$x = 2 \quad\quad x = 1$

Since $x = 1$ makes the denominator equal 0, the only solution is $x = 2$.

49. a) To solve equations that contain rational expressions, find the lowest common denominator. Then

multiply each term of the equation by the LCD. Solve the equation. Check all solutions to be sure none make any denominator equal to 0.

b) $\frac{1}{x - 1} - \frac{1}{x + 1} = \frac{3x}{x^2 - 1}$

$(x - 1)(x + 1)\left[\frac{1}{x - 1} - \frac{1}{x + 1}\right] = \frac{3x}{(x - 1)(x + 1)} \cdot (x - 1)(x + 1)$

$(\cancel{x - 1})(x + 1) \cdot \frac{1}{(\cancel{x - 1})} - (x - 1)(\cancel{x + 1}) \cdot \frac{1}{(\cancel{x + 1})} = \frac{3x}{(\cancel{x - 1})(\cancel{x + 1})} \cdot (\cancel{x - 1})(\cancel{x + 1})$

$$x + 1 - x + 1 = 3x$$
$$2 = 3x$$
$$\frac{2}{3} = x$$

51\. a) The problem on the right is an equation while the one on the left is not an equation.

b) To solve the problem on the left, write each fraction with the common denominator 12(x - 1). Then combine numerators over the common denominator.

To solve the problem on the right, multiply both sides of the equation by the LCD 12(x - 1) to eliminate fractions. Then solve the equation.

c) $\frac{x}{3} - \frac{x}{4} + \frac{1}{x - 1}$

$= \frac{x \cdot 4(x - 1)}{12(x - 1)} - \frac{x \cdot 3(x - 1)}{12(x - 1)} + \frac{12}{12(x - 1)}$

$= \frac{4x^2 - 4x - 3x^2 + 3x + 12}{12(x - 1)}$

$= \frac{x^2 - x + 12}{12(x - 1)}$

$= \frac{(x - 4)(x + 3)}{12(x - 1)}$

$\frac{x}{3} - \frac{x}{4} = \frac{1}{x - 1}$

$12(x - 1)\left[\frac{x}{3} - \frac{x}{4}\right] = \frac{1}{(x - 1)} \cdot 12(x - 1)$

$\overset{4}{\cancel{12}}(x - 1) \cdot \frac{x}{\cancel{3}} - \overset{3}{\cancel{12}}(x - 1) \cdot \frac{x}{\cancel{4}} = \frac{1}{(\cancel{x - 1})} \cdot 12(\cancel{x - 1})$

$$4x(x - 1) - 3x(x - 1) = 12$$
$$4x^2 - 4x - 3x^2 + 3x = 12$$
$$x^2 - x = 12$$
$$x^2 - x - 12 = 0$$
$$(x - 4)(x + 3) = 0$$
$$x - 4 = 0 \quad x + 3 = 0$$

$x = 4 \qquad x = -3$

Cumulative Review Exercises

53. Let x = smaller angle.
$2x + 30$ = larger angle.

$$x + 2x + 30 = 180$$
$$3x + 30 = 180$$
$$3x = 150$$
$$x = 50$$

Smaller angle: 50°
Larger angle:
$$2x + 30 = 2(50) + 30 = 100 + 30 = 130°$$

55. A linear equation is a first degree equation of the form $ax + b = 0$, $a \neq 0$. A quadratic equation is a second degree equation of the form $ax^2 + bx + c = 0$, $a \neq 0$.

Just for Fun

1. $$\frac{1}{p} + \frac{1}{q} = \frac{1}{f}$$

$$\frac{1}{30} + \frac{1}{q} = \frac{1}{10}$$

$$30q \cdot [\frac{1}{30} + \frac{1}{q}] = \frac{1}{10} \cdot 30q$$

$$30q \cdot \frac{1}{30} + 30q \cdot \frac{1}{q} = \frac{1}{10} \cdot 30q$$

$$q + 30 = 3q$$
$$30 = 2q$$
$$15 = q$$
$$q = 15 \text{ cm}$$

2. a) $$\frac{1}{R_T} = \frac{1}{R_1} + \frac{1}{R_2}$$

$R_1 = 200$ ohms;

$R_2 = 300$ ohms.

$$\frac{1}{R_T} = \frac{1}{200} + \frac{1}{300}$$

$$600R_T[\frac{1}{R_T}] = [\frac{1}{200} + \frac{1}{300}]600R_T$$

$$600R_T[\frac{1}{R_T}] = [\frac{1}{200}]600R_T + [\frac{1}{300}]600R_T$$

$$600 = 3R_T + 2R_T$$

$$600 = 5R_T$$

$$R_T = 120 \text{ ohms}$$

b) $$\frac{1}{R_T} = \frac{1}{R_1} + \frac{1}{R_2} + \frac{1}{R_3}$$

Since $R_1 = R_2 = R_3$, let R_1, R_2, and $R_3 = R$.

$R_T = 300$

$$\frac{1}{300} = \frac{1}{R} + \frac{1}{R} + \frac{1}{R}$$

$$\frac{1}{300} = \frac{3}{R}$$

$$300R[\frac{1}{300}] = [\frac{3}{R}]300R$$

$$R = 900 \text{ ohms}$$

Exercise Set 6.8

1. $x + 6 = \text{base}$
   $x = \text{height}$

   $80\text{cm}^2 = \text{area}$

   $$A = \frac{1}{2} \cdot b \cdot h$$

   $$80 = \frac{1}{2} \cdot (x + 6)x$$

   $$2(80) = \frac{2}{1} \cdot \left[\frac{1}{2}\right] \cdot (x + 6)x$$

   $$160 = (x + 6)x$$

   $$160 = x^2 + 6x$$

   $$0 = x^2 + 6x - 160$$
   $$0 = (x + 16)(x - 10)$$
   $$x + 16 = 0 \qquad x - 10 = 0$$
   $$x = -16 \qquad x = 10$$

   Since height cannot be a negative number, x cannot equal -16.
   Height = 10 cm.
   Base = $x + 6 = 10 + 6$
   = 16 cm.

3. x = a number.
   $3x$ = another number.

   $$\frac{1}{x} + \frac{1}{3x} = \frac{4}{3}$$

   $$3x\left[\frac{1}{x} + \frac{1}{3x}\right] = \frac{4}{3} \cdot \frac{3x}{1}$$
   $$3 + 1 = 4x$$
   $$4 = 4x$$
   $$1 = x$$
   A number: $x = 1$
   Another number:
   $3x = 3(1) = 3$

5. x = a number.

   $$\frac{1}{3} + \frac{1}{5} = \frac{1}{x}$$

   $$\frac{15x}{1} \cdot \left[\frac{1}{3} + \frac{1}{5}\right] = \frac{15x}{1} \cdot \frac{1}{x}$$
   $$5x + 3x = 15$$

   $$8x = 15$$

   $$x = \frac{15}{8}$$

   The number: $x = \frac{15}{8}$

7. Let x = one positive number.
   $x + 4$ = other positive number.

   $$\frac{1}{x} + \frac{1}{x + 4} = \frac{2}{3}$$

   $$3x(x + 4)\left[\frac{1}{x} + \frac{1}{x + 4}\right]$$

   $$= \frac{2}{3} \cdot 3x(x + 4)$$

   $$3x(x + 4) \cdot \frac{1}{x} + 3x(x + 4) \cdot \frac{1}{(x + 4)}$$

   $$= \frac{2}{3} \cdot 3x(x + 4)$$

   $$3x + 12 + 3x = 2x(x + 4)$$

   $$6x + 12 = 2x^2 + 8x$$

   $$0 = 2x^2 + 2x - 12$$

   $$0 = 2(x^2 + x - 6)$$
   $$0 = 2(x + 3)(x - 2)$$

   $x + 3 = 0$
   $x = -3$
   Since x is a positive number, $x \neq -3$.

   $x - 2 = 0$
   $x = 2$
   $x + 4$
   $= 2 + 4$
   $= 6$

   The numbers are 2 and 6.

9. x = speed of boat in still water.

$d = r \quad \frac{d}{r} = t$

	d	r	$t = \frac{d}{r}$
upstream	9	x-2	$\frac{9}{x - 2}$
downstream	11	x+2	$\frac{11}{x + 2}$

Time to travel 9 miles upstream = time to travel 11 miles downstream.

$$\frac{9}{x - 2} = \frac{11}{x + 2}$$

$$(x - 2)(x + 2)\left[\frac{9}{x - 2}\right]$$

$$= \left[\frac{11}{x + 2}\right](x - 2)(x + 2)$$

$$9(x + 2) = 11(x - 2)$$
$$9x + 18 = 11x - 22$$
$$18 = 2x - 22$$
$$40 = 2x$$
$$20 = x$$

The speed of the boat in still water: x = 20 mph.

11. r - rate of the propeller plane.
4r = rate of the jet.

$d = rt \quad \frac{d}{r} = t$

	d	r	$t = \frac{d}{r}$
jet	1800	4r	$\frac{1800}{4r}$
propeller	300	r	$\frac{300}{r}$

Time traveled in the jet + time traveled in the propellered plane = 5.

$$\frac{1800}{4x} + \frac{300}{x} = 5$$

$$4x\left[\frac{1800}{4x} + \frac{300}{x}\right] = 5(4x)$$
$$1800 + 1200 = 20x$$
$$300 = 20x$$
$$150 = x$$

Speed of propellered plane:
x = 150 mph.
Speed of jet:
4x = 4(150) = 600 mph.

13. d = distance walked.
d + 3 = distance jogged.

$d = rt \quad \frac{d}{r} = t$

	d	r	$t = \frac{d}{r}$
walked	d	2	$\frac{x}{2}$
jogged	d+3	4	$\frac{x + 3}{4}$

Time walked + time jogged = 3 hours.

$$\frac{d}{2} + \frac{d + 3}{4} = 3$$

$$4\left[\frac{d}{2} + \frac{d + 3}{4}\right] = 3(4)$$

$$2d + d + 3 = 12$$
$$3x + 3 = 12$$
$$3d = 9$$
$$d = 3$$

Distance walked:
  d = 3 miles
Distance jogged:
  d + 3 = 3 + 3 = 6 miles

15.

	Distance	Rate	Time
jog	d	8	$\frac{d}{8}$
walk	6-d	4	$\frac{6 - d}{4}$

$$\frac{d}{8} + \frac{6 - d}{4} = 1.2$$

$$8 \cdot \left(\frac{d}{8}\right) + 8\left(\frac{6 - d}{4}\right)$$

$$= (1.2)(8)$$
$$d + 2(6 - d) = 9.6$$
$$d + 12 - 2d = 9.6$$
$$12 - d = 9.6$$
$$-d = -2.4$$
$$d = 2.4 \text{ miles}$$

Jogs:
  2.4 miles
walks:
  6 - d = 6 - 2.4
    = 3.6 miles.

17.

	Distance	Rate	Time
Tailwind	d	600	$\frac{d}{600}$
Headwind	2800-d	500	$\frac{2800-d}{500}$

$$\frac{d}{600} + \frac{2800 - d}{500} = 5$$

$$\overset{5}{\cancel{3000}}\left(\frac{d}{\cancel{600}}\right)$$

$$+ \overset{6}{\cancel{3000}}\left(\frac{2800 - d}{\cancel{500}}\right)$$

$$= (5)(3000)$$
$$5d + 6(2800 - d) = 15{,}000$$
$$5d + 16{,}800 - 6d = 15{,}000$$
$$16{,}800 - d = 15{,}000$$
$$-d = -1800$$
$$d = 1800 \text{ miles}$$

Time with tailwind:

$$\frac{d}{600} = \frac{1800}{600} = 3 \text{ hours}$$

Time with headwind:

$$\frac{2800 - d}{500} = \frac{2800 - 1800}{500}$$

$$= \frac{1000}{500} = 2 \text{ hours}$$

19.

	Distance	Rate	Time
Freestyle	d	40	$\frac{d}{40}$
Breast stroke	d	30	$\frac{d}{30}$

$$\frac{d}{30} = \frac{d}{40} + 20$$

$$\overset{4}{120}\left(\frac{d}{30}\right) = \overset{3}{120}\left(\frac{d}{40}\right)$$

$$+ (120)(20)$$

$$4d = 3d + 2400$$

$$d = 2400 \text{ m}$$

21. t = time required to complete the job together:

$$\frac{t}{6} + \frac{t}{7} = 1$$

$$42\left[\frac{t}{6} + \frac{t}{7}\right] = 1(42)$$

$$7t + 6t = 42$$

$$13t = 42$$

$$t = \frac{42}{13}$$

Time required to complete the job:

$$t = 3\ \frac{3}{13} \text{ hours}$$

23. t = time required to fill tank.

$$\frac{t}{3} - \frac{t}{4} = 1$$

$$12\left[\frac{t}{3} - \frac{t}{4}\right] = 12(1)$$

$$4t - 3t = 12$$

$$t = 12$$

Time required to fill tank:
x = 12 hours.

25.

	Rate of work	time	Part of task
1st machine	$\frac{1}{8}$	3	$\frac{3}{8}$
2nd machine	$\frac{1}{x}$	3	$\frac{3}{x}$

$$\frac{3}{8} + \frac{3}{x} = 1$$

$$8x\left(\frac{3}{8}\right) + 8x\left(\frac{3}{x}\right) = 8x$$

$$3x + 24 = 8x$$

$$24 = 5x$$

$$\frac{24}{5} = x$$

$$x = \frac{4}{5} \text{ hours}$$

27.

	Rate of Work	Time	Part of Task
Large tractor	$\frac{1}{6}$	4	$\frac{4}{6}$
Small tractor	$\frac{1}{10}$	t	$\frac{t}{10}$

$$\frac{4}{6} + \frac{t}{10} = 1$$

$$\overset{5}{30}\left(\frac{4}{\underset{1}{6}}\right) + \overset{3}{30}\left(\frac{t}{\underset{1}{10}}\right) = 30$$

$$20 + 3t = 30$$

$$3t = 10$$

$$t = \frac{10}{3} = 3\frac{1}{3} \text{ days}$$

29.

	Rate of work	time	Part of Task
1st skimmer	$\frac{1}{60}$	t	$\frac{t}{60}$
2nd skimmer	$\frac{1}{50}$	t	$\frac{t}{50}$
valve	$\frac{1}{30}$	t	$\frac{t}{30}$

$$\frac{t}{60} + \frac{t}{50} - \frac{t}{30} = 1$$

$$\overset{5}{\cancel{300}}\left(\frac{t}{\underset{1}{\cancel{60}}}\right) + \overset{6}{\cancel{300}}\left(\frac{t}{\underset{1}{\cancel{50}}}\right) - \overset{10}{\cancel{300}}\left(\frac{t}{\underset{1}{\cancel{30}}}\right) = 300$$

$$5t + 6t - 10t = 300$$

$$t = 300 \text{ hours}$$

31.

	Rate of work	Time	Part of Task
Susan	$\frac{1}{20}$	t+11	$\frac{t+11}{20}$
Patty	$\frac{1}{25}$	t	$\frac{t}{25}$

$$\frac{t}{25} + \frac{t + 11}{20} = 1$$

$$\overset{4}{\cancel{100}}\left(\frac{t}{\underset{1}{\cancel{25}}}\right) + \overset{5}{\cancel{100}}\left(\frac{t + 11}{\underset{1}{\cancel{20}}}\right) = 100$$

$$4t + 5(t + 11) = 100$$

$$4t + 5t + 55 = 100$$

$$9t + 55 = 100$$

$$9t = 45$$

$$t = 5 \text{ hours}$$

Cumulative Review Exercises

33. $\frac{1}{2}(x + 3) - (2x + 6)$

$$= \frac{1}{2}x + \frac{3}{2} - 2x - 6$$

$$= \frac{1}{2}x - 2x + \frac{3}{2} - 6$$

$$= -\frac{3}{2}x - \frac{9}{2}$$

35. $\frac{x}{6x^2 - x - 15} - \frac{5}{9x^2 - 12x - 5}$

$$= \frac{x}{(2x + 3)(3x - 5)} - \frac{5}{(3x - 5)(3x + 1)}$$

$$= \frac{x(3x + 1)}{(2x + 3)(3x - 5)(3x + 1)} - \frac{5(2x + 3)}{(2x + 3)(3x - 5)(3x + 1)}$$

$$= \frac{3x^2 + x - 10x - 15}{(2x + 3)(3x - 5)(3x + 1)}$$

$$= \frac{3x^2 - 9x - 15}{(2x + 3)(3x - 5)(3x + 1)}$$

$$= \frac{3(x^2 - 3x - 5)}{(2x + 3)(3x - 5)(3x + 1)}$$

Just for fun

1. x = a number.
$x - 3$ = three less than a number.
$2x - 6$ = six less than twice a number.

$$\frac{1}{x - 3} = 2\left[\frac{1}{2x - 6}\right]$$

$$\frac{1}{x - 3} = \frac{2}{2(x - 3)}$$

$$\frac{1}{x - 3} = \frac{1}{x - 3}$$

$$(x - 3) \cdot \frac{1}{(x - 3)}$$

$$= \frac{1}{(x - 3)} \cdot (x - 3)$$

$$1 = 1$$

All real numbers except $x = 3$.

2. x = a number.

$$3x + 2\left[\frac{1}{x}\right] = 5$$

$$3x + \frac{2}{x} = 5$$

$$x\left[3x + \frac{2}{x}\right] = 5x$$

$$3x^2 + 2 = 5x$$

$$3x^2 - 5x + 2 = 0$$

$$(3x - 2)(x - 1) = 0$$

$$3x - 2 = 0 \qquad x - 1 = 0$$

$$3x = 2 \qquad x = 1$$

$$x = \frac{2}{3}$$

The numbers are $\frac{2}{3}$ and 1.

3.

	Quantity	Rate	Time
Donald	x	6	$\frac{x}{6}$
Juniper	x	3	$\frac{x}{3}$

$$\frac{x}{3} = \frac{x}{6} + 1.5$$

$$\overset{2}{\cancel{6}}\left(\frac{x}{\underset{1}{\cancel{3}}}\right) = \overset{1}{\cancel{6}}\left(\frac{x}{\underset{1}{\cancel{6}}}\right) + (6)(1.5)$$

$$2x = x + 9$$

$$x = 9 \text{ buckets}$$

Review Exercises

1. $\frac{6}{2x - 8}$ is defined for all values of x except when $x = 4$.

$$2x - 8 = 0$$

$$2x = 8$$

$$x = 4$$

3. $$\frac{x}{x - xy} = \frac{x}{x(1 - y)}$$

$$= \frac{1}{1 - y}$$

5. $$\frac{9x^2 + 6xy}{3x} = \frac{\cancel{3x}(3x + 2y)}{\cancel{3x}}$$

$$= 3x + 2y$$

7. $$\frac{x^2 - 4}{x - 2} = \frac{\cancel{(x - 2)}(x + 2)}{\cancel{x - 2}}$$

$$= x + 2$$

9. $$\frac{x^2 - 2x - 24}{x^2 + 6x + 8}$$

$$= \frac{(x + 4)(x - 6)}{(x + 4)(x + 2)}$$

$$= \frac{x - 6}{x + 2}$$

11. $\frac{4y}{3x} \cdot \frac{4x^2y}{2} = \frac{4y}{3x} \cdot \frac{4x^2y}{2}$

$$= \frac{8xy^2}{3}$$

13. $\frac{40a^3b^4}{7c^3} \cdot \frac{14c^5}{5a^5b}$

$$= \frac{40a^3b^4}{7c^3} \cdot \frac{14c^5}{5a^5b}$$

$$= \frac{16b^3c^2}{a^2}$$

15. $\frac{-x + 2}{3} \cdot \frac{6x}{x - 2}$

$$= \frac{-1(x - 2)}{3} \cdot \frac{6x}{x - 2}$$

$$= -2x$$

17. $\frac{a - 2}{a + 3} \cdot \frac{a^2 + 4a + 3}{a^2 - a - 2}$

$$= \frac{a - 2}{a + 3} \cdot \frac{(a + 3)(a + 1)}{(a - 2)(a + 1)}$$

$$= 1$$

19. $\frac{6y^3}{x} \div \frac{y^3}{6x} = \frac{6y^3}{x} \cdot \frac{6x}{y^3} = 36$

21. $\frac{3x + 3y}{x^2} \div \frac{x^2 - y^2}{x^2}$

$$= \frac{3x + 3y}{x^2} \cdot \frac{x^2}{x^2 - y^2}$$

$$= \frac{3(x + y)}{x^2} \cdot \frac{x^2}{(x - y)(x + y)}$$

$$= \frac{3}{x - y}$$

23. $\frac{4x}{a + 2} \div \frac{8x^2}{a - 2}$

$$= \frac{4x}{a + 2} \cdot \frac{a - 2}{8x^2}$$

$$= \frac{a - 2}{2x(a + 2)}$$

25. $\frac{x^2 - 3xy - 10y^2}{6x}$

$$\div \frac{x + 2y}{12x^2}$$

$$= \frac{(x - 5y)(x + 2y)}{6x}$$

$$\cdot \frac{12x^2}{x + 2y}$$

$$= 2x(x - 5y)$$

27. $\frac{x}{x + 2} + \frac{2}{x + 2}$ LCD $= x + 2$

$$\frac{x}{x + 2} + \frac{2}{x + 2}$$

$$= \frac{x + 2}{x + 2} = 1$$

29. $\frac{4x}{x + 2} + \frac{8}{x + 2}$ LCD $= x + 2$

$$\frac{4x}{x + 2} + \frac{8}{x + 2}$$

$$= \frac{4x + 8}{x + 2} + \frac{4(x + 2)}{x + 2}$$

$$= 4$$

31. $\frac{9x - 4}{x + 8} + \frac{76}{x + 8}$ LCD $= x + 8$

$$\frac{9x - 4}{x + 8} + \frac{76}{x + 8}$$

$$= \frac{9x - 4 + 76}{x + 8}$$

$= \dfrac{9x + 72}{x + 8}$

$= \dfrac{9(x + 8)}{x + 8}$

$= 9$

33. $\dfrac{4x^2 - 11x + 4}{x - 3}$

$- \dfrac{x^2 - 4x + 10}{x - 3}$

LCD = x - 3

$\dfrac{4x^2 - 11x + 4}{x - 3}$

$- \dfrac{x^2 - 4x + 10}{x - 3}$

$= \dfrac{4x^2 - 1x + 4 - [x^2 - 4x + 10]}{x - 3}$

$= \dfrac{4x^2 - 11x + 4 - x^2 + 4x - 10}{x - 3}$

$= \dfrac{3x^2 - 7x - 6}{x - 3}$

$= \dfrac{(3x + 2)(x - 3)}{x - 3}$

$= 3x + 2$

35. $\dfrac{x}{3} + \dfrac{5x}{8}$ LCD = 3 · 8 = 24

37. $\dfrac{6}{x + 1} - \dfrac{3x}{x}$ LCD = x(x + 1)

39. $\dfrac{7x - 12}{x^2 + x} - \dfrac{4}{x + 1}$

$= \dfrac{7x - 12}{x(x + 1)} - \dfrac{4}{x + 1}$

LCD = x(x + 1)

41. $\dfrac{4x^2}{x - 7} + 8x^2$

$= \dfrac{4x^2}{x - 7} + \dfrac{8x^2}{1}$

LCD = 1(x - 7) = x - 7

43. $\dfrac{4}{2x} + \dfrac{x}{x^2}$ LCD = $2x^2$

$\dfrac{4}{2x} + \dfrac{x}{x^2}$

$= [\dfrac{x}{x}] \cdot \dfrac{4}{2x} + \dfrac{x}{x^2} \cdot [\dfrac{2}{2}]$

$= \dfrac{4x + 2x}{2x^2}$

$= \dfrac{6x}{2x^2}$

$= \dfrac{3}{x}$

45. $\dfrac{5x}{3xy} - \dfrac{4}{x^2}$ LCD = $3x^2y$

$\dfrac{5x}{3xy} - \dfrac{4}{x^2}$

$= [\dfrac{x}{x}] \cdot \dfrac{5x}{3xy} - \dfrac{4}{x^2} [\dfrac{3y}{3y}]$

$= \dfrac{5x^2 - 12y}{3x^2y}$

47. $5 - \dfrac{3}{x + 3}$ LCD = x + 3

$5 - \dfrac{3}{x + 3}$

$= [\dfrac{x + 3}{x + 3}] \cdot \dfrac{5}{1} - \dfrac{3}{x + 3}$

$= \dfrac{5(x + 3) - 3}{x + 3}$

$= \dfrac{5x + 15 - 3}{x + 3}$

$$= \frac{5x + 12}{x + 3}$$

49. $\frac{3}{x + 3} + \frac{4}{x}$ LCD = $x(x + 3)$

$$\frac{3}{x + 3} + \frac{4}{x}$$

$$= [\frac{x}{x}] \cdot \frac{3}{(x + 3)} + \frac{4}{x} \cdot \frac{(x + 3)}{(x + 3)}$$

$$= \frac{3x + 4(x + 3)}{x(x + 3)}$$

$$= \frac{3x + 4x + 12}{x(x + 3)}$$

$$= \frac{7x + 12}{x(x + 3)}$$

51. $\frac{x + 4}{x + 3} - \frac{x - 3}{x + 4}$

LCD = $(x + 3)(x + 4)$

$$\frac{x + 4}{x + 3} - \frac{x - 3}{x + 4}$$

$$= \frac{(x + 4)}{(x + 4)} \cdot \frac{(x + 4)}{(x + 3)} - \frac{(x - 3)}{(x + 4)} \cdot \frac{(x + 3)}{(x + 3)}$$

$$= \frac{x^2 + 8x + 16 - [x^2 - 9]}{(x + 4)(x + 3)}$$

$$= \frac{x^2 + 8x + 16 - x^2 + 9}{(x + 4)(x + 3)}$$

$$= \frac{8x + 25}{(x + 3)(x + 4)}$$

53. $\frac{x + 3}{x^2 - 9} + \frac{2}{x + 3}$

LCD = $(x - 3)(x + 3)$

$$\frac{x + 3}{x^2 - 9} + \frac{2}{x + 3}$$

$$= \frac{x + 3}{(x + 3)(x - 3)} + \frac{2}{(x + 3)} \cdot \frac{(x - 3)}{(x - 3)}$$

$$= \frac{(x + 3) + 2(x - 3)}{(x + 3)(x - 3)}$$

$$= \frac{x + 3 + 2x - 6}{(x + 3)(x - 3)}$$

$$= \frac{3x - 3}{(x + 3)(x - 3)}$$

$$= \frac{3(x - 1)}{(x + 3)(x - 3)}$$

55. $\frac{x + 2}{x^2 - x - 6} + \frac{x - 3}{x^2 - 8x + 15}$

$$= \frac{x + 2}{(x - 3)(x + 2)} + \frac{x - 3}{(x - 5)(x - 3)}$$

LCD = $(x - 3)(x - 5)(x + 2)$

$$\frac{x + 2}{x^2 - x - 6} + \frac{x - 3}{x^2 - 8x + 15}$$

$$= \frac{x + 2}{(x - 3)(x + 2)} + \frac{x - 3}{(x - 5)(x - 3)}$$

$$= \frac{(x - 5)}{(x - 5)} \cdot \frac{(x + 2)}{(x - 3)(x + 2)} + \frac{(x - 3)}{(x - 5)(x - 3)} \cdot \frac{(x + 2)}{(x + 2)}$$

$$= \frac{[x^2-3x-10] + [x^2-x-6]}{(x - 5)((x - 3)(x + 2)}$$

$$= \frac{2x^2 - 4x - 16}{(x - 5)(x - 3)(x + 2)}$$

$$= \frac{2[x^2-2x-8]}{(x - 5)(x - 3)(x + 2)}$$

$$= \frac{2(x - 4)(x + 2)}{(x - 5)(x - 3)(x + 2)}$$

$$= \frac{2(x - 4)}{(x - 5)(x - 3)}$$

57. $$\frac{1 + \frac{5}{12}}{\frac{3}{8}} = \frac{\frac{12}{12} + \frac{5}{12}}{\frac{3}{8}}$$

$$= \frac{\frac{12 + 5}{12}}{\frac{3}{8}} = \frac{\frac{17}{12}}{\frac{3}{8}}$$

$$= \frac{17}{12} \cdot \frac{8}{3} = \frac{34}{9}$$

59. $$\frac{\frac{15xy}{6z}}{\frac{3x}{z^2}} = [\frac{15xy}{6z}]$$

$$[\frac{z^2}{3x}] = \frac{5yz}{6}$$

61. $$\frac{x + \frac{1}{y}}{y^2} = \frac{\frac{y}{y} \cdot \frac{x}{1} + \frac{1}{y}}{\frac{y^2}{1}}$$

$$= \frac{\frac{xy + 1}{y}}{\frac{y^2}{1}} = \frac{(xy + 1)}{y}$$

$$[\frac{1}{y^2}] = \frac{xy + 1}{y^3}$$

63. $$\frac{\frac{4}{x} + \frac{2}{x^2}}{6 - \frac{1}{x}} = \frac{\frac{x}{x} \cdot \frac{4}{x} + \frac{2}{x^2}}{\frac{x}{x} \cdot \frac{6}{1} - \frac{1}{x}}$$

$$= \frac{\frac{4x + 2}{x^2}}{\frac{6x - 1}{x}}$$

$$= \frac{(4x + 2)}{x^2} \cdot \frac{x}{(6x - 1)}$$

$$= \frac{4x + 2}{x(6x - 1)}$$

65. $$\frac{a^{-1}}{a^{-2}} = \frac{a^2}{a^1} = a^{2-1} = a^1 \text{ or } a$$

$$\frac{\frac{1}{a}}{\frac{1}{a^2}} = \frac{\frac{a^2}{1} \cdot \frac{1}{a}}{\frac{a^2}{1} \cdot \frac{1}{a^2}} = a$$

67. $$\frac{\frac{1}{x^2} + \frac{1}{x}}{\frac{1}{x^2} - \frac{1}{x}} = \frac{\frac{1}{x^2} + \frac{1}{x} \cdot \frac{1}{x}}{\frac{1}{x^2} - \frac{1}{x} \cdot \frac{1}{x}}$$

$$= \frac{\frac{1+x}{x^2}}{\frac{1-x}{x^2}} = \frac{(1+x)}{x^2}$$

$$\frac{x^2}{(1-x)} = \frac{1+x}{1-x}$$

69. $\frac{3}{x} = \frac{8}{24}$

$\frac{3}{x} = \frac{1}{3}$

$3x[\frac{3}{x}] = \frac{1}{3} \cdot 3x$

$9 = x$

71. $\frac{x+3}{5} = \frac{9}{5}$

$\frac{5(x+3)}{5} = [\frac{9}{5}] \cdot 5$

$x + 3 = 9$

$x = 6$

73. $\frac{3x+4}{5} = \frac{2x-8}{3}$

$15 \cdot [\frac{3x+4}{5}] = [\frac{2x-8}{3}]$

$\cdot 15$

$3(3x + 4) = 5(2x - 8)$

$9x + 12 = 10x - 40$

$12 = x - 40$

$52 = x$

75. $4 - \frac{5}{x+5} = \frac{x}{x+5}$

$(x + 5)[4 - \frac{5}{x+5}]$

$= \frac{x}{(x+5)} \cdot (x + 5)$

$(x + 5) \quad 4$

$- (x + 5)(\frac{5}{x+5})$

$= \frac{x}{(x+5)} \cdot (x + 5)$

$4x + 20 - 5 = x$

$4x + 15 = x$

$3x + 15 = 0$

$3x = -15$

$x = -5$

Since x = -5 makes the denominator equal 0, there is no solution.

77. $\frac{1}{x-2} + \frac{1}{x+2}$

$= \frac{1}{x^2 - 4}$

$\frac{1}{x-2} + \frac{1}{x+2}$

$= \frac{1}{(x-2)(x+2)}$

$(x - 2)(x + 2)$

$\cdot [\frac{1}{x-2} + \frac{1}{x+2}]$

$= [\frac{1}{(x-2)(x+2)}]$

$\cdot (x - 2)(x + 2)$

$(x + 2) + (x - 2) = 1$

$2x = 1$

$x = \frac{1}{2}$

79. $\frac{x}{x^2 - 9} + \frac{2}{x+3}$

$= \frac{4}{x-3}$

$\frac{x}{(x-3)(x+3)} + \frac{2}{x+3}$

$$= \frac{4}{x - 3}$$

$$(x - 3)(x + 3)$$

$$[\frac{x}{x - 3)(x + 3)}$$

$$+ \frac{2}{x + 3}]$$

$$= [\frac{4}{x - 3}]$$

$$(x - 3)(x + 3)$$

$$x + 2(x - 3) = 4(x + 3)$$
$$x + 2x - 6 = 4x + 12$$
$$3x - 6 = 4x + 12$$
$$-6 = x + 12$$
$$-18 = x$$

81. t = time it takes to fill the pool.

$$\frac{x}{7} - \frac{x}{12} = 1$$

$$84[\frac{x}{7} - \frac{x}{12}] = 1(84)$$

$$12x - 7x = 84$$
$$5x = 84$$
$$x = \frac{84}{5}$$

It will take 16 4/5 hours to fill the pool.

83. r = rate of the bus.
r + 40 = rate of the train.

$$d = rt \quad \frac{d}{r} = t$$

	d	r	$t = \frac{d}{t}$
bus	400	r	$\frac{400}{r}$
train	600	r+40	$\frac{600}{r+40}$

Amount of time for bus to travel 400 miltes = amount of time for train to travel 600 miles.

$$\frac{400}{r} = \frac{600}{r + 40}$$

$$r(r + 40) \text{ ' } [\frac{400}{r}]$$

$$= [\frac{600}{r + 40}] \text{ ' } x(x + 40)$$

$$400(r + 40) = 600r$$
$$400r + 16000 = 600r$$
$$16000 = 200r$$
$$80 = r$$

Bus speed: 80 kph
Train speed:
r + 40 = 80 + 40 = 120

Practice Test

1. $\frac{3x^2y}{4z^2} \cdot \frac{8xz^3}{9y^4} = \frac{3x^2y}{4z^2} \cdot \frac{8xz^3}{9y^4}$ (with cancellations: 3, 4, $8 \to 2$, $z^3 \to z$, $9 \to 3$, $y^4 \to y^3$)

$$= \frac{2x^3z}{3y^3}$$

2. $\dfrac{a^2 - 9a + 14}{a - 2}$

$\dfrac{a^2 - 4a - 21}{(a - 7)^2}$

$= \dfrac{(a - 7)(a - 2)}{a - 2}$

$\dfrac{(a - 7)(a + 3)}{(a - 7)(a - 7)}$

$= a + 3$

3. $\dfrac{x^2 - 9y^2}{3x + 6y} + \dfrac{x + 3y}{x + 2y}$

$= \dfrac{x^2 - 9y^2}{3x + 6y} \cdot \dfrac{x + 2y}{x + 3y}$

$= \dfrac{(x - 3y)(x + 3y)}{3(x + 2y)}$

$\dfrac{x + 2y}{x + 3y} = \dfrac{x - 3y}{3}$

4. $\dfrac{16}{y^2 + 2y - 15}$

$\div \dfrac{4y}{y - 3}$

$= \dfrac{16}{(y + 5)(y - 3)}$

$\dfrac{y - 3}{4y} = \dfrac{4}{y(y + 5)}$

5. $\dfrac{6x + 3}{2y} + \dfrac{x - 5}{2y}$ LCD $= 2y$

$\dfrac{6x + 3}{2y} + \dfrac{x - 5}{2y}$

$= \dfrac{6x + 3 + x - 5}{2y}$

$= \dfrac{7x - 2}{2y}$

6. $\dfrac{7x^2 - 4}{x + 3} - \dfrac{6x + 7}{x + 3}$

LCD $= x + 3$

$\dfrac{7x^2 - 4}{x + 3} - \dfrac{6x + 7}{x + 3}$

$= \dfrac{7x^2 - 4 - (6x+7)}{x + 3}$

$= \dfrac{7x^2 - 4 - 6x - 7}{x + 3}$

$= \dfrac{7x^2 - 6x - 11}{x + 3}$

7. $\dfrac{5}{x} + \dfrac{3}{2x^2}$ LCD $= 2x^2$

$\dfrac{5}{x} + \dfrac{3}{2x^2} = [\dfrac{2x}{2x} \cdot \dfrac{5}{x} + \dfrac{3}{2x^2}]$

$= \dfrac{10x + 3}{2x^2}$

8. $5 - \dfrac{6x}{x + 2} = \dfrac{5}{1} - \dfrac{6x}{x + 2}$

LCD $= x + 2$

$5 - \dfrac{6x}{x + 2} = \dfrac{5}{1} - \dfrac{6x}{x + 2}$

$= \dfrac{(x + 2)}{(x + 2)} \cdot \dfrac{5}{1} - \dfrac{6x}{(x + 2)}$

$= \dfrac{5(x + 2) - 6x}{x + 2}$

$= \dfrac{5x + 10 - 6x}{x + 2}$

$= \dfrac{-x + 10}{x + 2}$

9. $\dfrac{x - 5}{x^2 - 16} - \dfrac{x - 2}{x^2 + 2x - 8}$

$= \dfrac{x - 5}{(x - 4)(x + 4)}$

$- \dfrac{x - 2}{(x + 4)(x - 2)}$

LCD $= (x - 4)(x + 4)(x - 2)$

$$\frac{x-5}{x^2-16} - \frac{x-2}{x^2+2x-8}$$

$$= \frac{x-5}{(x-4)(x+4)}$$

$$- \frac{x-2}{(x+4)(x-2)}$$

$$= \frac{(x-2)}{(x-2)} \cdot \frac{(x-5)}{(x-4)(x+4)}$$

$$- \frac{(x-2)}{(x+4)(x-2)}$$

$$\cdot \frac{(x-4)}{(x-4)}$$

$$= \frac{x^2-7x+10-[x^2-6x+8]}{(x+4)(x-2)(x-4)}$$

$$= \frac{x^2-7x+10-x^2+6x-8}{(x+4)(x-2)(x-4)}$$

$$= \frac{-x+2}{(x+4)(x-2)(x-4)}$$

$$= \frac{-1(x-2)}{(x+4)(x-2)(x-4)}$$

$$= \frac{-1}{(x+4)(x-4)}$$

10. $$\frac{3+\frac{5}{8}}{2-\frac{3}{4}} = \frac{\frac{8}{8}\cdot\frac{3}{1}+\frac{5}{8}}{\frac{4}{4}\cdot\frac{2}{1}-\frac{3}{4}}$$

$$= \frac{\frac{24+5}{8}}{\frac{8-3}{4}} = \frac{\frac{29}{8}}{\frac{5}{4}}$$

$$= \frac{29}{8} \cdot \frac{4}{5} = \frac{29}{10}$$

11. $$\frac{x+\frac{x}{y}}{\frac{1}{x}} = \frac{\frac{y}{y}\cdot\frac{x}{1}+\frac{x}{y}}{\frac{1}{x}}$$

$$= \frac{\frac{xy+x}{y}}{\frac{1}{x}}$$

$$= \left[\frac{xy+x}{y}\right] \cdot \frac{x}{1}$$

$$= \frac{x^2y+x^2}{y} = \frac{x^2(y+1)}{y}$$

12. $$\frac{x}{3} - \frac{x}{4} = 5$$

$$12\left[\frac{x}{3} - \frac{x}{4}\right] = 5(12)$$

$$4x - 3x = 60$$

$$x = 60$$

13. $$\frac{x}{x-8} + \frac{6}{x-2}$$

$$= \frac{x^2}{x^2-10x+16}$$

$$\frac{x}{x-8} + \frac{6}{x-2}$$

$$= \frac{x^2}{(x-8)(x-2)}$$

$$(x-8)(x-2)$$

$$\cdot \left[\frac{x}{x-8} + \frac{6}{x-2}\right]$$

$$= \left[\frac{x^2}{(x-8)(x-2)}\right]$$

$$\cdot (x-8)(x-2)$$

$$(x-2)x + 6(x-8) = x^2$$

$$x^2 - 2x + 6x - 48 = x^2$$

$$x^2 + 4x - 48 = x^2$$
$$4x - 48 = 0$$
$$4x = 48$$
$$x = 12$$

14. Let t = time they work together.

$$\frac{1}{8} + \frac{1}{5} = \frac{1}{t}$$

$$40t[\frac{1}{8} + \frac{1}{5}] = [\frac{1}{t}]40t$$

$$5t + 8t = 40$$
$$13t = 40$$

$$t = \frac{40}{13}$$

$$t = 3\frac{1}{13}$$

Cumulative Review Test

1. $3x^2 - 2xy - 7$

$$= 3(-3)^2 - 2(-3)(5) - 7$$
$$= 3(9) - 2(-15) - 7$$
$$= 27 + 30 - 7$$
$$= 50$$

2. $-4 - [2(-6 \div 3)^2)] \div 2$

$$= -4 - [2(-2)^2] \div 2$$
$$= -4 - [2(4)] \div 2$$
$$= -4 - [8] \div 2$$
$$= -4 - 4$$
$$= -8$$

3. $4y + 3 = -2(y + 6)$

$$4y + 3 = -2y - 12$$
$$6y + 3 = -12$$
$$6y = -15$$

$$y = -\frac{15}{6} = -\frac{5}{2}$$

4. $(\frac{6x^2y^3}{2x^5y})^3 = (\frac{3y^2}{x^3})^3$

$$= \frac{27y^6}{x^9}$$

5. $P = 2E + 3R$

$$P - 2E = 3R$$

$$\frac{P - 2E}{3} = R$$

6. $(6x^2 - 3x - 5)$

$$- (3x^2 + 8x - 9)$$

$$= 6x^2 - 3x - 5 - 3x^2 - 8x + 9$$

$$= 6x^2 - 3x^2 - 3x - 8x - 5 + 9$$

$$= 3x^2 - 11x + 4$$

7.
$$\begin{array}{r} 4x^2 - 6x + 3 \\ 3x - 5 \\ \hline 12x^3 - 18x^2 + 9x \\ - 20x^2 + 30x - 15 \\ \hline 12x^3 - 38x^2 + 39x - 15 \end{array}$$

8. $6a^2 - 6a - 5a + 5$

$$= 6a(a - 1) - 5(a - 1)$$
$$= (6a - 5)(a - 1)$$

9. $10x^2 - 5x + 5$

$$= 5(2x^2 - x + 1)$$

10. $x^2 - 10x + 24$

$$= (x - 6)(x - 4)$$

11. $6x^2 - 11x - 10$

$$= (3x + 2)(2x - 5)$$

12. $2x^2 = 11x - 12$

$$2x^2 - 11x + 12 = 0$$
$$(2x - 3)(x - 4) = 0$$
$$2x - 3 = 0 \quad x - 4 = 0$$

$$2x = 3 \qquad x = 4$$
$$x = \frac{3}{2}$$

13. $$\frac{x^2 - 9}{x^2 - x - 6} \cdot \frac{x^2 - 2x - 8}{2x^2 - 7x - 4}$$
$$= \frac{(x - 3)(x + 3)}{(x - 3)(x + 2)} \cdot \frac{(x - 4)(x + 2)}{(2x + 1)(x - 4)}$$
$$= \frac{(x + 3)}{(2x + 1)}$$

14. $$\frac{x}{x + 4} - \frac{3}{x - 5}$$
$$= \frac{x(x - 5)}{(x + 4)(x - 5)} - \frac{3(x + 4)}{(x + 4)(x - 5)}$$
$$= \frac{x^2 - 5x - 3x - 12}{(x + 4)(x - 5)}$$
$$= \frac{x^2 - 8x - 12}{(x + 4)(x - 5)}$$

15. $$\frac{4}{x^2 - 3x - 10} + \frac{2}{x^2 + 5x + 6}$$
$$= \frac{4}{(x - 5)(x + 2)} + \frac{2}{(x + 2)(x + 3)}$$
$$= \frac{4(x + 3)}{(x - 5)(x + 2)(x + 3)} + \frac{2(x - 5)}{(x - 5)(x + 2)(x + 3)}$$
$$= \frac{4x + 12 + 2x - 10}{(x - 5)(x + 2)(x + 3)}$$
$$= \frac{6x + 2}{(x - 5)(x + 2)(x + 3)}$$

16. $$\frac{x}{6} - \frac{x}{4} = \frac{1}{8}$$
$$\overset{4}{24}\left(\frac{x}{\underset{1}{6}}\right) - \overset{6}{24}\left(\frac{x}{\underset{1}{4}}\right) = \overset{3}{24}\left(\frac{1}{\underset{1}{8}}\right)$$
$$4x - 6x = 3$$
$$-2x = 3$$
$$x = -\frac{3}{2}$$

17. $$\frac{1}{x - 4} + \frac{2}{x - 3} = \frac{4}{x^2 - 7x + 12}$$
$$(x - 4)(x - 3) \cdot \left[\frac{1}{x - 4} + \frac{2}{x - 3}\right] = \frac{4}{(x - 3)(x - 4)} \cdot (x - 4)(x - 3)$$
$$(x - 4)(x - 3)\left(\frac{1}{x - 4}\right) + (x - 4)(x - 3)\left(\frac{2}{x - 3}\right) = \frac{4}{(x - 3)(x - 4)} \cdot (x - 4)(x - 3)$$
$$x - 3 + 2(x - 4) = 4$$
$$x - 3 + 2x - 8 = 4$$
$$3x - 11 = 4$$
$$3x = 15$$
$$x = 5$$

18. Let x = amount of medical bills.

$$0.10x = 100 + 0.05x$$
$$0.05x = 100$$

$x = \$2000$

19. Let x = amount of Hippy dog food.

$$3(6) + 4x = 3.2(6 + x)$$
$$18 + 4x = 19.2 + 3.2x$$
$$18 + 0.8x = 19.2$$
$$0.8x = 1.2$$
$$x = 1.5 \text{ pounds}$$

20.

	Distance	Rate	Time
Up hill	x	4	$\frac{x}{4}$
Down hill	6-x	12	$\frac{6-x}{12}$

$$\frac{x}{4} + \frac{6 - x}{12} = 1$$

$$\overset{3}{\cancel{12}}\left(\frac{x}{\underset{1}{\cancel{4}}}\right) + \overset{1}{\cancel{12}}\left(\frac{6 - x}{\underset{1}{\cancel{12}}}\right) = 12$$

$$3x + 6 - x = 12$$
$$2x + 6 = 12$$
$$2x = 6$$
$$x = 3 \text{ miles at each rate}$$

CHAPTER SEVEN

Exercise Set 7.1

1. $A(3,1)$; $B(-3,0)$; $C(1,-3)$;
   $D(-2,-3)$; $E(0,3)$; $F(\frac{-3}{2},-1)$

3. $A(4,2)$; $B(-3,2)$; $C(0,-3)$;
   $D(-2,0)$; $E(-3,-4)$; $F(-4,-2)$

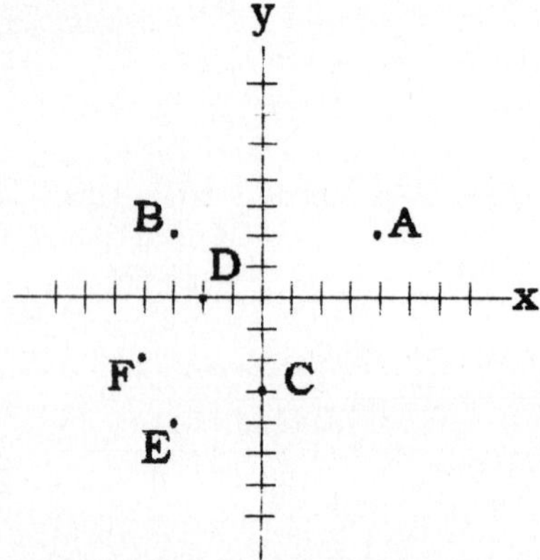

5. $A(1,-1)$; $B(5,3)$; $C(-3,-5)$;
   $D(0,-2)$; $E(2,0)$

   Yes, these points are collinear.

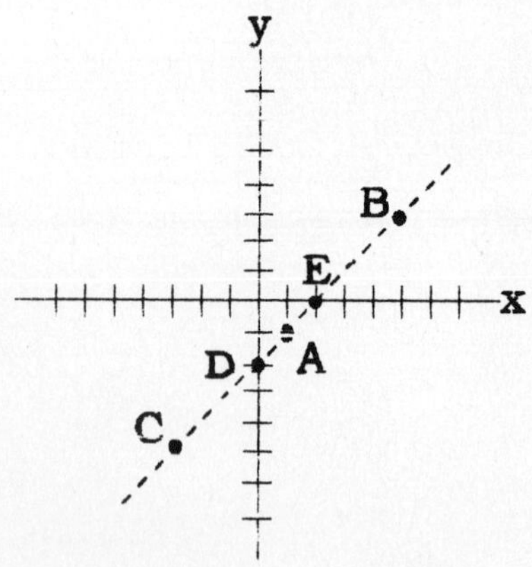

7. The x-coordinate is always listed first.

Cumulative Review Exercises

9. $(-2x^2 - 5x + 9)$
   $- (6x^2 - 4x + 5)$
   $= -2x^2 - 5x + 9 - 6x^2 + 4x - 5$
   $= -2x^2 - 6x^2 - 5x + 4x + 9 - 5$
   $= -8x^2 - x + 4$

11. $x^2 - 2x + 3xy - 6y$
    $= x(x - 2) + 3y(x - 2)$
    $= (x - 2)(x + 3y)$

Exercise Set 7.2

1. $y = 2x - 1$

   a) $(0,-1)$
      $-1 = 2(0) - 1$
      $-1 = 0 - 1$
      $-1 = -1$
      $(0,-1)$ satisfies the equation.

   b) $(-1,0)$
      $0 = 2(-1) - 1$
      $0 = -2 - 1$
      $0 \neq -3$
      $(-1,0)$ does not satisfy the equation.

   c) $(5,3)$
      $3 = 2(5) - 1$
      $3 = 10 - 1$
      $3 \neq 9$
      $(5,3)$ does not satisfy the equation.

   d) $(6,11)$
      $11 = 2(6) - 1$

$11 = 12 - 1$
$11 = 11$
(6,11) satisfies the equation.

e) $(\frac{1}{2},0)$

$0 = 2(\frac{1}{2}) - 1$

$0 = \frac{2}{2} - 1$

$0 = 1 - 1$
$0 = 0$

$(\frac{1}{2},0)$ satisfies the equation.

3. $5x - 6 = 2y$

a) (9,20)
$5(9) - 6 = 2(20)$
$45 - 6 = 40$
$39 \neq 40$
(9,20) does not satisfy the equation.

b) (-2,-8)
$5(-2) - 6 = 2(-8)$
$-10 - 6 = -16$
$-16 = -16$
(-2,-8) satisfies the equation.

c) (0,-3)
$5(0) - 6 = 2(-3)$
$-6 = -6$
(0,-3) satisfies the equation.

d) $(\frac{6}{5},0)$

$5(\frac{6}{5}) - 6 = 12$

$6 - 6 = 0$
$0 = 0$

$(\frac{6}{5},0)$ satisfies the equation.

e) $(\frac{-3}{8},6)$

$5(\frac{-3}{8}) - 6 = 2(6)$

$\frac{-15}{8} - 6 = 12$

$\frac{-15}{8} - \frac{48}{8} = 12$

$\frac{-63}{8} = 12$

$-7\frac{7}{8} \neq 12$

$(\frac{-3}{8},6)$ does not satisfy the equation.

5. a) $2x + y = 6$
$2(2) + y = 6$
$4 + y = 6$
$y = 2$

b) $2x + y = 6$
$2(-1) + y = 6$
$-2 + y = 6$
$y = 8$

c) $2x + y = 6$
$2x + (-5) = 6$
$2x = 11$
$x = \frac{11}{2}$

d) $2x + y = 6$
$2x + (-3) = 6$
$2x = 9$
$x = \frac{9}{2}$

e) $2x + y = 6$
$2x + 0 = 6$
$2x = 6$
$x = 3$

d) $2x + y = 6$

$$2\left(\frac{1}{2}\right) + y = 6$$

$$1 + y = 6$$
$$y = 5$$

7. $y = 6$

x	y	Ordered Pairs
-1	6	(-1,6)
3	6	(3,6)
4	6	(4,6)

This equation states that y has to be six, however x may be any real number.

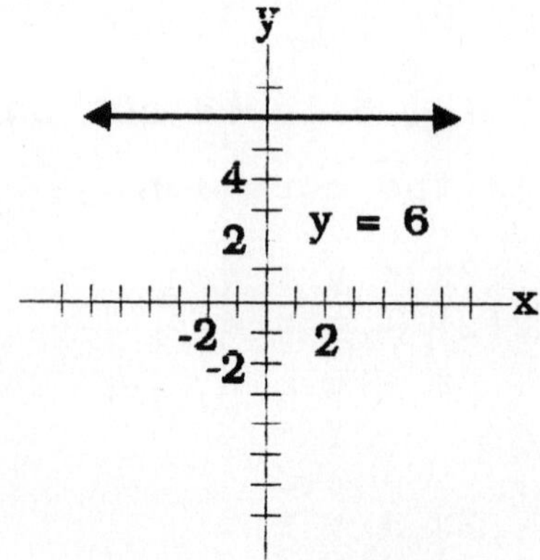

9. $x = 3$

x	y	Ordered Pairs
3	-1	(3,-1)
3	0	(3,0)
3	1	(3,1)

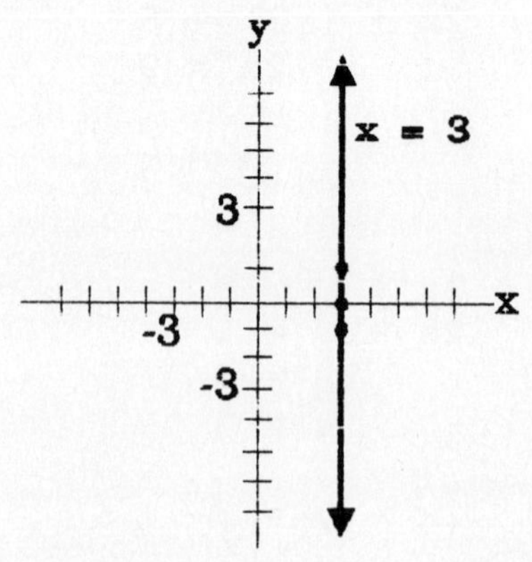

11. $y = 4x - 2$

x	y	Ordered Pairs
-2	-10	(-2,-10)
0	-2	(0,-2)
1	2	(1,2)

Let $x = -2$.
$y = 4(-2) - 2 = -8 - 2$
$= -10$
Let $x = 0$.
$y = 4(0) - 2 = -2$
Let $x = 1$.
$y = 4(1) - 2 = 4 - 2 = 2$

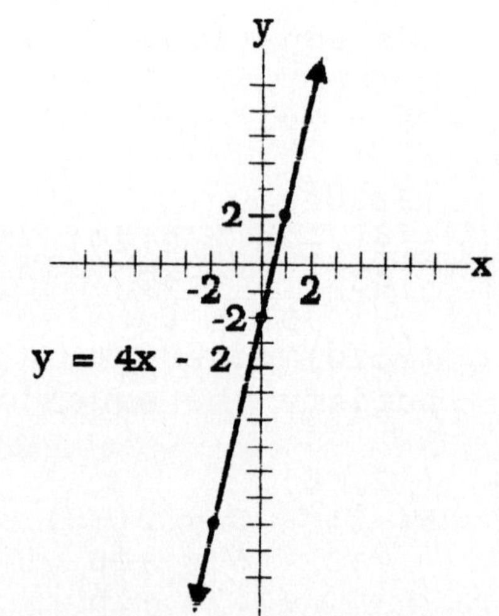

13. $y = 6x + 2$

x	y	Ordered Pairs
-1	-4	(-1,-4)
0	2	(0,2)
1	8	(1,8)

Let $x = -1$.
$y = 6(-1) + 2$
$= -6 + 2 = -4$
Let $x = 0$.
$y = 6(0) + 2$
$= 0 + 2 = 2$
Let $x = 3$.
$y = 6(1) + 2$
$= 6 + 2 = 8$

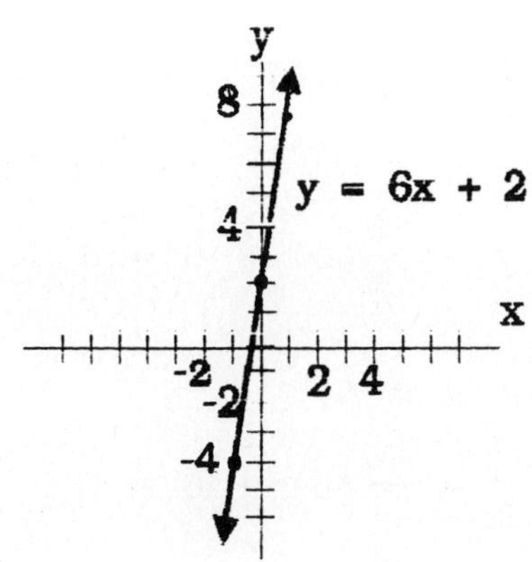

15. $y = \frac{-1}{2}x + 3$

x	y	Ordered Pairs
-2	4	(-2,4)
0	3	(0,3)
2	2	(2,2)

Let x = -2.

$$y = \frac{-1}{2}(-2) + 3$$
$$= 1 + 3 = 4$$

Let x = 0.

$$y = \frac{-1}{2}(0) + 3$$
$$= 0 + 3 = 3$$

Let x = 2.

$$y = \frac{-1}{2}(2) + 3$$
$$= -1 + 3 = 2$$

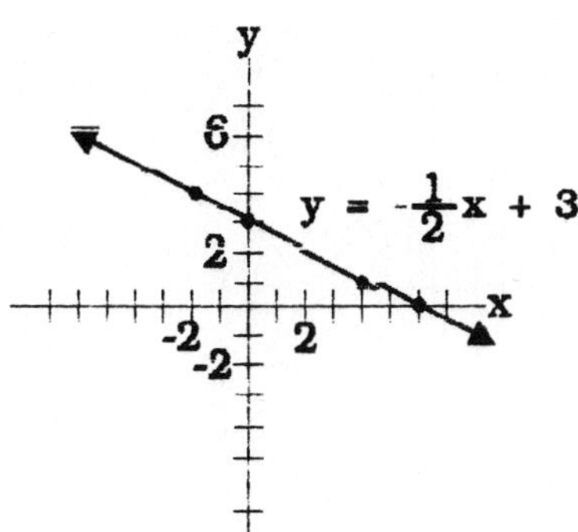

17. $6x - 2y = 4$

$$6x - 6x - 2y = 4 - 6x$$
$$-2y = 4 - 6x$$
$$\frac{-2y}{-2} = \frac{(4 - 6x)}{-2}$$
$$y = -2 + 3x$$
$$y = 3x - 2$$

x	y	Ordered Pairs
-1	-5	(-1,-5)
0	-2	(0,-2)
2	4	(2,4)

Let x = -1.
$y = 3(-1) - 2 = -5$
Let x = 0.
$y = 3(0) - 2 = -2$
Let x = 2.
$y = 3(2) - 2 = 4$

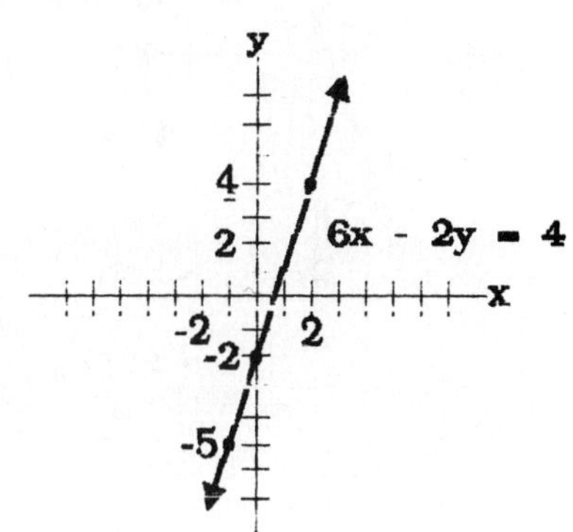

19. $5x - 2y = 8$

$$5x - 8 = 2y$$
$$\frac{5x}{2} - \frac{8}{2} = y$$
$$\frac{5x}{2} - 4 = y$$

x	y	Ordered Pairs
0	-4	(0,-4)
2	1	(2,1)
4	6	(4,6)

Let x = 0.

$y = \frac{5}{2}(0) - 4 = -4$

Let x = 2.

$y = \frac{5}{2}(2) - 4 = 5(1) - 4$

$= 5 - 4 = 1$

Let x - 4.

$y = \frac{5}{2}(4) - 4 = 5(2) - 4$

$= 10 - 4 = 6$

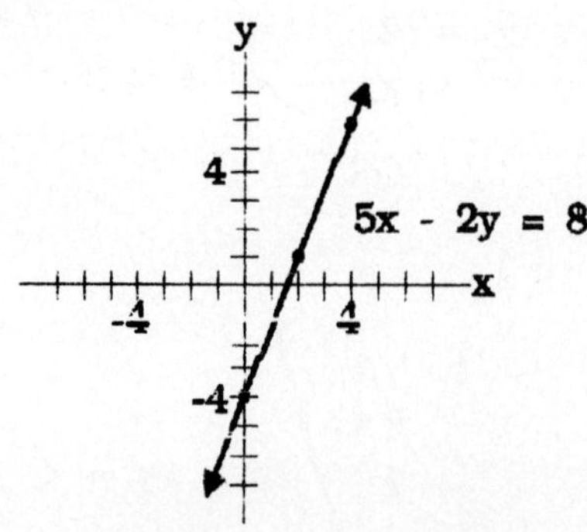

21.
$$\begin{aligned} 6x + 5y &= 30 \\ 6x - 6x + 5y &= -6x + 30 \\ 5y &= -6x + 30 \\ \frac{5y}{5} &= \frac{(-6x + 30)}{5} \\ y &= \frac{-6x}{5} + 6 \end{aligned}$$

x	y	Ordered Pairs
-5	12	(-5,12)
0	6	(0,6)
5	0	(5,0)

Let x = -5.

$y = \frac{-6}{5}(-5) + 6 = 12$

Let x = 0.

$y = \frac{-6}{5}(0) + 6 = 6$

Let x = 5.

$y = \frac{-6}{5}(5) + 6 = 0$

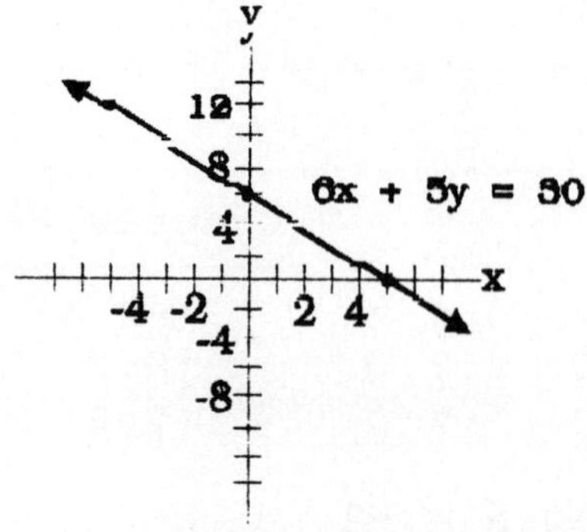

23.
$$\begin{aligned} -4x - y &= -2 \\ -4x + 4x - 1y &= 4x - 2 \\ \frac{-1y}{-1} &= \frac{4x}{-1} \frac{-2}{-1} \\ y &= -4x + 2 \end{aligned}$$

x	y	Ordered Pairs
-2	10	(-2,10)
0	2	(0,2)
3	-10	(3,-10)

Let x = -2.
$y = -4(-2) + 2 = 10$
Let x = 0.
$y = -4(0) + 2 = 2$
Let x = 3.
$y = -4(3) + 2 = -10$

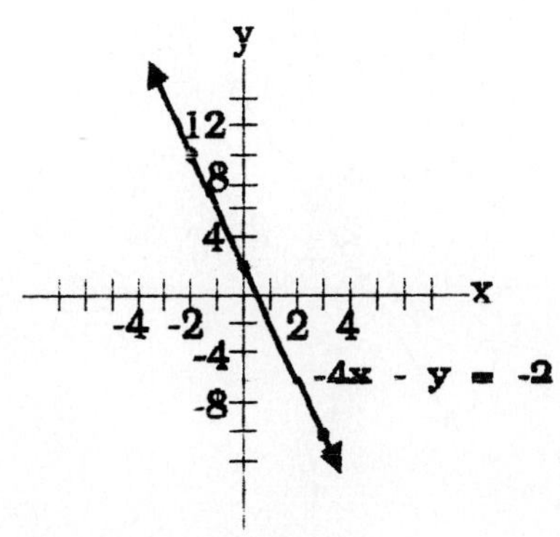

25. $y = 20x + 40$

x	y	Ordered Pairs
-2	0	(-2,0)
0	40	(0,40)
2	80	(2,80)

Let $x = -2$.
$y = 20(-2) + 40$
$= -40 + 40 = 0$
Let $x = 0$.
$y = 20(0) + 40$
$= 0 + 40 = 40$
Let $x = 2$.
$y = 20(2) + 40$
$= 40 + 40 = 80$

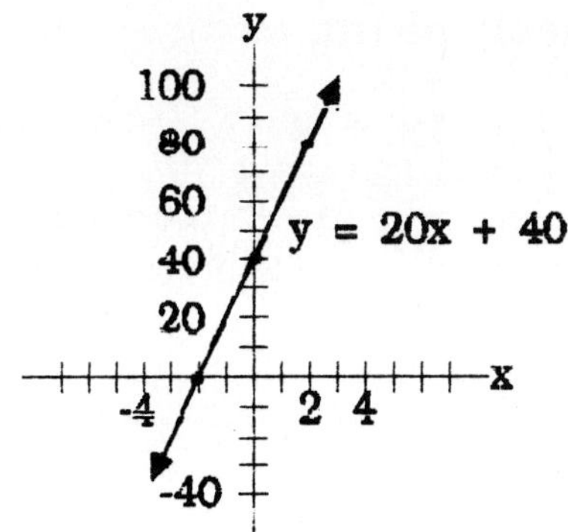

27. $y = \frac{2}{3}x$

x	y	Ordered Pairs
-3	-2	(-3,-2)
0	0	(0,0)
3	2	(3,2)

Let $x = -3$.

$y = \frac{2}{3}(-3) = -2$

Let $x = 0$.

$y = \frac{2}{3}(0) = 0$

Let $x = 3$.

$y = \frac{2}{3}(3) = 2$

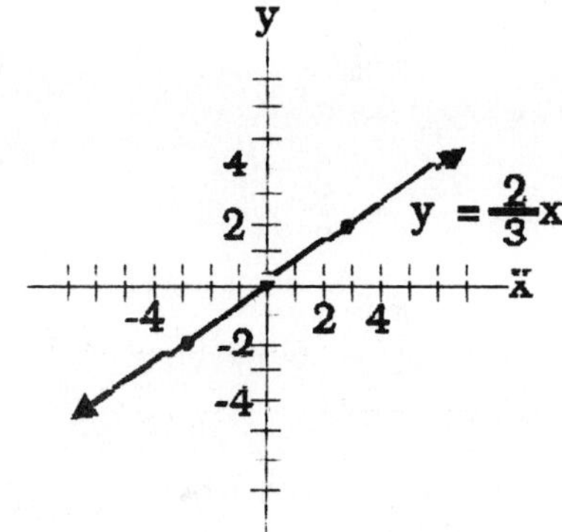

29. $y = \frac{1}{2}x + 4$

x	y	Ordered Pairs
-2	3	(-2,3)
0	4	(0,4)
2	5	(2,5)

Let $x = -2$.

$y = \frac{1}{2}(-2) + 4 = 3$

Let $x = 0$.

$y = \frac{1}{2}(0) + 4 = 4$

Let $x = 2$.

$$y = \frac{1}{2}(2) + 4 = 5$$

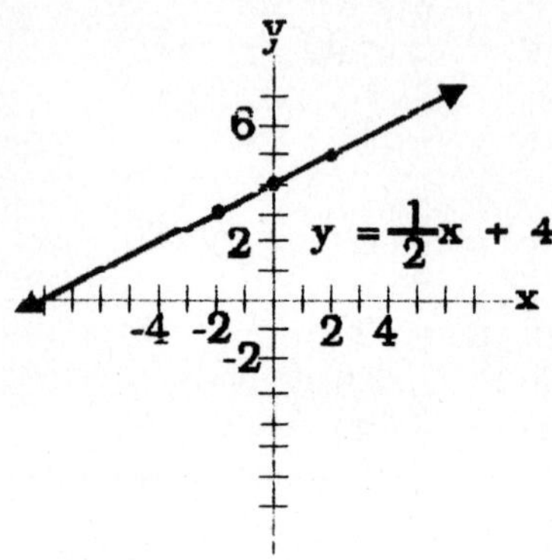

31. $2y = 3x + 6$

$$\frac{2y}{2} = \frac{3x}{2} + \frac{6}{2}$$

$$y = \frac{3}{2}x + 3$$

x	y	Ordered Pairs
-2	0	(-2,0)
0	3	(0,3)
2	6	(2,6)

Let $x = -2$.

$$y = \frac{3}{2}(-2) + 3 = 0$$

Let $x = 0$.

$$y = \frac{3}{2}(0) + 3 = 3$$

Let $x = 2$.

$$y = \frac{3}{2}(2) + 3 = 6$$

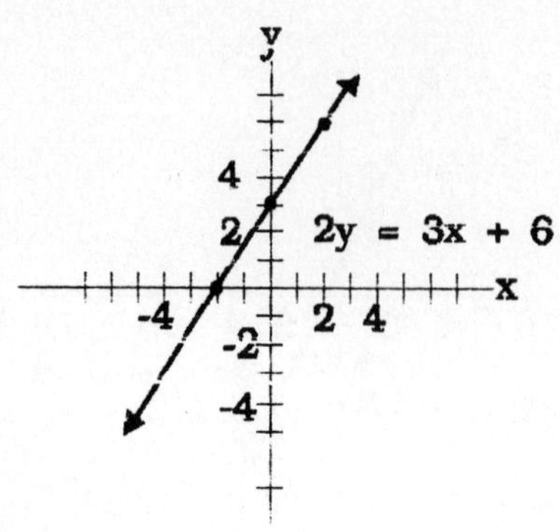

33. $y = 2x + 4$

Find y-intercept.
Let $x = 0$.

$y = 2x + 4$
$y = 2(0) + 4$
$y = 0 + 4$
$y = 4 \quad (0,4)$

Find x-intercept.
Let $y = 0$.

$y = 2x + 4$
$0 = 2x + 4$
$0 - 4 = 2x + 4 - 4$
$-4 = 2x$

$$\frac{-4}{2} = \frac{2x}{2}$$

$-2 = x \quad (-2,0)$

Check point
Let $x = 2$.

$y = 2x + 4$
$y = 2(2) + 4$
$y = 4 + 4$
$y = 8 \quad (2,8)$

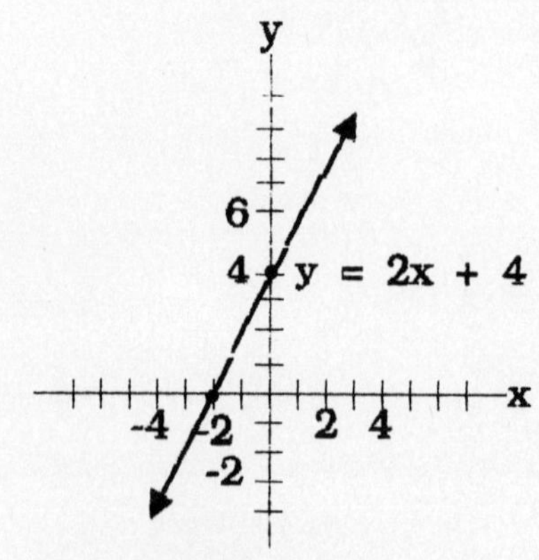

35. $y = 4x - 3$

Find the y-intercept.
Let $x = 0$.

$y = 4x - 3$
$y = 4(0) - 3$
$y = 0 - 3$
$y = -3 \quad (0,-3)$

Find the x-intercept.
Let $y = 0$.

$y = 4x - 3$
$0 = 4x - 3$
$0 + 3 = 4x - 3 + 3$
$3 = 4x$
$\frac{3}{4} = \frac{4x}{4}$
$\frac{3}{4} = x \quad (\frac{3}{4},0)$

Check point
Let $x = 2$.

$y = 4(2) - 3$
$y = 8 - 3$
$y = 5 \quad (2,5)$

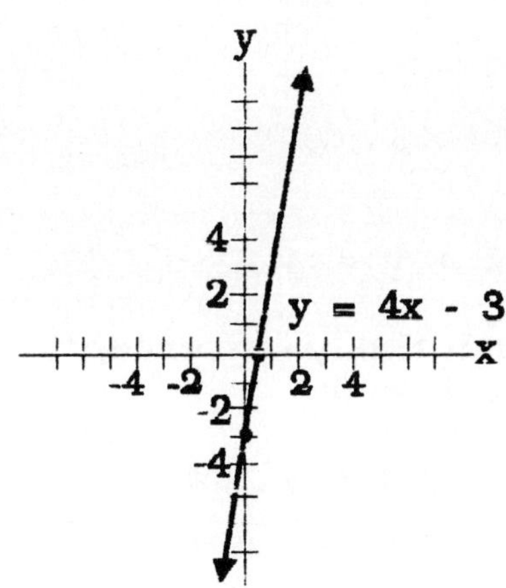

37. $y = -6x + 5$

Find the y-intercept.
Let $x = 0$.

$y = -6x + 5$
$y = -6(0) + 5$
$y = 0 + 5$
$y = 5 \quad (0,5)$

Find the x-intercept.
Let $y = 0$.

$y = -6x + 5$
$0 = -6x + 5$
$0 - 5 = -6x + 5 - 5$
$-5 = -6x$
$\frac{-5}{-6} = \frac{-6x}{-6}$
$\frac{5}{6} = x \quad (\frac{5}{6},0)$

Check point.
Let $x = 1$.

$y = -6x + 5$
$y = -6(1) + 5$
$y = -6 + 5$
$y = -1 \quad (1,-1)$

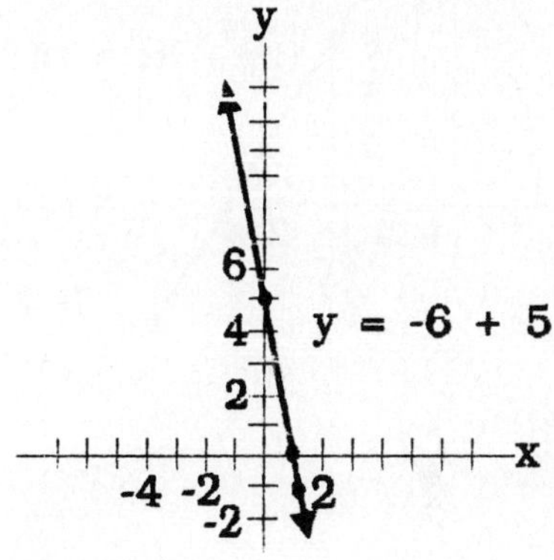

39. $2y + 3x = 12$

Find the y-intercept.
Let $x = 0$.

$2y + 3x = 12$
$2y + 3(0) = 12$
$2y + 0 = 12$
$\frac{2y}{2} = \frac{12}{2}$
$y = 6 \quad (0,6)$

Find the x-intercept.
Let $y = 0$.

$2y + 3x = 12$

$$2(0) + 3x = 12$$
$$0 + 3x = 12$$
$$3x = 12$$
$$\frac{3x}{3} = \frac{12}{3}$$
$$x = 4 \quad (4,0)$$

Check point.
Let x = 2.

$$2y + 3x = 12$$
$$2y + 3(2) = 12$$
$$2y + 6 = 12$$
$$2y + 6 - 6 = 12 - 6$$
$$2y = 6$$
$$\frac{2y}{2} = \frac{6}{2}$$
$$y = 3 \quad (2,3)$$

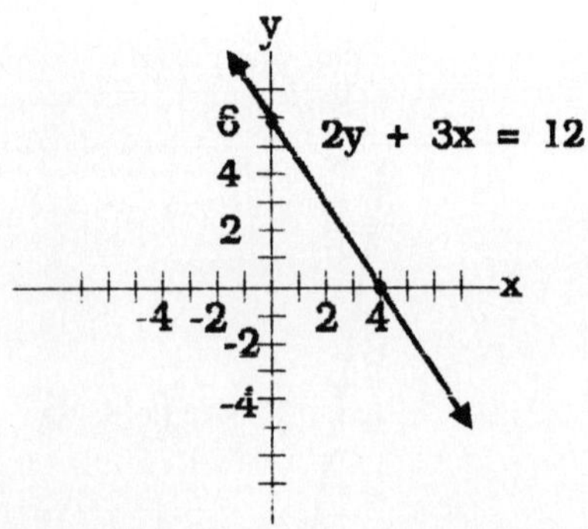

41. $4x = 3y - 9$

Find y-intercept.
Let x = 0.

$$4x = 3y - 9$$
$$4(0) = 3y - 9$$
$$0 + 9 = 3y - 9 + 9$$
$$9 = 3y$$
$$\frac{9y}{3} = \frac{3y}{3}$$
$$3 = y \quad (0,3)$$

Find the x-intercept.
Let y = 0.

$$4x = 3y - 9$$
$$4x = 3(0) - 9$$

$$4x = -9$$
$$\frac{4x}{4} = \frac{-9}{4}$$
$$x = \frac{-9}{4} \quad (\frac{-9}{4},0)$$

Check point.
Let x = 3.

$$4x = 3y - 9$$
$$4(3) = 3y - 9$$
$$12 = 3y - 9$$
$$12 + 9 = 3y - 9 + 9$$
$$21 = 3y$$
$$\frac{21}{3} = \frac{3y}{3}$$
$$7 = y \quad (3,7)$$

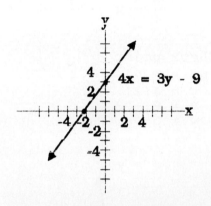

43. $\frac{1}{2} x + y = 4$

Find the y-intercept.
Let x = 0.

$$\frac{1}{2} x + y = 4$$
$$\frac{1}{2}(0) + y = 4$$
$$0 + y = 4$$
$$y = 4 \quad (0,4)$$

Find the x-intercept.
Let y = 0.

$$\frac{1}{2}x + y = 4$$

$$\frac{1}{2}x + 0 = 4$$

$$\frac{1}{2}x = 4$$

$$2\left(\frac{1}{2}x\right) = 2(4)$$

$$x = 8 \quad (8,0)$$

Check point.
Let x = 2.

$$\frac{1}{2}x + y = 4$$

$$\frac{1}{2}(2) + y = 4$$

$$1 + y = 4$$
$$1 - 1 + y = 4 - 1$$
$$y = 3 \quad (2,3)$$

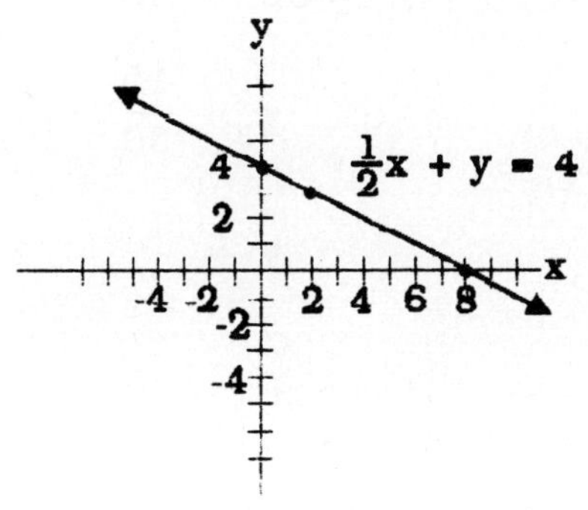

$$\frac{-12y}{-12} = \frac{24}{-12}$$

$$y = -2 \quad (0,-2)$$

Find the x-intercept.
Let y = 0.

$$6x - 12y = 24$$
$$6x - 12(0) = 24$$
$$6x - 0 = 24$$
$$6x = 24$$
$$\frac{6x}{6} = \frac{24}{6}$$
$$x = 4 \quad (4,0)$$

Check point.
Let x = 12.

$$6x - 12y = 24$$
$$6(12) - 12y = 24$$
$$72 - 12y = 24$$
$$72 - 72 - 12y = 24 - 72$$
$$-12y = -48$$
$$\frac{-12y}{-12} = \frac{-48}{-12}$$
$$y = 4 \quad (12,4)$$

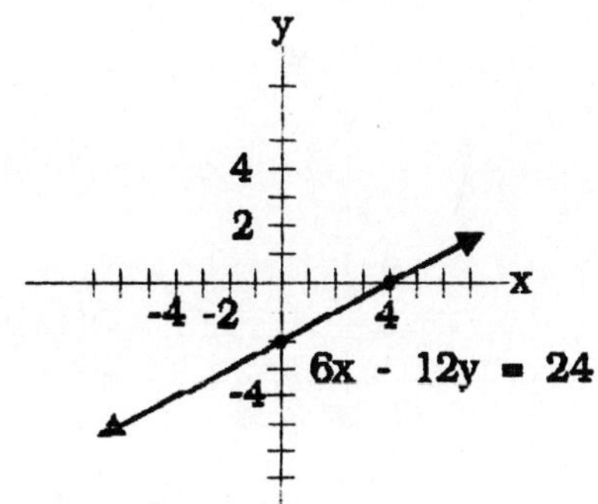

45. $6x - 12y = 24$

Find the y-intercept.
Let x = 0.

$$6x - 12y = 24$$
$$6(0) - 12y = 24$$
$$0 - 12y = 24$$
$$-12y = 24$$

47. $8y = 6x - 12$

Find the y-intercept.
Let x = 0.

$$8y = 6x - 12$$
$$8y = 6(0) - 12$$
$$8y = 0 - 12$$
$$8y = -12$$

$$\frac{8y}{8} = \frac{-12}{8}$$

$$y = \frac{-3}{2} \quad (0, \frac{-3}{2})$$

Find the x-intercept.
Let y = 0.

$$8y = 6x - 12$$
$$8(0) = 6x - 12$$
$$0 = 6x - 12$$
$$0 + 12 = 6x - 12 + 12$$
$$12 = 6x$$
$$\frac{12}{6} = \frac{6x}{6}$$
$$2 = x \quad (2,0)$$

Check point.
Let x = 2.

$$8y = 6x - 12$$
$$8y = 6(2) - 12$$
$$8y = 12 - 12$$
$$8y = 0$$
$$\frac{8y}{8} = \frac{0}{8}$$
$$y = 0 \quad (2,0)$$

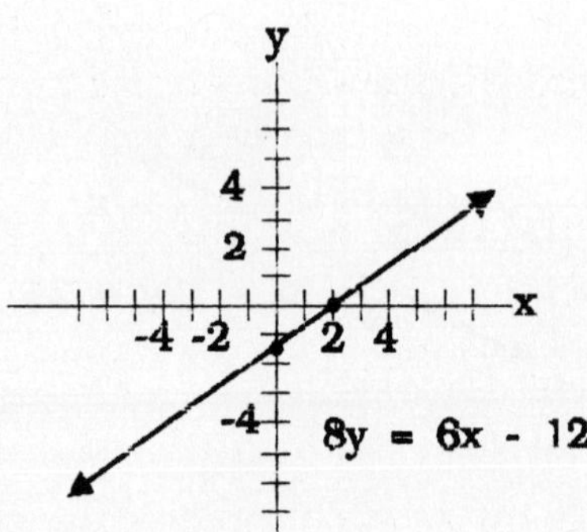

49. $30y + 10x = 45$

Find the y-intercept.
Let x = 0.

$$30y + 10x = 45$$
$$30y + 10(0) = 45$$
$$30y + 0 = 45$$
$$30y = 45$$
$$\frac{30y}{30} = \frac{45}{30}$$
$$y = \frac{3}{2} \quad (0, \frac{3}{2})$$

Find the x-intercept.
Let y = 0.

$$30y + 10x = 45$$
$$30(0) + 10x = 45$$
$$0 + 10x = 45$$
$$10x = 45$$
$$\frac{10x}{10} = \frac{45}{10}$$
$$x = \frac{9}{2} \quad (\frac{9}{2}, 0)$$

Check point.
Let x = 3.

$$30y + 10x = 45$$
$$30y + 10(3) = 45$$
$$30y + 30 = 45$$
$$30y + 30 - 30 = 45 - 30$$
$$30y = 15$$
$$\frac{30y}{30} = \frac{15}{30}$$
$$y = \frac{15}{30}$$
$$y = \frac{1}{2} \quad (3, \frac{1}{2})$$

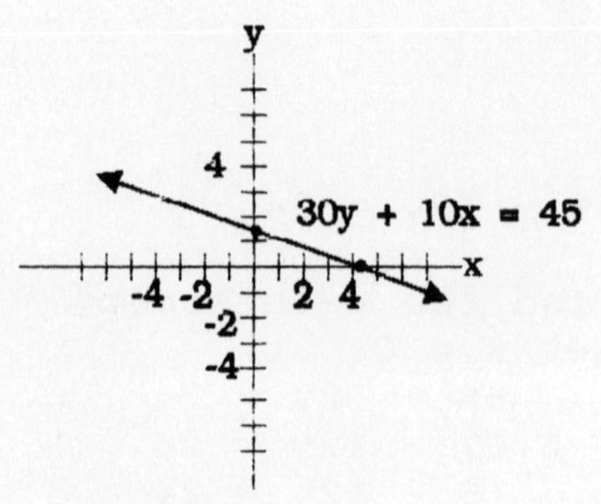

51. $40x + 6y = 40$

Find the y-intercept.
Let x = 0.

$$40(x) + 6y = 40$$
$$40(0) + 6y = 40$$
$$0 + 6y = 40$$
$$6y = 40$$
$$\frac{6y}{6} = \frac{40}{6}$$
$$y = \frac{20}{3} \quad (0, \frac{20}{3})$$

Find the x-intercept.
Let y = 0.

$$40x + 6y = 40$$
$$40x + 6(0) = 40(0)$$
$$40x + 0 = 40$$
$$40x = 40$$
$$\frac{40x}{40} = \frac{40}{40}$$
$$x = 1 \quad (1,0)$$

Check point.

Let $x = \frac{-1}{2}$.

$$40x + 6y = 40$$
$$40(\frac{-1}{2}) + 6y = 40$$
$$\frac{-40}{2} + 6y = 40$$
$$-20 + 6y = 40$$
$$-20 + 20 + 6y = 40 + 20$$
$$6y = 60$$
$$\frac{6y}{6} = \frac{60}{6}$$
$$y = 10 \quad (\frac{-1}{2}, 10)$$

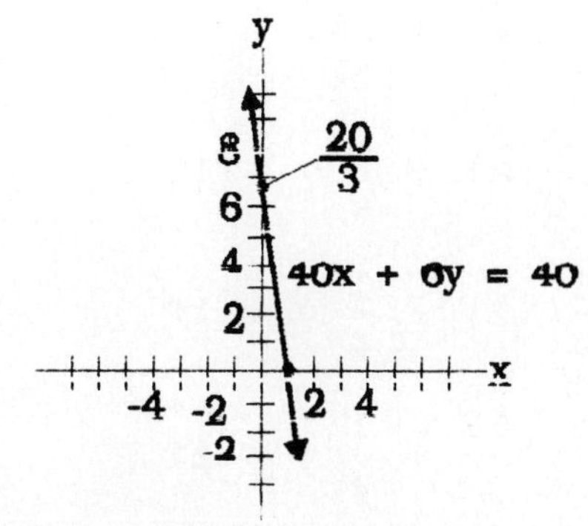

53. $\frac{1}{3}x + \frac{1}{4}y = 12$

Find the y-intercept.
Let x = 0

$$\frac{1}{3}(x) + \frac{1}{4}y = 12$$
$$\frac{1}{3}(0) + \frac{1}{4}y = 12$$
$$0 + \frac{1}{4}y = 12$$
$$\frac{1}{4}y = 12$$
$$4(\frac{1}{4}y) = 4(12)$$
$$y = 48 \quad (0,48)$$

Find the x-intercept.
Let y = 0.

$$\frac{1}{3}x + \frac{1}{4}y = 12$$
$$\frac{1}{3}x + \frac{1}{4}(0) = 12$$
$$\frac{1}{3}x + 0 = 12$$
$$\frac{1}{3}x = 12$$
$$3(\frac{1}{3}x) = 3(12)$$
$$x = 36 \quad (36,0)$$

Check point.
Let x = 12.

$$\frac{1}{3}(x) + \frac{1}{4}y = 12$$

$$\frac{1}{3}(12) + \frac{1}{4}y = 12$$

$$\frac{12}{3} + \frac{1}{4}y = 12$$

$$4 + \frac{1}{4}y = 12$$

$$4 - 4 + \frac{1}{4}y = 12 - 4$$

$$\frac{1}{4}y = 8$$

$$4(\frac{1}{4}y) = 8(4)$$

$$y = 32 \quad (12,32)$$

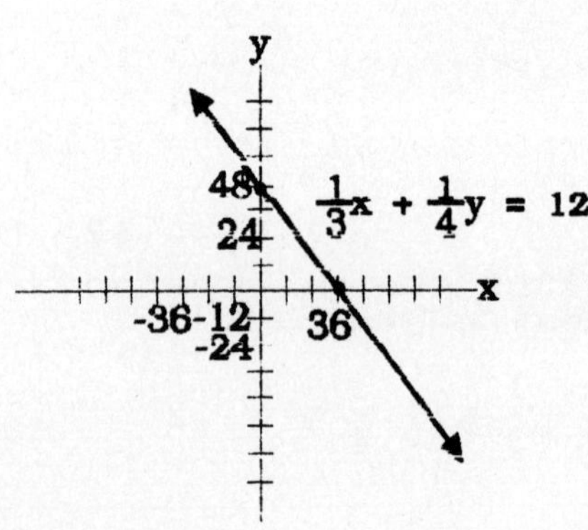

55. $\frac{1}{2}x = \frac{2}{5}y - 80$

Find the y-intercept.
Let x = 0.

$$\frac{1}{2}x = \frac{2}{5}y - 80$$

$$\frac{1}{2}(0) = \frac{2}{5}y - 80$$

$$0 = \frac{2}{5}y - 80$$

$$0 + 80 = \frac{2}{5}y - 80 + 80$$

$$80 = \frac{2}{5}y$$

$$\frac{5}{2}(80) = \frac{5}{2}(\frac{2}{5}y)$$

$$200 = y \quad (0,200)$$

Find the x-intercept.
Let y = 0.

$$\frac{1}{2}x = \frac{2}{5}y - 80$$

$$\frac{1}{2}x = \frac{2}{5}(0) - 80$$

$$\frac{1}{2}x = 0 - 80$$

$$\frac{1}{2}x = -80$$

$$\frac{2}{1}(\frac{1}{2})x = \frac{2}{1}(-80)$$

$$x = -160 \quad (-160,0)$$

Check point.
Let x = 20

$$\frac{1}{2}x = \frac{2}{5}y - 80$$

$$\frac{1}{2}(20) = \frac{2}{5}y - 80$$

$$\frac{20}{2} = \frac{2}{5}y - 80$$

$$10 = \frac{2}{5}y - 80$$

$$10 + 80 = \frac{2}{5}y - 80 + 80$$

$$90 = \frac{2}{5}y$$

$$\frac{5}{2}(90) = \frac{5}{2}\left(\frac{2}{5}y\right)$$

$$225 = y \quad (20, 225)$$

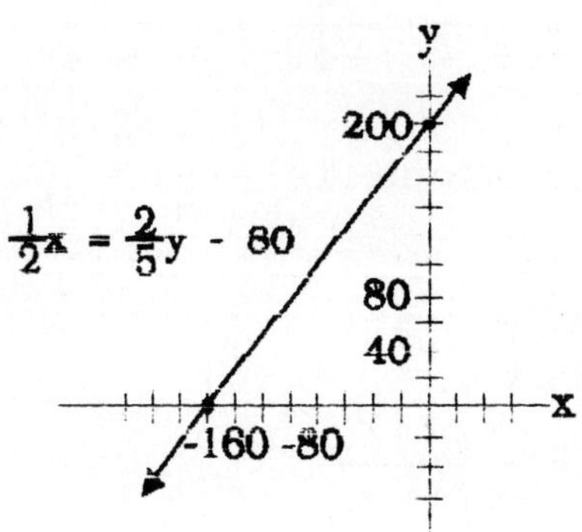

57. $x = -3$

59. $y = 3$

61. The set of points whose coordinates satisfy the equation.

63. Two points are needed to graph a straight line. Three points should be used: 2 to graph the line and 1 check point.

65. A vertical line.

67.
$$\begin{aligned} ax + by &= c & ax + by &= c \\ a(-3)+b(0) &= 18 & a(0)+b(6) &= 18 \\ 3a + 0 &= 18 & 0 + 6b &= 18 \\ -3a &= 18 & 6b &= 18 \\ a &= -6 & b &= 3 \end{aligned}$$
$$-6x + by = 18$$

69.
$$\begin{aligned} ax - by &= c & ax - by &= c \\ a(-5)-b(0) &= 30 & a(0)-b(-15) &= 30 \\ -5a &= 30 & 15b &= 30 \\ a &= -6 & b &= 2 \end{aligned}$$
$$-6x + 2y = 30$$

Cumulative Review Exercises

71.

	d	r	t
Faster rider	1.5(x+3)	x+3	1.5
Slower rider	1.5x	x	1.5

$$\begin{aligned} 1.5x + 1.5(x + 3) &= 18 \\ 1.5x + 1.5x + 4.5 &= 18 \\ 3.0x + 4.5 &= 18 \\ 3.0x &= 13.5 \\ x &= 4.5 \end{aligned}$$

Slow rider: 4.5 mph.
Faster rider: 4.5 + 3 = 7.5 mph

73.
$$x - 14 = \frac{-48}{x}$$

$$x(x - 14) = \left(\frac{-48}{x}\right) \cdot x$$

$$\begin{aligned} x^2 - 14x &= -48 \\ x^2 - 14x + 48 &= 0 \\ (x - 6)(x - 8) &= 0 \end{aligned}$$
$$x - 6 = 0 \qquad x - 8 = 0$$
$$x = 6 \qquad x = 8$$

Exercise Set 7.3

1. $(x_1, y_1) = (4, 1)$
$(x_2, y_2) = (5, 6)$

$$m = \frac{y_2 - y_1}{x_2 - x_1}$$

$$m = \frac{6 - 1}{5 - 4} = \frac{5}{1} = 5$$

The slope is 5.

3. $(x_1, y_1) = (9, 0)$
$(x_2, y_2) = (5, -2)$

$$m = \frac{y_2 - y_1}{x_2 - x_1}$$

$$m = \frac{-2 - 0}{5 - 9} = \frac{-2}{-4} = \frac{1}{2}$$

The slope is $\frac{1}{2}$.

5. $(x_1, y_1) = (3, 8)$
$(x_2, y_2) = (-3, 8)$

$$m = \frac{y_2 - y_1}{x_2 - x_1}$$

$$m = \frac{8 - 8}{-3 - 3} = \frac{0}{-6} = 0$$

The slope is 0.

7. $(x_1, y_1) = (-4, 6)$
$(x_2, y_2) = (-2, 6)$

$$m = \frac{y_2 - y_1}{x_2 - x_1}$$

$$m = \frac{6 - 6}{-2 - (-4)} = \frac{0}{-2 + 4}$$

$$= \frac{0}{2} = 0$$

The slope is 0.

9. $(x_1, y_1) = (3, 4)$
$(x_2, y_2) = (3, -2)$

$$m = \frac{y_2 - y_1}{x_2 - x_1}$$

$$m = \frac{-2 - 4}{3 - 3} = \frac{-6}{0}$$

The slope is undefined.

11. $(x_2, y_1) = (-4, 2)$
$(x_1, y_2) = (5, -3)$

$$m = \frac{y_2 - y_1}{x_2 - x_1}$$

$$m = \frac{-3 - 2}{5 - (-4)} = \frac{-5}{5 + 4} = \frac{-5}{9}$$

The slope is $-\frac{5}{9}$.

13. $(x_1, y_1) = (-1, 7)$
$(x_2, y_2) = (4, -3)$

$$m = \frac{y_2 - y_1}{x_2 - x_1}$$

$$m = \frac{-3 - 7}{4 - (-1)} = \frac{-10}{4 + 1}$$

$$= \frac{-10}{5} = -2$$

The slope is -2.

15. $m = \frac{\text{vertical change}}{\text{horizontal change}} = \frac{4}{2}$
$= 2$

The slope is 2.

17. $m = \frac{\text{vertical change}}{\text{horizontal change}} = \frac{-3}{2}$

The slope is $-\frac{3}{2}$.

19. $m = \frac{\text{vertical change}}{\text{horizontal change}} = \frac{-1}{7}$

The slope is $-\frac{1}{7}$.

21. $m = \frac{\text{vertical change}}{\text{horizontal change}} = \frac{7}{4}$

The slope is $\frac{7}{4}$.

23. $m = \dfrac{\text{vertical change}}{\text{horizontal change}} = \dfrac{0}{3}$

$= 0$

The slope is 0.

25. $m = \dfrac{\text{vertical change}}{\text{horizontal change}} = \dfrac{\;}{0}$

The slope is undefined.

27. The slope of a line is the ratio of the vertical change to the horizontal change between any two points on a line.

29. Lines that rise from left to right have a positive slope; lines that fall from left to right have a negative slope.

31. A vertical line has no slope. Since we cannot divide by 0, we say that the slope of a vertical line is undefined.

Cumulative Review Exercises

33. a) A quadratic equation in one variable is an equation that contains only one variable, and the greatest exponent on that variable is 2.

b) $x^2 + 2x - 3 = 0$
(answers will vary)

35. a) A linear equation in two variables is an equation that contains two variables where the exponent on both variables is 1.

b) $y = 3x - 2$
(answers will vary)

Just for fun

1. $(x_1, y_1) = (\frac{1}{2}, -\frac{3}{8})$

$(x_2, y_2) = (-\frac{4}{9}, -\frac{7}{2})$

$m = \dfrac{y_2 - y_1}{x_2 - x_1}$

$m = \dfrac{-\frac{7}{2} - (-\frac{3}{8})}{-\frac{4}{9} - (\frac{1}{2})} = \dfrac{-\frac{28}{8} + \frac{3}{8}}{-\frac{8}{18} - \frac{9}{18}}$

$= \dfrac{\frac{-25}{8}}{\frac{-17}{18}} = \dfrac{-25}{\cancel{8}_{4}} \cdot \dfrac{-\cancel{18}^{9}}{17} = \dfrac{225}{68}$

The slope is $\frac{225}{68}$.

2. $(x_1, y_1) = (6, -4)$

$m = -\frac{5}{3}$

$m = \dfrac{y_2 - y_1}{x_2 - x_1}$

$-\frac{5}{3} = \dfrac{y_2 - (-4)}{x_2 - 6}$

$-\frac{5}{3} = \dfrac{y_2 + 4}{x_2 - 6}$

$-5(x_2 - 6) = 3(y_2 + 4)$

$-5x_2 + 30 = 3y_2 + 12$

$-5x_2 - 3y_2 = -18$

$5x_2 + 3y_2 = 18$

All points on the line 5x + 3y = 18 will work, like (3,1), (0,6), etc.

3. $(x_1, y_1) = (-5, 2)$
$(x_2, y_2) = (x_2, -7)$

$m = -\frac{3}{4}$

$m = \frac{y_2 - y_1}{x_2 - x_1}$

$-\frac{3}{4} = \frac{-7 - 2}{x_2 - (-5)}$

$-\frac{3}{4} = \frac{-9}{x_2 + 5}$

$-3(x_2 + 5) = 4(-9)$
$-3x_2 - 15 = -36$
$-3x = -21$
$x_2 = 7$

The x-coordinate is 7.

4. a) The angle (less than or equal to 90°) that the hill makes with the horizontal is measured to determine the slope.

b) The slope of a line has no specific units. The slope of a hill is measured in degrees.

Exercise Set 7.4.

1. y = 2x - 1

slope is 2
y-intercept is -1

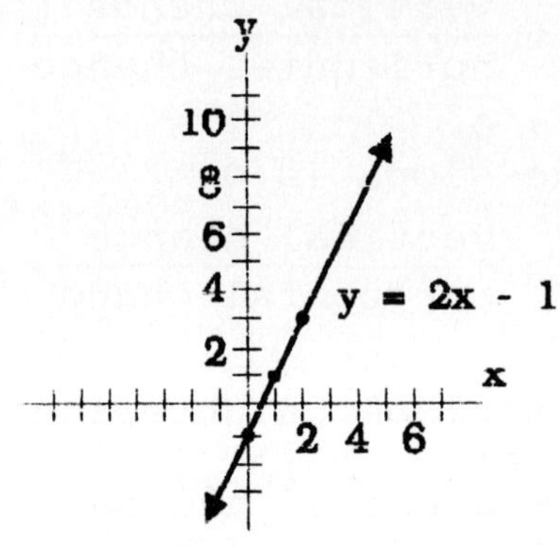

3. y = -x + 5
y = -1(x) + 5

slope is -1
y-intercept is 5

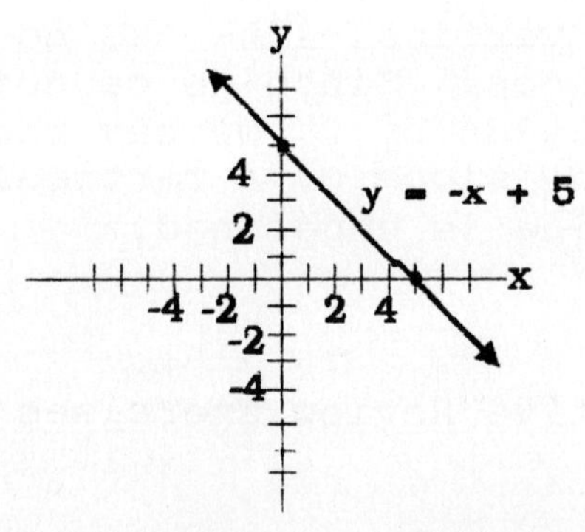

5. y = -4x
y = -4x + 0

Slope is -4
y-intercept is 0
m = -4, b = 0

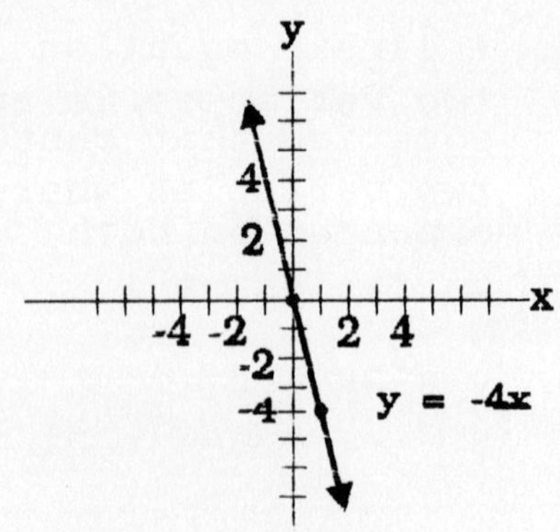

7. $$-2x + y = -3$$
$$-2x + 2x + y = 2x - 3$$
$$y = 2x - 3$$

slope is 2
y-intercept is -3
m = 2, b = -3

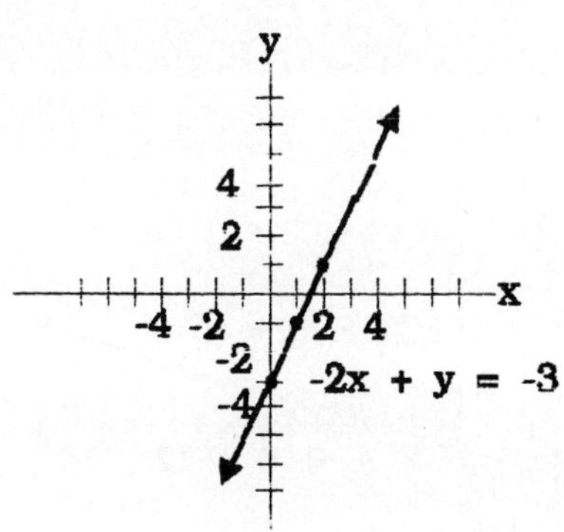

9. $$3x + 3y = 9$$
$$3x - 3x + 3y = -3x + 9$$
$$3y = -3x + 9$$
$$\frac{3y}{3} = \frac{-3x}{3} + \frac{9}{3}$$
$$y = -1x + 3$$

slope is -1
y-intercept is 3
m = -1, b = 3

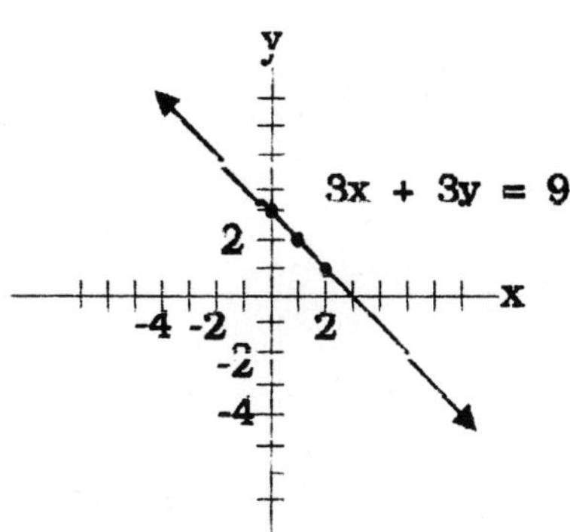

11. $$-x + 2y = 8$$
$$-x + x + 2y = x + 8$$
$$2y = x + 8$$
$$\frac{2y}{2} = \frac{x}{2} + \frac{8}{2}$$
$$y = \frac{1}{2} x + 4$$

slope is $\frac{1}{2}$

y-intercept is 4

$m = \frac{1}{2}$, b = 4

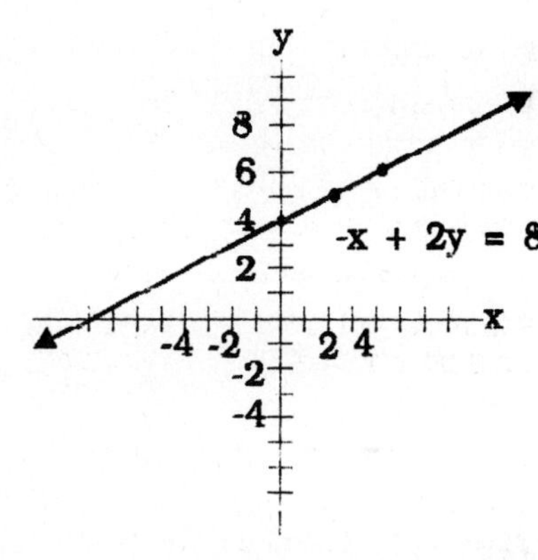

13. $$4x = 6y + 9$$
$$4x - 9 = 6y + 9 - 9$$
$$4x - 9 = 6y$$
$$\frac{4x - 9}{6} = y$$
$$\frac{4x}{6} - \frac{9}{6} = y$$
$$\frac{2x}{3} - \frac{3}{2} = y$$

slope is $\frac{2}{3}$

y-intercept is $\frac{-3}{2}$

$m = \frac{2}{3},\ b = \frac{-3}{2}$

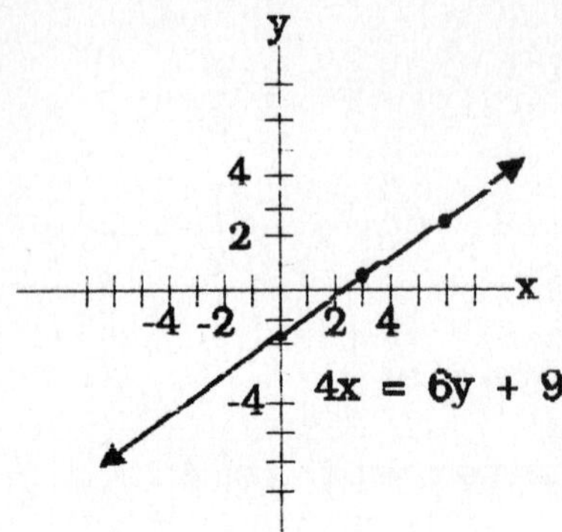

15. $$-6x = -2y + 8$$
$$-6x - 8 = -2y + 8 - 8$$
$$-6x - 8 = -2y$$
$$\frac{-6x}{-2} - \frac{8}{-2} = \frac{-2y}{-2}$$
$$3x + 4 = y$$

slope is 3
y-intercept is 4
$m = 3,\ b = 4$

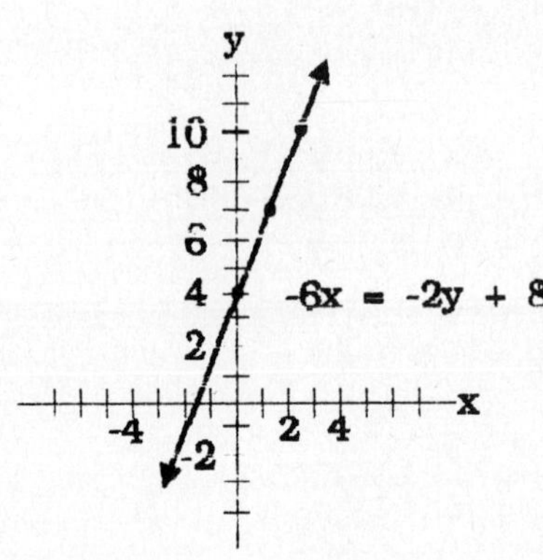

17. $$-3x + 8y = -8$$
$$-3x + 3x + 8y = 3x - 8$$
$$8y = 3x - 8$$

$$\frac{8y}{8} = \frac{3x}{8} - \frac{8}{8}$$

$$y = \frac{3}{8}x - 1$$

slope is $\frac{3}{8}$

y-intercept is -1

$m = \frac{3}{8},\ b = -1$

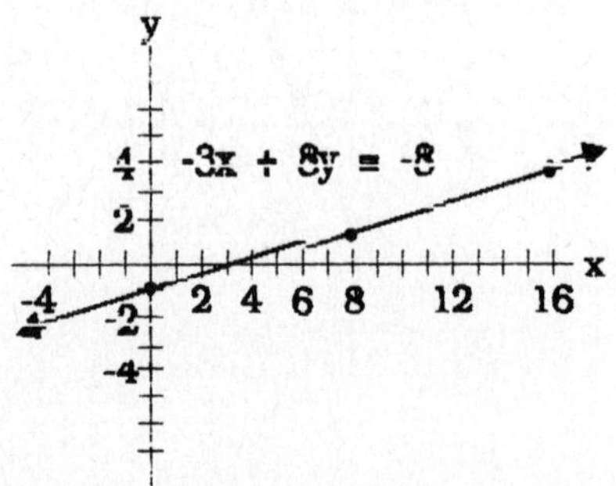

19. $$3x = 2y - 4$$
$$3x + 4 = 2y$$
$$\frac{3x + 4}{2} = y$$
$$\frac{3x}{2} + \frac{4}{2} = y$$
$$\frac{3x}{2} + 2 = y$$

slope is $\frac{3}{2}$

y-intercept is 2

$m = \frac{3}{2},\ b = 2$

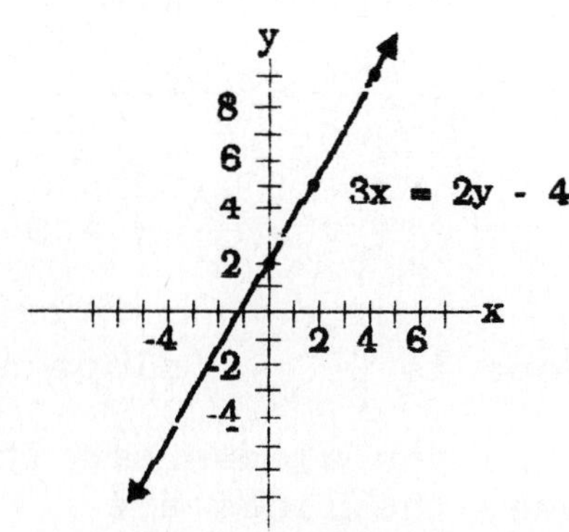

21. $$20x = 80y + 40$$

$$20x - 40 = 80y$$

$$\frac{20x - 40}{80} = \frac{80y}{80}$$

$$\frac{20x}{80} - \frac{40}{80} = y$$

$$\frac{1}{4}x - \frac{1}{2} = y$$

slope is $\frac{1}{4}$ y-intercept is

$-\frac{1}{2}$

$m = \frac{1}{4}$, $b = -\frac{1}{2}$

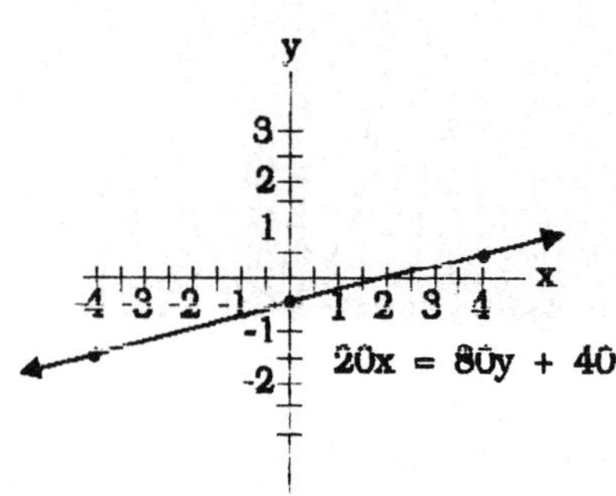

23. The graph goes through the points (-2,0) and (0,2).

$$m = \frac{y_2 - y_1}{x_2 - x_1} = \frac{2 - 0}{0 - (-2)}$$

$$= \frac{2}{2} = 1$$

y-intercept is 2

$y = mx + b$ $y = 1x + 2$

$y = x + 2$

25. The graph goes through the points (0,2) and (3,1).

$$m = \frac{y_2 - y_1}{x_2 - x_1} = \frac{2 - 1}{0 - 3}$$

$$= \frac{1}{-3} = -\frac{1}{3}$$

y-intercept is 2

$y = mx + b$ $y = \frac{-1}{3}x + 2$

27. The graph goes through the points (0,15) and (10,0).

$$m = \frac{y_2 - y_1}{x_2 - x_1} = \frac{15 - 0}{0 - 10}$$

$$= \frac{15}{10} = -\frac{3}{2}$$

y-intercept is 15

$y = mx + b$ $y = \frac{-3}{2}x + 15$

29. The graph goes through the points (-4,0) and (0,-2).

$$m = \frac{y_2 - y_1}{x_2 - x_1} = \frac{-2 - 0}{0 - (-4)}$$

$$= \frac{-2}{4} = \frac{-1}{2}$$

y-intercept is -2

$y = mx + b$ $y = \frac{-1}{2}x - 2$

31. $3x + 3y = 8$ $y = \frac{-2}{3}x + 5$

$$2x - 2x + 3y = -2x + 8$$

$$\frac{3y}{3} = \frac{-2x}{3} + \frac{8}{3}$$

$$y = \frac{-2}{3}x + \frac{8}{3}$$

slope is $\frac{-2}{3}$ slope is $\frac{-2}{3}$

Since both lines have the same slope, they are parallel.

33. $3x - 5y = 7$

$$3x - 3x - 5y = -3x + 7$$

$$-5y = -3x + 7$$

$$\frac{-5}{-5} = \frac{-3x}{-5} + \frac{7}{-5}$$

$$y = \frac{3}{5}x - \frac{7}{5}$$

slope is $\frac{3}{5}$

$$5y + 3x = 2$$

$$5y + 3x - 3x = -3x + 2$$

$$\frac{5y}{5} = \frac{-3x}{5} + \frac{2}{5}$$

$$y = \frac{-3}{5}x + \frac{2}{5}$$

slope is $\frac{-3}{5}$

Since slopes are not the same, the lines are not parallel.

35. $6x + 2y = 8$ $4x - 9 = -y$

$2y = -6x + 8$ $-4x + 9 = y$

$$y = \frac{-6x + 8}{2}$$

$$y = -3x + 4$$

slope is -3 slope is -4

Since the slopes are not the same, the lines are not parallel.

37. $2y - 6 = -5x$ $y = \frac{-5}{2}x - 2$

$$2y - 6 + 6 = -5x + 6$$

$$2y = -5x + 6$$

$$\frac{2y}{2} = \frac{-5x}{2} + \frac{6}{2}$$

$$y = \frac{-5}{2}x + 3$$

slope is $\frac{-5}{2}$ slope is $\frac{-5}{2}$

Since the slopes are the same, the lines are parallel.

39. a) Plotting points:

$4x + 3y = 6$

x	y
0	2
3	-2
-3	6

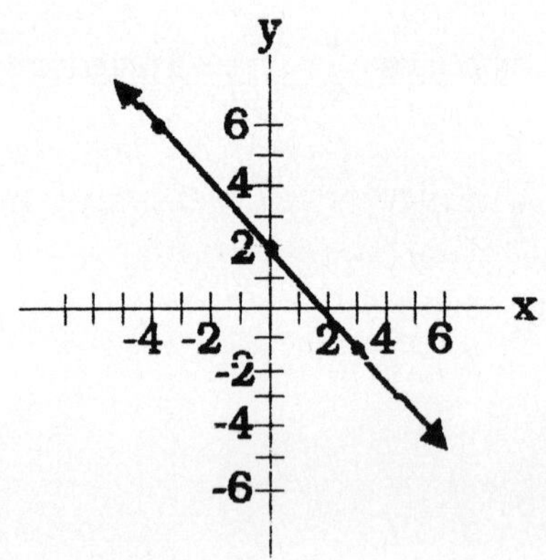

b) Intercepts:

$4x + 3y = 6$

x-intercept
Let $y = 0$

$$4x + 3(0) = 6$$

$$4x = 6$$

$$x = \frac{6}{4} = \frac{3}{2}$$

y-intercept
Let $x = 0$

$$(0) + 3y = 6$$

$$3y = 6$$

$$y = 2$$

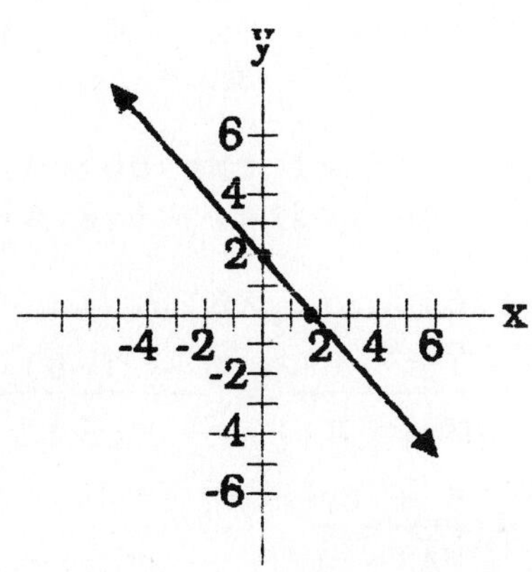

c) Slope and y-intercept:

$$4x + 3y = 6$$
$$3y = -4x + 6$$
$$y = \frac{-4x + 6}{3}$$
$$y = -\frac{4}{3}x + 2$$

Plot the y-intercept 2. Since the slope is - 4/3, obtain the second point by moving down 4 units and to the right 3 units.

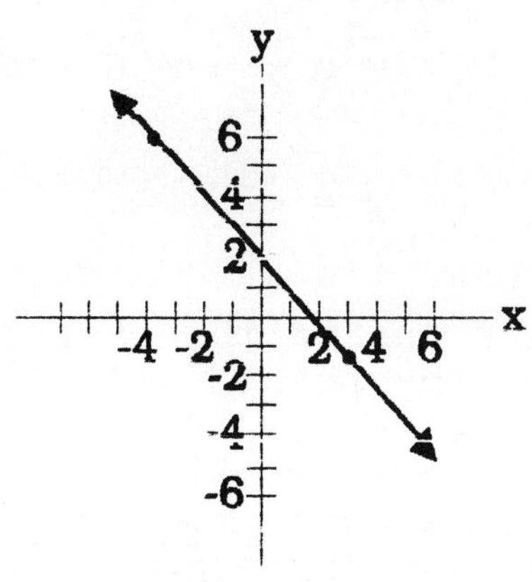

41. A negative slope means the values of y decrease as the values of x increase.

43. Compare their slopes. If the slope are the same and their y-intercepts are different, the lines are parallel.

Cumulative Review Exercises

45. $\frac{250 \text{ rpm}}{20 \text{ mph}} = \frac{x \text{ rpm}}{30 \text{ mph}}$

$$\frac{250}{20} = \frac{x}{30}$$
$$20x = (30)(250)$$
$$20x = 7500$$
$$x = 375 \text{ rpm}$$

47. $4x^2 - 16y^4 = 4(x^2 - 4y^4)$

$$= 4(x - 2y^2)(x + 2y^2)$$

Exercise Set 7.5

1. Slope = 5, through (0,4)
Point-slope form of a line
$y - y_1 = m(x - x_1)$

$$y - 4 = 5(x - 0)$$
$$y - 4 = 5x$$
$$y = 5x + 4$$

3. Slope = -2 through (-4,5)
Point-slope form of a line.
$(y - y_1 = m(x - x_1))$

$$y - (5) = -2(x + 4)$$
$$y - 5 = -2(x + 4)$$
$$y - 5 = -2x - 8$$
$$y - 5 + 5 = -2x - 8 + 5$$
$$y = -2x - 3$$

5. Slope = $\frac{1}{2}$ through (-1,-5)

$(y - y_1 = m(x - x_1)$

$$y - (-5) = (\frac{1}{2})[x - (-1)]$$

$$y + 5 = \frac{1}{2}(x + 1)$$

$$y + 5 = \frac{1}{2}x + \frac{1}{2}$$

$$y = \frac{1}{2}x + \frac{1}{2} - 5$$

$$y = \frac{1}{2}x + \frac{1}{2} - \frac{10}{2}$$

$$y = \frac{1}{2}x - \frac{9}{2}$$

7. Slope $= \frac{3}{5}$ through (4,-2)

$$y - y_1 = m(x - x_1)$$

$$y - (-2) = \frac{3}{5}(x - 4)$$

$$y + 2 = \frac{3}{5}x - \frac{12}{5}$$

$$y + 2 - 2 = \frac{3}{5}x - \frac{12}{5} - 2$$

$$y = \frac{3}{5}x - \frac{12}{5} - \frac{10}{5}$$

$$y = \frac{3}{5}x - \frac{22}{5}$$

9. (-4,-2) (-2,4)

$$m = \frac{y_2 - y_1}{x_2 - x_1} = \frac{4 - (-2)}{-2 - (-4)}$$

$$= \frac{4 + 2}{-2 + 4}$$

$$= \frac{6}{2}$$

$$= 3$$

$$y - y_1 = m(x - x_1)$$

$$y - 4 = 3[x - (-2)]$$

$$y - 4 = 3x + 6$$

$$y - 4 + 4 = 3x + 6 + 4$$

$$y = 3x + 10$$

Pick one of the coordinates for the point. (-2,4)

11. (-4,6) (4,-6)

$$m = \frac{y_2 - y_1}{x_2 - x_1} = \frac{6 - (-6)}{-4 - (4)}$$

$$= \frac{6 + 6}{-4 - 4}$$

$$= \frac{12}{-8}$$

$$= \frac{-3}{2}$$

$$y - y_1 = m(x - x_1)$$

$$y - (-6) = \frac{-3}{2}[x - (4)]$$

$$y + 6 = \frac{-3}{2}(x - 4)$$

$$y + 6 = \frac{-3x}{2} + 6$$

$$y = \frac{-3x}{2} + 6 - 6$$

$$y = \frac{-3x}{2}$$

Pick one of the coordinates for the point. (4,-6)

13. (10,3) (0,-2)

$$m = \frac{y_2 - y_1}{x_2 - x_1} = \frac{-2 - 3}{0 - 10}$$

$$= \frac{-5}{-10}$$

$$= \frac{1}{2}$$

$$y - y_1 = m(x - x_1)$$

$$y - (-2) = \frac{1}{2}(x - 0)$$

$$y + 2 = \frac{1}{2}x$$

$$y + 2 - 2 = \frac{1}{2}x - 2$$

$$y = \frac{1}{2}x - 2$$

Pick one of the coordinates for the point. (0,-2)

15. Standard form, slope-intercept form, point-slope form, standard form:

$$4 - 2y = 6x$$
$$4 - 2y + 2y = 6x + 2y$$
$$4 = 6x + 2y$$
$$6x + 2y = 4$$

Slope intercept form

$$4 - 2y = 6x$$
$$4 - 4 - 2y = 6x - 4$$
$$-2y = 6x - 4$$
$$\frac{-2y}{-2} = \frac{6x}{-2} - \frac{4}{-2}$$
$$y = -3x + 2$$

Point-slope form

$$4 - 2y = 6x$$
$$\frac{4}{-2} - \frac{-2y}{-2} = \frac{6x}{-2}$$
$$-2 + y = -3x$$
$$y - 2 = -3(x + 0)$$

17. Write the equation.
Standard form:
$5x - 3y = 8$

Slope-intercept form:

$$y = \frac{5}{3}x - \frac{8}{3}$$

Point-slope form:

$$y + \frac{8}{3} = \frac{5}{3}(x - 0)$$

Cumulative Review Exercises

19. $$\frac{x^2 + 2x - 8}{x^2 - 16} \div \frac{2x^2 - 5x - 3}{x^2 - 7x + 12}$$

$$= \frac{(x + 4)(x - 2)}{(x - 4)(x + 4)}$$

$$\div \frac{(2x + 1)(x - 3)}{(x - 4)(x - 3)}$$

$$= \frac{(x + 4)(x - 2)}{(x + 4)(x + 4)}$$

$$\cdot \frac{(x + 4)(x + 3)}{(2x + 1)(x + 3)}$$

$$= \frac{(x - 2)}{(2x + 1)}$$

21. Let b = base.
2b − 7 = height.

$$A = \frac{1}{2}bh$$

$$36 = \frac{1}{2}b(2b - 7)$$

$$(2)(36) = 2\cdot\frac{1}{2}b(2b - 7)$$

$$72 = b(2b - 7)$$

$$72 = 2b^2 - 7b$$
$$0 = 2b^2 - 7b - 72$$
$$0 = (2b + 9)(b - 8)$$

$2b + 9 = 0$ $b - 8 = 0$

$2b = -9$ $b = 8$

$b = -\frac{9}{2}$

Base cannot be negative.

Base: 8 feet
Height:
$2b - 7$
$= 2(8) - 7$
$= 16 - 7$
$= 9$ feet

Just for fun

1. $3x - 4y = 6$

$$-4y = -3x + 6$$

$$y = \frac{-3x + 6}{-4}$$

$$y = \frac{3}{4} x - \frac{3}{2}$$

Slope: $\frac{3}{4}$

Line parallel to $3x - 4y = 6$ has same slope, $m = 3/4$. Through $(-4,-1)$,

$$y - y_1 = m(x - x_1)$$

$$y - (-1) = \frac{3}{4}[x - (-4)]$$

$$y + 1 = \frac{3}{4}(x + 4)$$

$$y + 1 = \frac{3}{4} x + 3$$

$$y = \frac{3}{4} x + 2$$

2. $-5x + 2y = -4$

$$2y = 5x - 4$$

$$y = \frac{5x - 4}{2}$$

$$y = \frac{5}{2} x - 2$$

Line perpendicular to $-5x + 2y = -4$ has slope of $-2/5$.

$$m = -\frac{2}{5} \text{ through } (2,\tfrac{1}{2})$$

$$y - y_1 = m(x - x_1)$$

$$y - \frac{1}{2} = -\frac{2}{5}(x - 2)$$

$$y - \frac{1}{2} = -\frac{2}{5} x + \frac{4}{5}$$

$$y = -\frac{2}{5} x + \frac{4}{5} + \frac{1}{2}$$

$$y = -\frac{2}{5} x + \frac{8 + 5}{10}$$

$$y = -\frac{2}{5} x + \frac{13}{10}$$

3. Factor 3/4 from the x term and constant. Express the left side of the equation as $y = 0$.

Exercise Set 7.6

1. $x > 3$
Draw $x = 3$.
(use a dashed line since $>$ is used in the inequality.)

Pick a point not on the line and determine if it satisfies the inequality, (0,0).

$x > 3 \quad 0 > 3 \quad$ False
Shade the side of the line that does not contain (0,0).

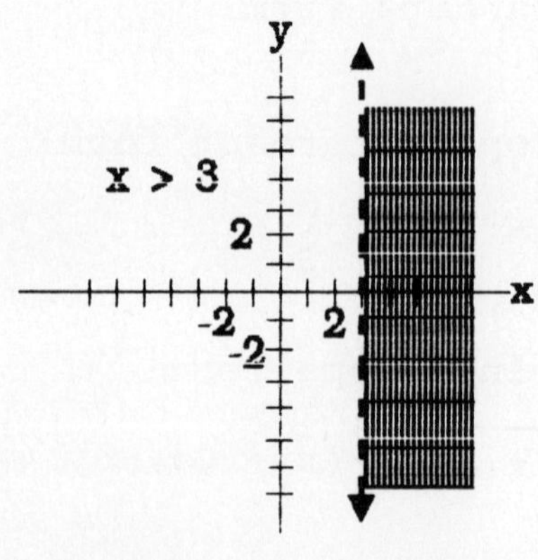

3. $x \geq \frac{5}{2}$

Draw $x = \frac{5}{2}$.
(Use a solid line since $\geq$ is used in the inequality.)

Pick a point not on the line and determine if it satisfies the inequality (0,0).

$x \geq \frac{5}{2}$ $\quad 0 \geq \frac{5}{2}$ $\quad$ False

Since $0 \geq 5/2$ is false, shade the point on the opposite side of the line from (0,0).

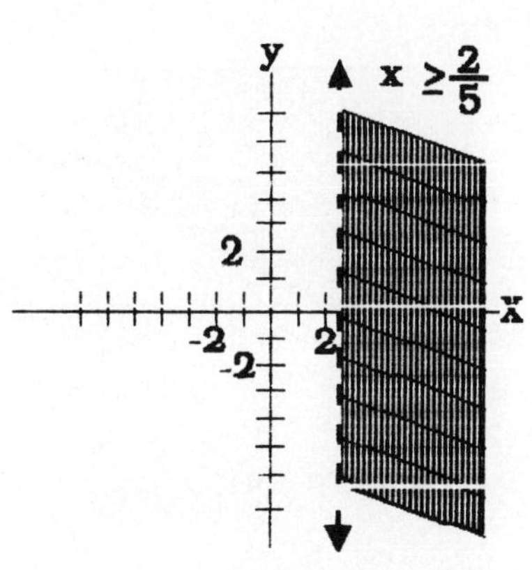

5. $y \geq 2x$

x	y
-2	-4
0	0
2	4

Pick a point not on the line and determine if it satisfies the inequality (4,1).

$y \geq 2x \quad 1 \geq 2(4) \quad 1 \geq 8$
false

Shade the side of the line that does not contain the point (4,1).

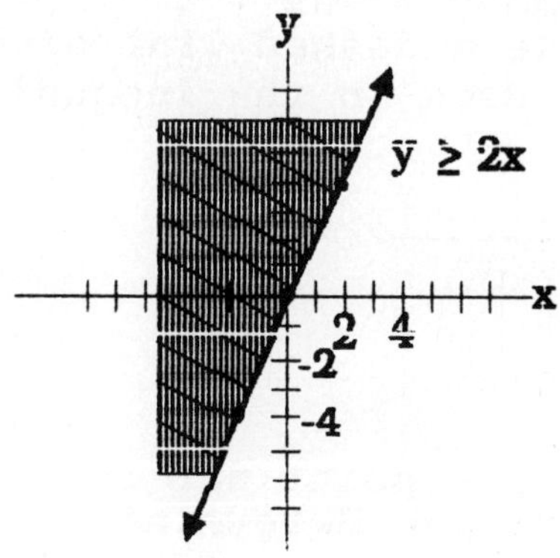

7. $y < 2x + 1$

x	y
0	1
-2	-3
1	3

Draw $y = 2x + 1$
(Use a dashed line.)

Pick a point not on the line and determine if it satisfies the inequaltiy (1,0).

$y < 2x + 1 \quad 0 < 2(1) + 1$
$0 < 3$ True

Shade the side of the line that contains the point (1,0).

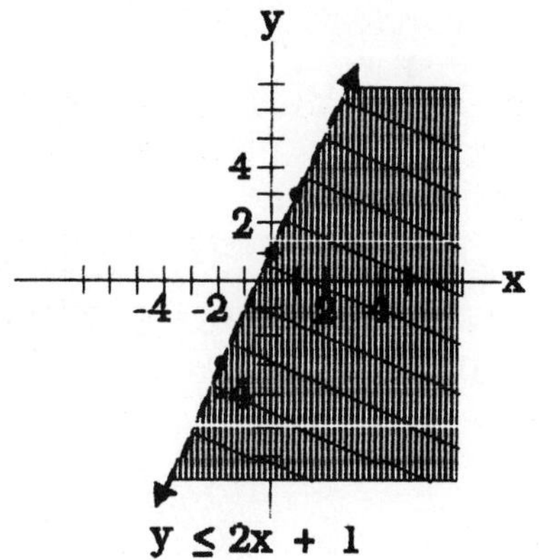

9. $y < -3x + 4$

Draw $y = -3x + 4$
(Use a dashed line since $<$ is used in the inequality.

x	y
-1	7
0	4
1	1

Pick a point not on the line and determine if it satisfies the inequality (0,0).

$y < -3x + 4 \quad 0 < -3(0) + 4$
$0 < 4$ True

Shade the side of the line that contains the point (0,0).

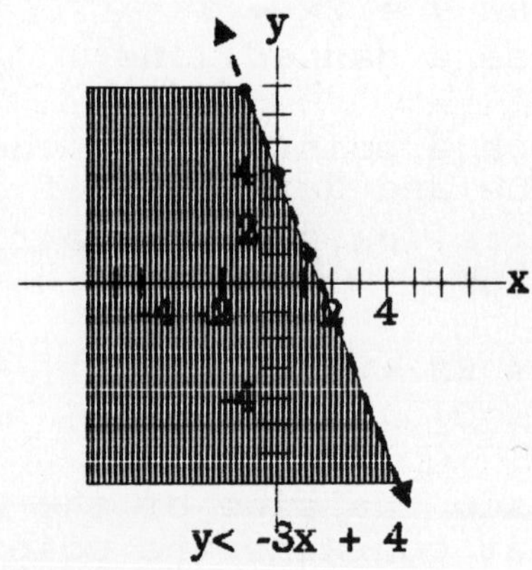

11. $y \geq \frac{1}{2}x - 4$

Draw $y = \frac{1}{2}x - 4$
(Use a solid line since $\geq$ is used in the original inequality.)

x	y
-2	-5
0	-4
2	-3

Pick a point not on the line and determine if it satisfies the inequality (0,0).

$y \geq \frac{1}{2}x - 4 \quad 0 \geq \frac{1}{2}(0) - 4$

$0 \geq -4$ True

Shade the side of the line that contains the point (0,0).

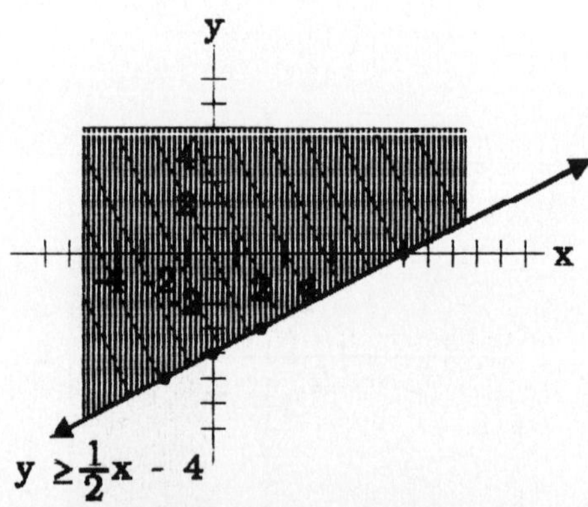

13. $y \leq - x + 6$

Draw $y = \frac{1}{3}x + 6$

(Use a solid line since $\leq$ is used in the inequality.)

x	y
-3	5
0	6
3	7

Pick a point not on the line and determine if it

satisfies the inequality (0,0).

$y \le \frac{1}{3}x + 6 \quad 0 \le \frac{1}{3}(0) + 6$

$0 \le 0 + 6 \quad 0 \le 6$ True

Shade the side of the line that contains the point (0,0).

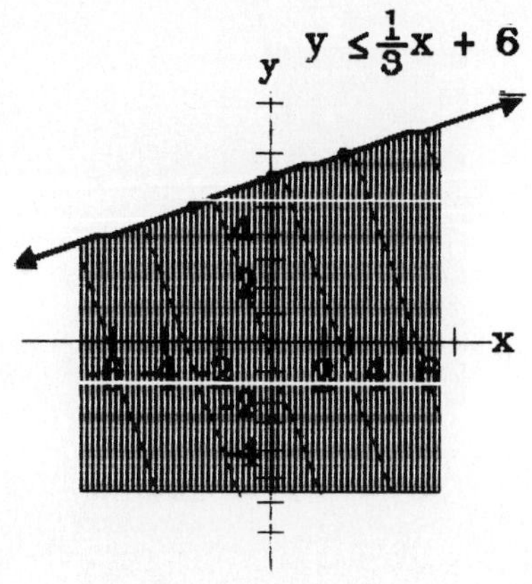

15. $y \le -3x + 5$

Draw $y = -3x + 5$
Use a solid line since ≤ is used in the inequality.

x	y
-1	8
0	5
1	2

Pick a point not on the line and determine if it satisfies the inequality (0,0).

$y \le -3x + 5 \quad 0 \le -3(0) + 5$
$0 \le 5$ True

Shade the side of the line that contains the point (0,0).

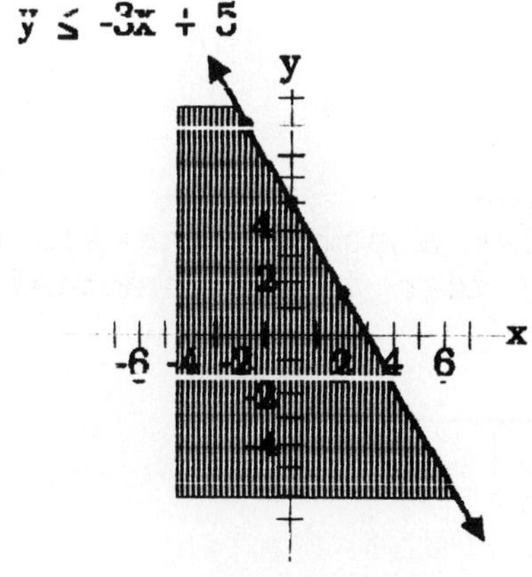

17. $y > 5x - 9$

Draw $y = 5x - 9$

(Use a dashed line since > is used in the inequality.)

x	y
-1	-14
0	-9
1	-4

Pick a point not on the line and determine if it satisfies the inequality (0,0).

$y > 5x - 9 \quad 0 > 5(0) - 9$
$0 > -9$ True

Shade the side of the line that contains the point (0,0).

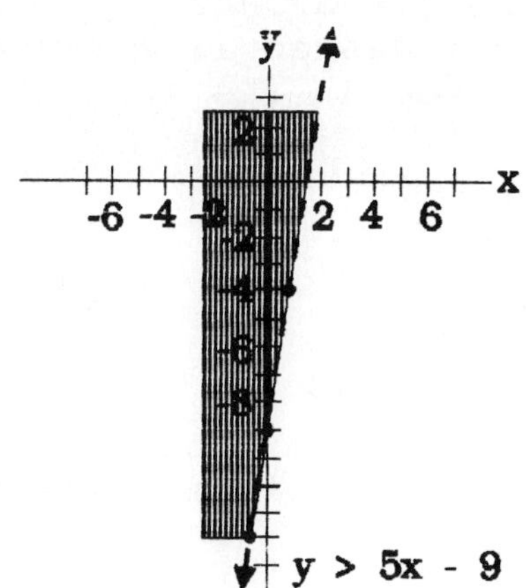

19. $y \le -x + 4$

Draw $y = -x + 4$
(Use a solid line since $\le$ is used in the inequality.)

x	y
-1	5
0	4
1	3

Pick a point not on the line and determine if it satisfies the inequality (0,0).

$y \le -x + 4 \quad 0 \le -0 + 4$
$0 \le 4$ True

Shade the side of the line that contains the point (0,0).

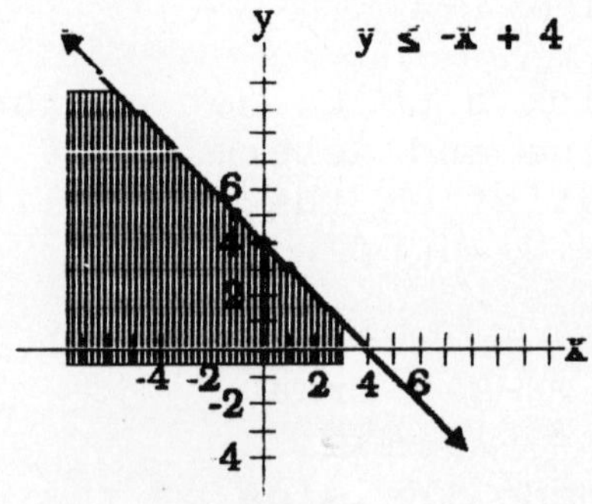

21. $y > 3x - 2$

Draw $y = 3x - 2$
Use a dashed line since $>$ is used in the inequality.

x	y
-2	-8
0	-2
1	1

Pick a point not on the line and determine if it satisfies the inequality (0,0).

$y > 3x - 2 \quad 0 > 3(0) - 2$
$0 > -2$ True

Shade the side of the line that contains the point (0,0).

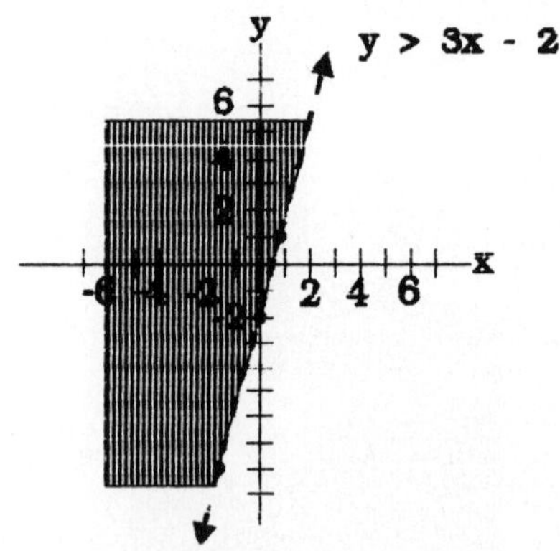

23. $y \ge -4x + 3$

Draw $y = -4x + 3$
(Use a solid line since $\ge$ is used in the inequality.)

x	y
-2	11
0	3
1	-1

Pick a point not on the line and determine if it satisfies the inequality (0,0).

$y \ge - 4x + 3 \quad 0 \ge -4(0) + 3$
$0 \ge 3$ False

Shade the side of the line that does not contain the point (0,0).

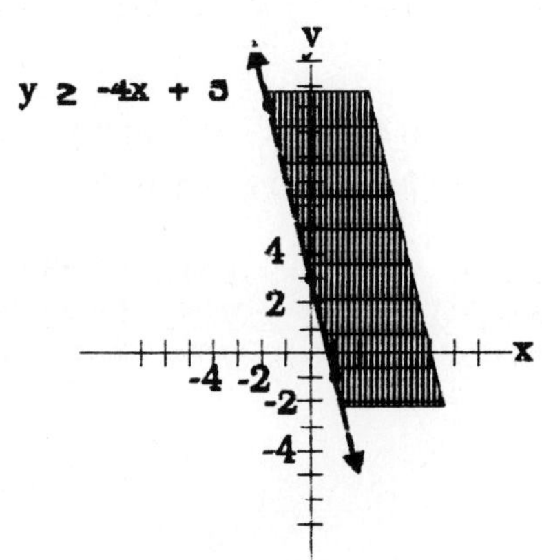

25. $y < \frac{-x}{3} - 2$

Draw $y = \frac{-x}{3} - 2$

(Use a dashed line since < is used in the inequality.)

x	y
-3	-1
0	-2
3	-3

Pick a point not on the line and determine if it satisfies the inequality (0,-5).

$y < \frac{-x}{3} - 2 \quad -5 < \frac{-0}{3} - 2$

$-5 < -2$ True

Shade the side of the line that contains the poinit (0,-5).

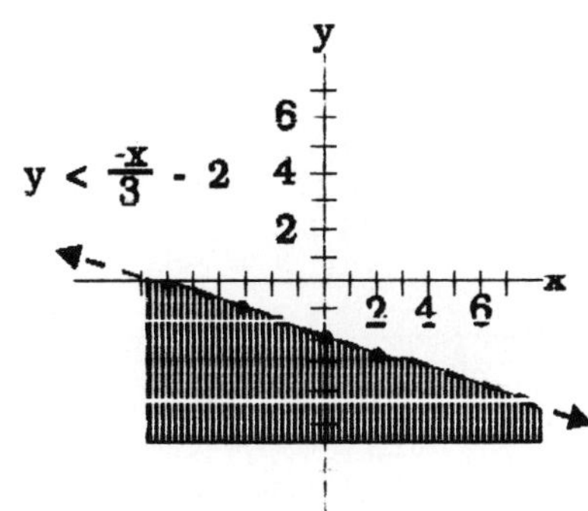

27. Points on the line satisfy an equation (=) but not an inequality that is strictly greater than or less than.

Cumulative Review Exercises

29. $i = prt$

$300 = p(0,08)(3)$

$300 = p(0.24)$

$\frac{300}{0.24} = \frac{p(0.24)}{(0.24)}$

$\$1250 = p$

31. $6x - 5y = 9$

$-5y = -6x + 9$

$y = \frac{-6x + 9}{-5}$

$y = \frac{6}{5} x - \frac{9}{5}$

$m = \frac{6}{5}$

$b = - \frac{9}{5}$

Review Exercises

1.

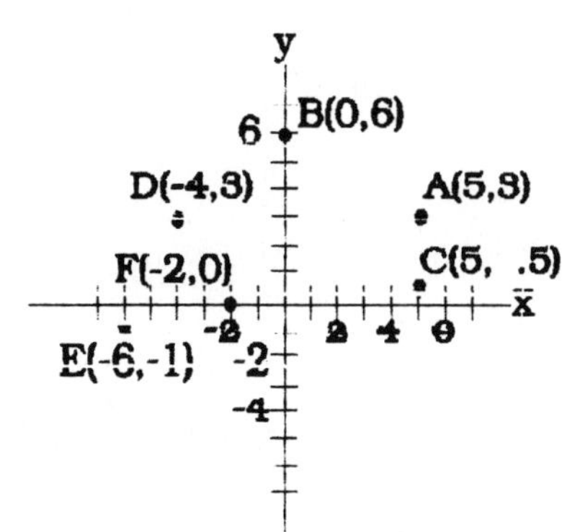

3. a) $2x + 3y = 9 (4,3)$

$2(4) + 3(3) = 9$

$8 + 9 = 9$

$17 \neq 9$

(4,3) is not a solution.

b) $2x + 3y = 9 (0,3)$

$2(0) + 3(3) = 9$

$0 + 9 = 9$

$9 = 9$

(0,3) is a solution.

c) $2x + 3y = 9 (-1,4)$

$2(-1) + 3(4) = 9$

$-2 + 12 = 9$

$10 \neq 9$

(-1,4) is not a solution.

d) $2x + 3y = 9 \ (2,\frac{5}{3})$

$2(2) + 3(\frac{5}{3}) = 9$

$4 + 5 = 9$

$9 = 9$

$(2,\frac{5}{3})$ is a solution.

5.

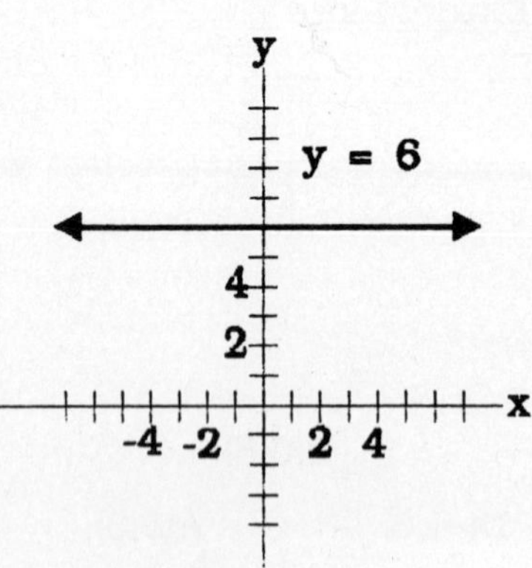

7. $y = 3x$

x	y	Ordered Pairs
-2	-6	(-2,-6)
0	0	(0,0)
1	3	(1,3)

Let $x = -2$: $y = 3(-2) = 6$

Let $x = 0$: $y = 3(0) = 0$

Let $x = 1$: $y = 3(1) = 3$

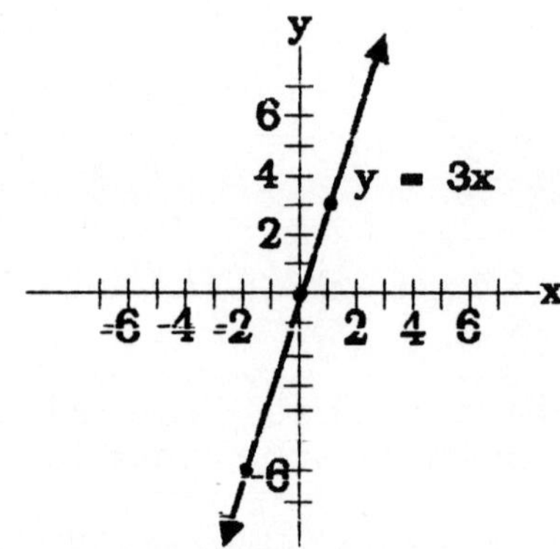

9. $y = -3x + 4$

x	y	Ordered Pairs
0	4	(0,4)
1	1	(1,1)
2	-2	(2,-2)

Let $x = 0$: $y = -3(0) + 4$
$= 4$

Let $x = 1$: $y = -3(1) + 4$
$= 1$

Let $x = 2$: $y = -3(2) + 4$
$= -2$

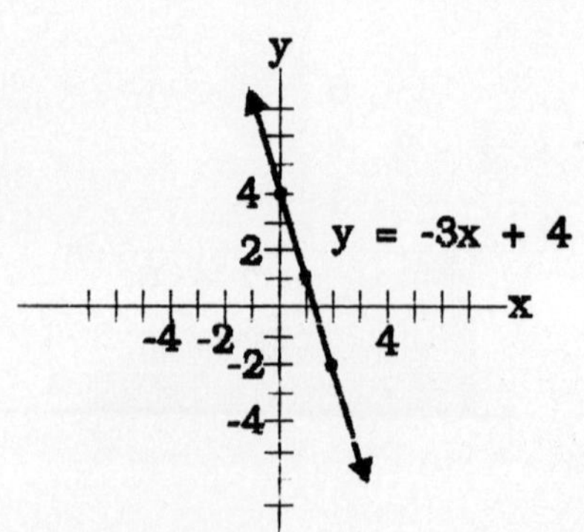

11. $2x + 3y = 6$

$$2x - 2x + 3y = -2x + 6$$
$$3y = -2x + 6$$
$$\frac{3y}{3} = \frac{-2x}{3} + \frac{6}{3}$$
$$y = \frac{-2x}{3} + 2$$

x	y	Ordered Pairs
0	2	(0,2)
3	0	(3,0)
6	−2	(6,−2)

Let $x = 0$: $y = \frac{-2}{3}(0) + 2 = 2$

Let $x = 3$: $y = \frac{-2}{3}(3) + 2 = 0$

Let $x = 6$: $y = \frac{-2}{3}(6) + 2 = -2$

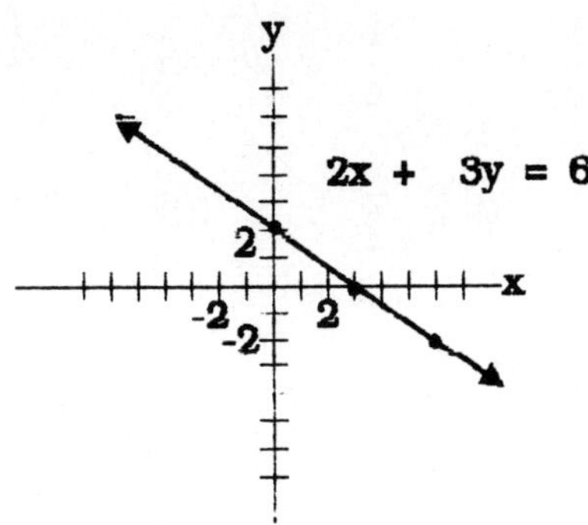

13. $2y = 3x - 6$

Find the y-intercept.
Let $x = 0$.
$$2y = 3(0) - 6$$
$$2y = -6$$
$$\frac{2y}{2} = \frac{-6}{2}$$
$$y = -3 \quad (0,-3)$$

Find the x-intercept.
Let $y = 0$.
$$2(0) = 3x - 6$$
$$0 = 3x - 6$$
$$0 + 6 = 3x - 6 + 6$$
$$6 = 3x$$
$$\frac{6}{3} = \frac{3x}{3}$$
$$2 = x \quad (2,0)$$

Check point:
Let $x = 6$.
$$2y = 3(6) - 6$$
$$2y = 18 - 6$$
$$2y = 12$$
$$2y = 12$$
$$\frac{2y}{2} = \frac{12}{2}$$
$$y = 6 \quad (6,6)$$

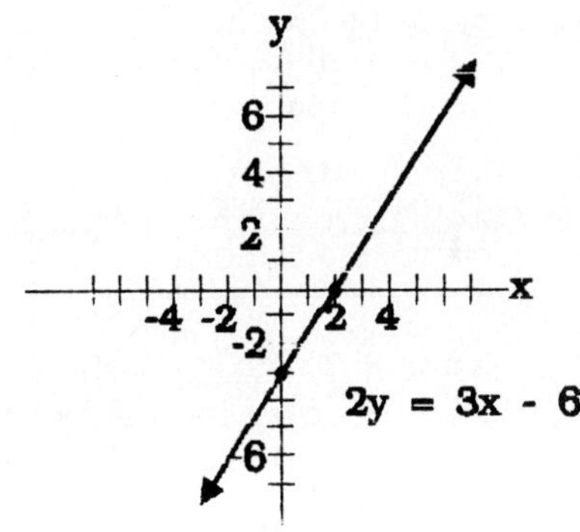

15. $-5x - 2y = 10$

Find the y-intercept.
Let $x = 0$.
$$-5(0) - 2y = 10$$
$$-2y = 10$$
$$\frac{-2y}{-2} = \frac{10}{-2}$$
$$y = -5 \quad (0,-5)$$

Find the x-intercept.
Let y = 0.

$$-5x - 2(0) = 10$$
$$-5x = 10$$
$$\frac{-5x}{-5} = \frac{10}{-5}$$
$$x = -2 \quad (-2,0)$$

Check point.
Let x = 2.

$$-5(2) - 2(y) = 10$$
$$-10 - 2y = 10$$
$$-10 + 10 - 2y = 10 + 10$$
$$-2y = 20$$
$$\frac{-2y}{-2} = \frac{20}{2}$$
$$y = -10 \quad (2,-10)$$

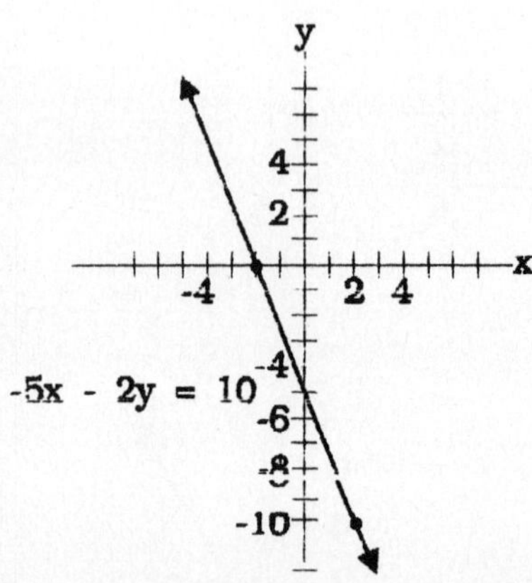

17.
$$25x + 50y = 100$$
$$25x - 25x + 50y = -25x + 100$$
$$50y = -25x + 100$$
$$\frac{50y}{50} = \frac{-25x}{50} + \frac{100}{50}$$
$$y = \frac{-1}{2}x + 2$$

x	y	Ordered Pairs
0	2	(0,2)
2	1	(2,1)
4	0	(4,0)

Let x = 0: $y = \frac{-1}{2}(0) + 2$
$= 2$

Let x = 2: $y = \frac{-1}{2}(2) + 2$
$= 1$

Let x = 4: $y = \frac{-1}{2}(4) + 2$
$= 0$

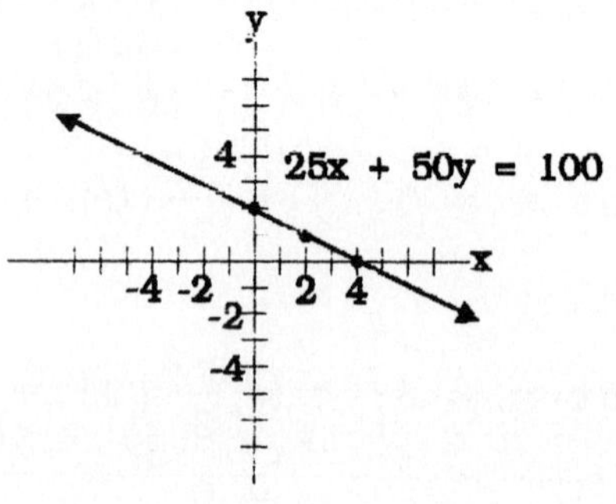

19. $\frac{2}{3}x = \frac{1}{4}y + 20$

Find the y-intercept.
Let x = 0.

$$\frac{2}{3}(0) = \frac{1}{4}y + 20$$
$$0 = \frac{1}{4}y + 20$$
$$0 - 20 = \frac{1}{4}y$$
$$-80 = y \quad (0,-80)$$

Find the x-intercept.
Let y = 0.

$$\frac{2}{3}x = \frac{1}{4}(0) + 20$$
$$\frac{2}{3}x = 20$$

$$\frac{3}{2}\cdot\frac{2}{3}x = \frac{3}{2}\cdot 20$$

$$x = 30 \quad (30,0)$$

Check point.
Let $y = 4$.

$$\frac{2}{3}x = \frac{1}{4}(4) + 20$$

$$\frac{2}{3}x = 21$$

$$\frac{3}{2}\cdot\frac{2}{3}x = \frac{3}{2}\cdot 21$$

$$x = \frac{63}{2} \quad (\frac{63}{2},4)$$

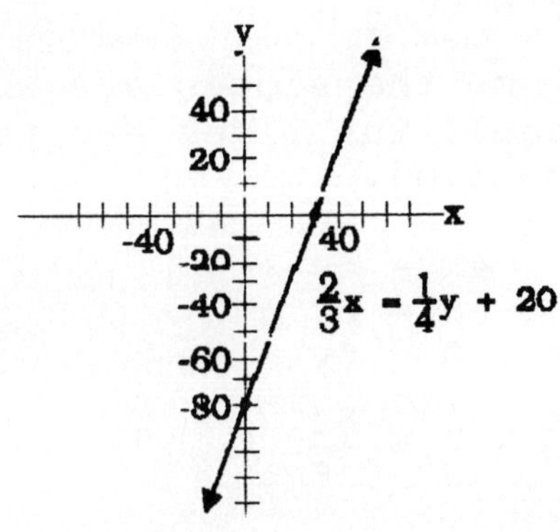

21. $(x_1,y_1) = (-4,-2)$

$(x_2,y_2) = (8,-3)$

$$m = \frac{y_2 - y_1}{x_2 - x_1}$$

$$m = \frac{-3 - (-2)}{8 - (-4)} = \frac{-3 + 2}{8 + 4}$$

$$= \frac{-1}{12}$$

The slope is $-\frac{1}{12}$.

23. The slope of a horizontal line is 0.

25. $m = \dfrac{\text{vertical change}}{\text{horizontal change}} = \dfrac{-5}{7}$

The slope is $-\frac{5}{7}$.

27. $m = \dfrac{\text{vertical change}}{\text{horizontal change}}$

$$= \frac{2}{8} = \frac{1}{4}$$

The slope is $\frac{1}{4}$.

29. $y = 3x + 5$
$m = 3$
$b = 5$

31.
$$2x + 3y = 8$$
$$2x - 2x + 3y = -2x + 8$$
$$3y = -2x + 8$$
$$\frac{3y}{3} = \frac{-2x}{3} + \frac{8}{3}$$
$$y = \frac{-2x}{3} + \frac{8}{3}$$

$$m = -\frac{2}{3}$$

$$b = \frac{8}{3}$$

33. $4y = 6x + 12$

$$\frac{4y}{4} = \frac{6x}{4} + \frac{12}{4}$$

$$y = \frac{3x}{2} + 3$$

$$m = \frac{3}{2}$$

$$b = 3$$

35.
$$9x + 7y = 15$$
$$9x - 9x + 7y = -9x + 15$$

$$7y = -9x + 15$$
$$\frac{7y}{7} = \frac{-9x}{7} + \frac{15}{7}$$
$$y = \frac{-9}{7}x + \frac{15}{7}$$

$m = \frac{-9}{7}$

$b = \frac{15}{7}$

37. $$4x - 8 = 0$$
$$4x - 8 + 8 = 0 + 8$$
$$4x = 8$$
$$\frac{4x}{4} = \frac{8}{4}$$
$$x = 2$$

$x = 2$ is a vertical line. Slope is undefined, no y-intercept.

39. The graph passes through the points (-1,0) and (1,4).

$$m = \frac{y_2 - y_1}{x_2 - x_1} = \frac{4 - 0}{1 - (-1)}$$
$$= \frac{4}{2} = 2$$

y-intercept = 2

$y = mx + b \quad y = 2x + 2$

41. The graph passes through the points (0,2) and (4,0).

$$m = \frac{y_2 - y_1}{x_2 - x_1} = \frac{0 - 2}{4 - 0} = \frac{-2}{4}$$
$$= \frac{-1}{2}$$

y intercept = 2

$y = mx + b$

$$y = \frac{-1}{2}x + 2$$

43. $$2x - 3y = 9$$
$$2x - 2x - 3y = -2x + 9$$
$$-3y = -2x + 9$$
$$\frac{-3y}{-3} = \frac{-2x}{-3} + \frac{9}{-3}$$
$$y = \frac{2}{3}x - 3$$

Slope is $\frac{2}{3}$

$$3x - 2y = 6$$
$$3x - 3x - 2y = -3x + 6$$
$$-2y = -3x + 6$$
$$\frac{-2y}{-2} = \frac{-3x}{-2} + \frac{6}{-2}$$
$$y = \frac{3}{2}x - 3$$

Slope = $\frac{3}{2}$

Since the slopes are not equal, the lines are not parallel.

45. $$4x = 6y + 3$$
$$4x - 3 = 6y + 3 - 3$$
$$4x - 3 = 6y$$
$$\frac{4x}{6} - \frac{3}{6} = \frac{6y}{6}$$
$$\frac{2}{3}x - \frac{1}{2} = y$$
$$y = \frac{2}{3}x - \frac{1}{2}$$

Slope is $\frac{2}{3}$

$$-2x = -3y + 10$$
$$-2x - 10 = -3y + 10 - 10$$
$$-2x - 10 = -3y$$
$$\frac{-2}{-3}x - \frac{10}{-3} = \frac{-3y}{-3}$$
$$\frac{2}{3}x + \frac{10}{3} = y$$

$$y = \frac{2}{3}x + \frac{10}{3}$$

Slope is $\frac{2}{3}$

Since both lines have the same slope, they are parallel.

47. Slope = -3 through (-1,5).
Point (-1,5) Slope is -3

$$y - y_1 = m(x - x_1)$$

$$y - 5 = -3[x - (-1)]$$
$$y - 5 = -3(x + 1)$$
$$y - 5 = -3x - 3$$
$$y - 5 + 5 = -3x - 3 + 5$$
$$y = -3x + 2$$

49. Slope = 0 through (4,2)
Point (4,2) Slope is 0

$$y - y_1 = m(x - x_1)$$

$$y - 2 = 0(x - 4)$$
$$y - 2 = 0$$
$$y - 2 + 2 = 0 + 2$$
$$y = 2$$

51. (4,3) (2,1)

$$m = \frac{y_2 - y_1}{x_2 - x_1} = \frac{3 - 1}{4 - 2} = \frac{2}{2} = 1$$

Pick one of the two coordinates for the point (2,1).

$$y - y_1 - m(x - x_1)$$

$$y - 1 = 1(x - 2)$$
$$y - 1 = x - 2$$
$$y = x - 2 + 1$$
$$y = x - 1$$

53. (-4,-2) (-4,3)

$$m = \frac{y_2 - y_1}{x_2 - x_1} = \frac{3 - (-2)}{-4 - (-4)}$$

$$= \frac{3 + 5}{-4 + 4} = \frac{8}{0}$$

Slope is undefined.
$x = -4$

55. $x < 4$

Draw $x = 4$.
(Use a dashed line since < is used in the inequality.)

Pick a point not on the line and determine if it satisfies the inequality (0,0).

$x < 4$ $0 < 4$ True

Shade the side of the line that contains (0,0).

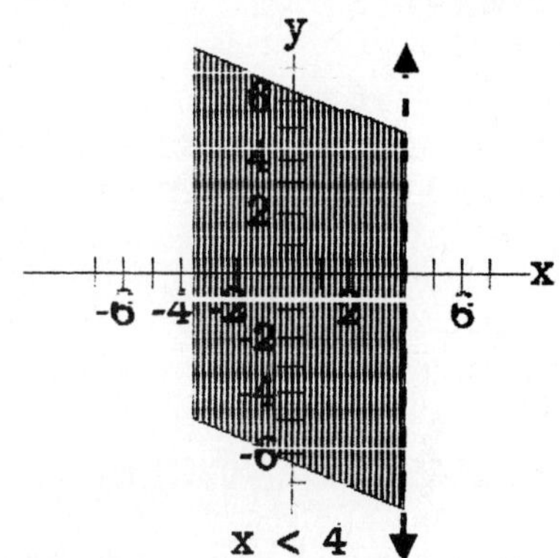

57. $y > 2x + 1$

Draw $y = 2x + 1$
(Use a broken line.)

x	y
-2	-3
0	1
1	3

Pick a point not on the line and determine if it satisfies the inequality (-1,0).

$y > 2x + 1 \quad 0 > 2(-1) + 1$
$0 > -2 + 1 \quad 0 > -1$ True

Shade the side of the line that contains the point $(-1,0)$.

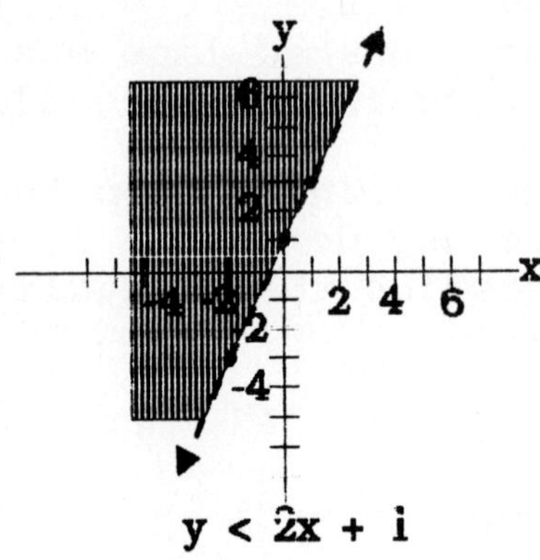

59. $y \geq 6x + 5$

Draw $y = 6x + 5$
(Use a solid line since $\geq$ is used in the inequality.)

x	y
-2	-7
0	5
1	11

Pick a point not on the line and determine if it satisfies the inequality $(-3,0)$.

$y \geq 6x + 5 \quad 0 \geq 6(-3) + 5$
$0 \geq -13$ True

Shade the side of the line containing $(-3,0)$.

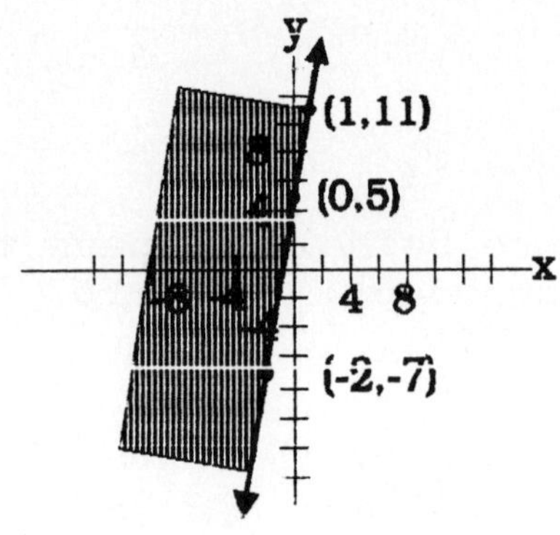

61. $y \leq \frac{1}{3}x - 2$

Draw $y = \frac{1}{3}x - 2$

(Use a solid line since $\leq$ is used in the inequality.)

x	y
0	-2
3	-1
6	0

Pick a point not on the line and determine if it satisfies the inequality $(12,0)$.

$y \leq \frac{1}{3}x - 2 \quad 0 \leq \frac{1}{3}(12) - 2$

$0 \leq 2$ True

Shade the side of the line that contains the point $(12,0)$.

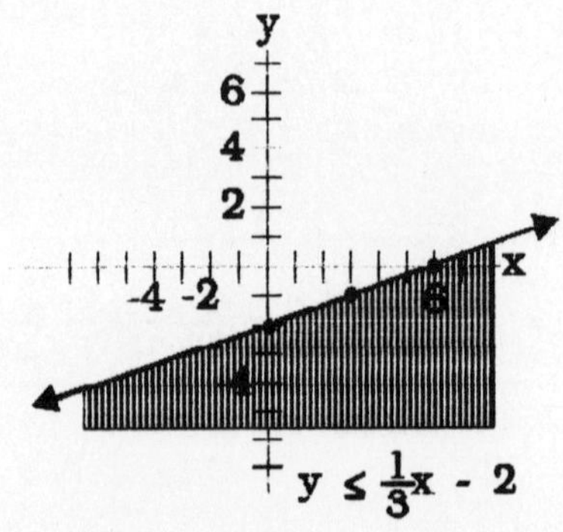

Practice Test

1. a) $3y = 5x - 9 \quad (3,2)$
$$3(2) = 5(3) - 9$$
$$6 = 15 - 9$$
$$6 = 6$$
(3,2) is a solution.

b) $3y = 5x - 9 \quad (\frac{9}{5},0)$
$$3(0) = 5(\frac{9}{5}) - 9$$
$$0 = 9 - 9$$
$$0 = 0$$
$(\frac{9}{5},0)$ is a solution.

c) $3y = 5x - 9 \quad (-2,-6)$
$$3(-6) = 5(-2) - 9$$
$$-18 = -10 - 9$$
$$-18 \neq -19$$
(-2,-6) is not a solution.

d) $3y = 5x - 9 \quad (0,3)$
$$3(3) = 5(0) - 9$$
$$9 = 0 - 9$$
$$9 \neq -9$$
(0,3) is not a solution.

2. $(x_1,y_1) = (-4,3)$
$(x_2,y_2) = (2,-5)$
$$m = \frac{y_2 - y_1}{x_2 - x_1}$$
$$m = \frac{-5 - 3}{2 - (-4)} = \frac{-5 - 3}{2 + 4}$$
$$= \frac{-8}{6} = \frac{-4}{3}$$

The slope is $-\frac{4}{3}$.

3. $4x - 9y = 15$
$$-9y = -4x + 15$$
$$y = \frac{-4 + 15}{-9}$$
$$y = \frac{-4x}{-9} + \frac{15}{-9}$$
$$y = \frac{4x}{9} - \frac{5}{3}$$

Slope is $\frac{4}{9}$.

y-intercept is $\frac{-5}{3}$

4. The line passes through the points (-1,0) and (0,-1).
$$m = \frac{y_2 - y_1}{x_2 - x_1} = \frac{0 - (-1)}{-1 - 0}$$
$$= \frac{1}{-1} = -1$$

y-intercept is -1.
The equation is $y = -x - 1$.

5. $y - y_1 = m(x - x_1)$
Point (1,3) $\quad m = 3$
$$y - (3) = 3[x - (1)]$$
$$y - 3 = 3(x - 1)$$
$$y - 3 = 3x - 3$$
$$y = 3x - 3 + 3$$
$$y = 3x$$

6. (3,-1) $\quad$ (-4,2)
$$m = \frac{y_2 - y_1}{x_2 - x_1} = \frac{2 - (-1)}{-4 - 3}$$
$$= \frac{2 + 1}{-4 - 3} = \frac{3}{-7} = \frac{-3}{7}$$

Pick one of the coordinates (3,-1).

$$y - y_1 = m(x - x_1)$$

$$y - (-1) = \frac{-3}{7}[x - (3)]$$

$$y + 1 = \frac{-3}{7}(x - 3)$$

$$y + 1 = \frac{-3x}{7} + \frac{9}{7}$$

$$y = \frac{-3x}{7} + \frac{9}{7} - 1$$

$$y = \frac{-3x}{7} + \frac{2}{7}$$

7. $2y = 3x - 6 \qquad y - \frac{3}{2}x = -5$

$$y = \frac{3x - 6}{2} \qquad y = \frac{3}{2}x - 5$$

$$y = \frac{3}{2}x - 3$$

The lines are parallel. Each equation was solved for y to find the slope and the y-intercept of the lines. Since their slopes are equal and their y-intercepts are different, the lines are parallel.

8. $x = -5$

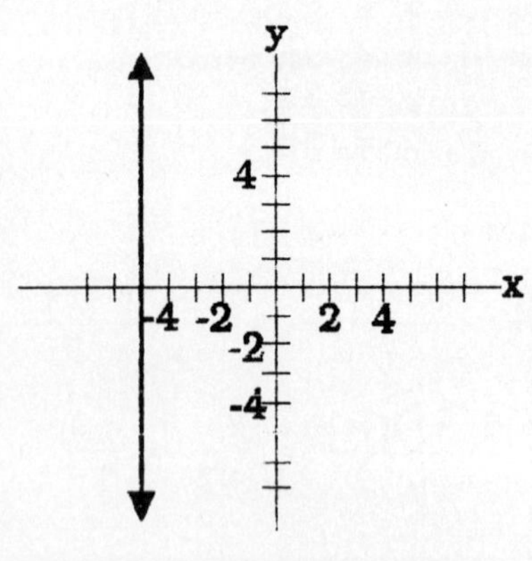

9. $y = 3x - 2$

x	y
0	-2
1	1
2	4

$y = 3(0) - 2 = -2$
$y = 3(1) - 2 = 1$
$y = 3(2) - 2 = 4$

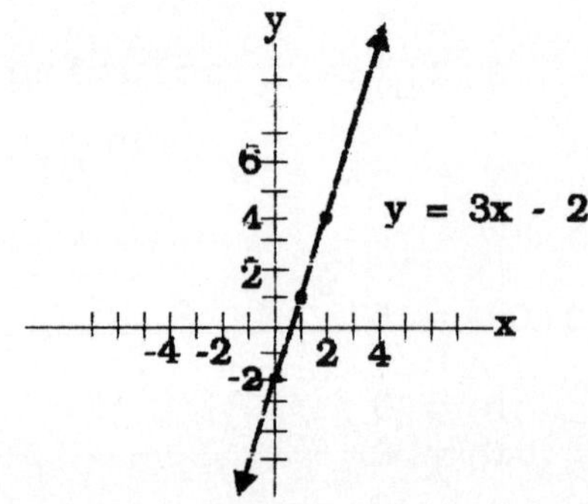

10. $3x + 5y = 15$

Find y-intercept.
Let $x = 0$.
$3(0) + 5y = 15$
$5y = 15$
$y = 3$
The y-intercept is (0,3).

Find x-intercept.
Let $y = 0$.
$3x + 5(0) = 15$
$3x = 15$
$x = 5$
The x-intercept is (5,0).

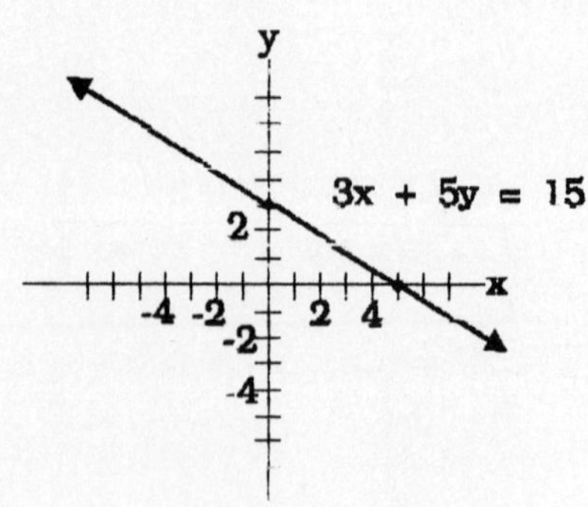

11. $4x = -y + 10$

x	y
0	10
2	2
4	-6

$4(0) = -y + 10 \quad y = 10$
$4(2) = -y + 10 \quad y = 2$
$4(4) = -y + 10 \quad y = -6$

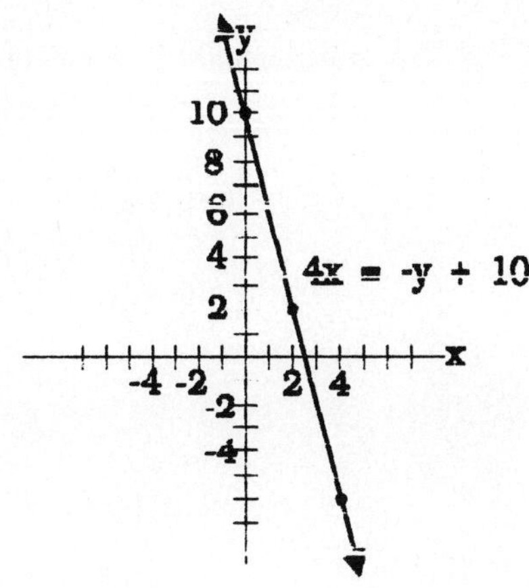

12. $3x - 2y = 8$

x	y
0	-4
2	-1
4	2

$3(0) - 2y = 8 \quad -2y = 8 \quad y = -4$
$3(2) - 2y = 8 \quad 6 - 2y = 8$
$-2y = 2 \quad y = -1$
$3(4) - 2y = 8 \quad 12 - 2y = 8$
$-2y = -4 \quad y = 2$

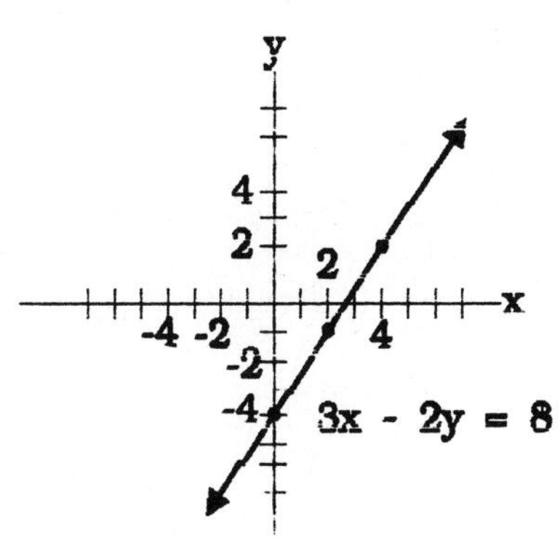

13. $y \geq -3x + 5$

Draw $y = -3x + 5$
(Use a solid line.)

x	y
-1	8
0	5
2	-1

Pick a point not on the line and determine if it satisfies the inequality (3,0).

$y \geq -3x + 5 \qquad 0 \geq -3(3) + 5$
$0 \geq -9 + 5 \qquad 0 \geq -4$ True

Shade the side of the line that contains the point (3,0).

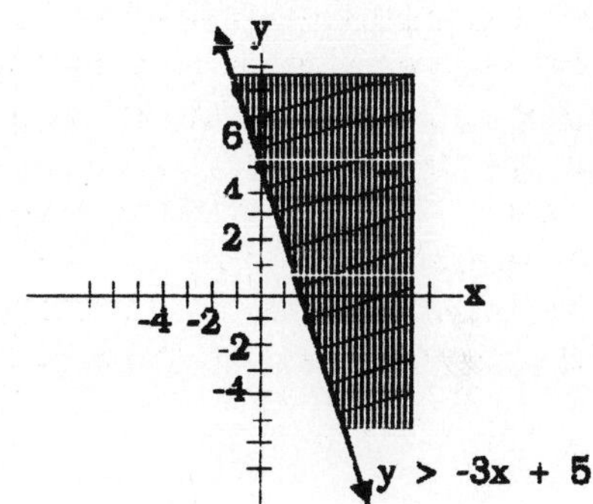

14. $y < 4x - 2$

Draw $y = 4x - 2$.
(Use a broken line.)

x	y
0	-2
1	2
2	6

Pick a point not on the

line and determine if it satisfies the inequality (2,0).

$y < 4x - 2 \quad 0 < 4(2) - 2$
$0 < 8 - 2 \quad 0 < 6$ True

Shade the side of the line that contains the point (2,0).

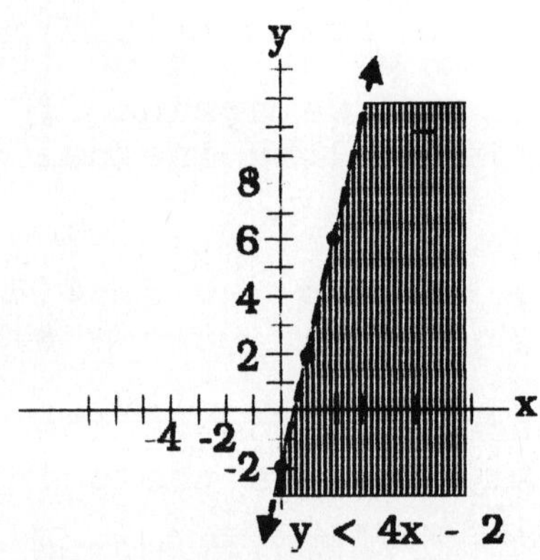

CHAPTER EIGHT

Exercise Set 8.1

1. $y = 3x - 4 \quad y = -x + 4$

a) $(-2,2)$

$2 = 3(-2) - 4 \quad 2 = -(-2) + 4$
$2 = -6 - 4 \quad 2 = 2 + 4$
$2 = -10 \quad 2 = 6$
False False
$(-2,2)$ is not a solution.

b) $(-4,-8)$

$y = 3x - 4 \quad y = -x + 4$
$-8 = 3(-4) - 4 \quad -8 = -(-4) + 4$
$-8 = -12 - 4 \quad -8 = 4 + 4$
$-8 = -16 \quad -8 = 8$
False False
$(-4,-8)$ is not a solution.

c) $(2,2)$

$y = 3x - 4 \quad y = -x + 4$
$2 = 3(2) - 4 \quad 2 = -(2) + 4$
$2 = 6 - 4 \quad 2 = -2 + 4$
$2 = 2 \quad 2 = 2$
True True
$(2,2)$ is a solution.

3. $y = 2x - 3 \quad y = x + 5$

a) $(8,13)$

$y = 2x - 3 \quad y = x + 5$
$13 = 2(8) - 3 \quad 13 = 8 + 5$
$13 = 16 - 3 \quad 13 = 13$
$13 = 13$
True True
$(8,13)$ is a solution.

b) $(4,5)$

$y = 2x - 3 \quad y = x + 5$
$5 = 2(4) - 3 \quad 5 = 4 + 5$
$5 = 8 - 3 \quad 5 = 9$
$5 = 5$
True False
$(4,5)$ is not a solution.

c) $(4,9)$

$y = 2x - 3 \quad y = x + 5$
$9 = 2(4) - 3 \quad 9 = 4 + 5$
$9 = 8 - 3 \quad 9 = 9$
$9 = 5$
False True
$(4,9)$ is not a solution.

5. $3x - y = 6 \quad 2x + y = 9$

a) $(3,3)$

$3(3) - 3 = 6 \quad 2(3) + 3 = 9$
$9 - 3 = 6 \quad 6 + 3 = 9$
$6 = 6 \quad 9 = 9$
True True
$(3,3)$ is a solution.

b) $(4,-2)$

$3x - y = 6 \quad 2x + y = 9$
$3(4) - (-2) = 6 \quad 2(4) - 2 = 9$
$12 + 2 = 6 \quad 8 - 2 = 9$
$14 = 6 \quad 6 = 9$
False False
$(4,-2)$ is not a solution.

c) $(-6,3)$

$3x - y = 6 \quad 2x + y = 9$
$3(-6) - 3 = 6 \quad 2(-6) + 3 = 9$
$-18 - 3 = 6 \quad -12 + 3 = 9$
$-21 = 6 \quad -9 = 9$
False False
$(-6,3)$ is not a solution.

7. $2x - 3y = 6 \quad y = \frac{2}{3}x - 2$

a) $(3,0)$

$2(3) - 3(0) = 6 \quad 0 = \frac{2}{3}(3) - 2$
$6 - 0 = 6 \quad 0 = 2 - 2$
$6 = 6 \quad 0 = 0$
True True
$(3,0)$ is a solution.

b) $(3,-2)$

$2x - 3y = 6 \quad y = \frac{2}{3}x - 2$

$2(3) - 3(-2) = 6 \quad -2 = \frac{2}{3}(3) - 2$

$6 + 6 = 6 \quad -2 = 2 - 2$

$12 = 6 \quad -2 = 0$

False False

$(3,-2)$ is not a solution.

c) $(1,\frac{-4}{3})$

$2x - 3y = 6 \quad y = \frac{2}{3}x - 2$

$2(1) - 3(\frac{-4}{3}) = 6 \quad \frac{-4}{3} = \frac{2}{3}(1) - 2$

$2 + 4 = 6 \quad \frac{-4}{3} = \frac{2}{3} - 2$

$6 = 6 \quad \frac{-4}{3} = \frac{2}{3} - \frac{6}{3}$

$\frac{-4}{3} = \frac{-4}{3}$

True True

$(1,\frac{-4}{3})$ is a solution.

9. $3x - 4y = 8 \quad 2y = \frac{3}{2}x - 4$

a) $(0,-2)$

$3(0)-4(-2) = 8 \quad 2(-2) = \frac{3}{2}(0)-4$

$0 + 8 = 8 \quad -4 = 0 - 4$

$8 = 8 \quad -4 = -4$

True True

$(0,-2)$ is a solution.

b) $(1,-6)$

$3x - 4y = 8 \quad 2y = \frac{3}{2}x - 4$

$3(1)-4(-6) = 8 \quad 2(-6) = \frac{3}{2}(1)-4$

$3 + 24 = 8 \quad -12 = \frac{3}{2} - \frac{8}{2}$

$27 = 8 \quad -12 = \frac{-5}{2}$

False False

$(1,-6)$ is not a solution.

c) $(\frac{-1}{3},\frac{-9}{4})$

$3x - 4y = 8 \quad 2y = \frac{3}{2}x - 4$

$3[\frac{-1}{3}]-4[\frac{-9}{4}] = 8 \quad 2[\frac{-9}{4}] = \frac{3}{2}\cdot[\frac{-1}{3}]-4$

$-1 + 9 = 8 \quad \frac{-9}{2} = \frac{-1}{2} - 4$

$8 = 8 \quad \frac{-9}{2} = \frac{-1}{2} - \frac{8}{2}$

$\frac{-9}{2} = \frac{-9}{2}$

True True

$(\frac{-1}{3},\frac{-9}{4})$ is a solution.

11. $y = 2x - 3 \quad 2x - 3y = 4$

a) $(\frac{1}{2},-2)$

$-2 = 2(\frac{1}{2}) - 3 \quad 2(\frac{1}{2}) - 3(-2) = 4$

$-2 = 1 - 3 \quad 1 + 6 = 4$

$-2 = -2 \quad 7 = 4$

True False

$(\frac{1}{2},-2)$ is not a solution.

b) $(\frac{5}{4},\frac{-1}{2})$

$y = 2x - 3 \quad 2x - 3y = 4$

$\frac{-1}{2} = 2(\frac{5}{4}) - 3 \quad 2(\frac{5}{4}) - 3(\frac{-1}{2}) = 4$

$\frac{-1}{2} = \frac{5}{2} - 3 \quad \frac{5}{2} + \frac{3}{2} = 4$

$\frac{-1}{2} = \frac{5}{2} - \frac{6}{2}$ $\qquad$ $\frac{8}{2} = 4$

$\frac{-1}{2} = \frac{-1}{2}$ $\qquad$ $4 = 4$

True $\qquad$ True

$(\frac{5}{4}, \frac{-1}{2})$ is a solution.

c) $(\frac{1}{5}, \frac{-10}{3})$

$y = 2x - 3$ $\qquad$ $2x - 3y = 4$

$\frac{-10}{3} = 2(\frac{1}{5}) - 3$ $\qquad$ $2(\frac{1}{5}) - 3(\frac{-10}{3}) = 4$

$\frac{-10}{3} = \frac{2}{5} - 3$ $\qquad$ $\frac{2}{5} + 10 = 4$

$\frac{-10}{3} = \frac{2}{5} - \frac{15}{5}$ $\qquad$ $\frac{2}{5} + \frac{50}{5} = 4$

$\frac{-10}{3} = \frac{-13}{5}$ $\qquad$ $\frac{52}{5} = 4$

$\frac{-50}{15} = \frac{-39}{15}$

False $\qquad$ False

$(\frac{1}{5}, \frac{-10}{3})$ is not a solution.

Cumulative Review Exercises

13. a) Natural numbers: $\{6\}$
    b) Whole numbers: $\{0,6\}$
    c) Integers: $\{-4,0,6\}$
    d) Rational numbers:

    $\{6,-4,0,2\frac{1}{2},-\frac{9}{5},4,22\}$

    e) Irrational numbers:

    $\{-\sqrt{7},\sqrt{3}\}$

    f) Real numbers:

    $\{6,-4,0,\sqrt{3},2\frac{1}{2},-\frac{9}{5},4.22,-\sqrt{7}\}$

15. $|-6| > |-2|$

17. $-3^3 = -27$

19. $-3^4 = -81$

21. $(\frac{3x^2y^4}{x^3y^2})^2 = (\frac{3y^2}{x})^2 = \frac{9y^4}{x^2}$

Exercise Set 8.2

1. Consistent, one solution.

3. Dependent, infinite number of solutions.

5. Consistent, one solution.

7. Dependent, infinite number of solutions.

9. $y = 3x - 2$ $\qquad$ $2y = 4x - 6$

   $\frac{1}{2}(2y) = (4x - 6)\frac{1}{2}$

   $y = 2x - 3$

   One solution

11. $3y = 2x + 3$ $\qquad$ $y = \frac{2}{3}x - 2$

    $\frac{1}{3}(3y) = (2x + 3)\frac{1}{3}$

    $y = \frac{2}{3}x + 1$

    No solution.

13. $4x = y - 6$ $\qquad$ $3x = 4y + 5$

    $4x + 6 = y$ $\qquad$ $3x - 5 = 4y$

    $\frac{1}{4}(3x - 5) = (4y)\frac{1}{4}$

$$\frac{3}{4}x - \frac{5}{4} = y$$

One solution

15. $2x = 3y + 4 \quad 6x - 9y = 12$

$-3y = -2x + 4 \quad -9y = -6x + 12$

$\frac{-1}{3}(-3y) \quad \frac{-1}{9}(-9y)$

$= (-2x+4)\frac{-1}{3} \quad = (-6x+12)\frac{-1}{9}$

$y = \frac{2}{3}x - \frac{4}{3} \quad y = \frac{2}{3}x - \frac{4}{3}$

Infinite number of solutions.

17. $y = \frac{3}{2}x + \frac{1}{2} \quad 3x - 2y = \frac{-1}{2}$

$\frac{-1}{2}(-2y)$

$= (-3x - \frac{1}{2})(\frac{-1}{2})$

$y = \frac{3}{2}x + \frac{1}{4}$

No solution

19. $y = x + 2$

$m = 1 \quad b = 2$

$y = -x + 2$

$m = -1 \quad b = 2$

Solution: (0,2)

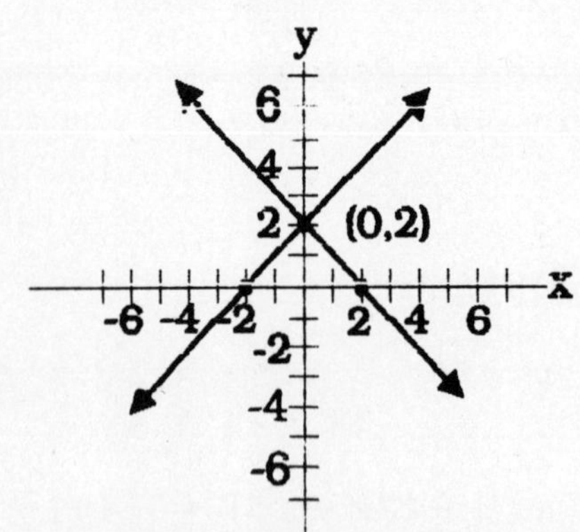

21. $y = 3x - 6$

$m = 3 \quad b = -6$

$y = -x + 6$

$m = -1 \quad b = 6$

Solution: (3,3)

23. $2x = 4 \quad y = -3$

$\frac{1}{2}(2x) = (4)\frac{1}{2} \quad m = 0 \quad b = -3$

$x = 2$

Vertical line crossing the x axis at 2.

Solution: (2,-3)

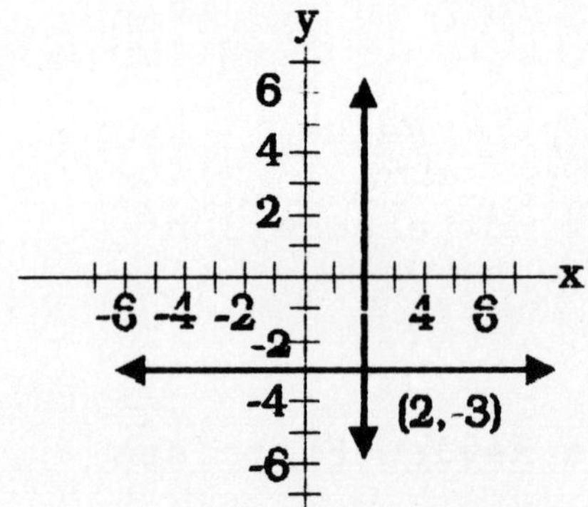

25. $y = x + 2 \quad x + y = 4$

$m = 1 \quad b = 2 \quad x - x + y = -x + 4$

$y = -x + 4$

$m = -1 \quad b = 4$

Solution: (1,3)

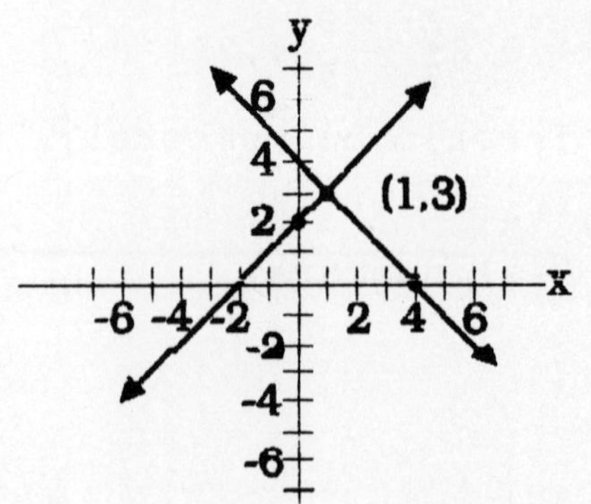

27. $y = \frac{-1}{2}x + 4$

$m = \frac{-1}{2} \quad b = 4$

$$x + 2y = 6$$
$$-x + x + 2y = -x + 6$$
$$2y = -x + 6$$
$$\frac{1}{2}(2y) = (-x + 6)\frac{1}{2}$$

$$y = \frac{-1}{2}x + 3$$

$$m = \frac{-1}{2} \quad b = 3$$

Inconsistent

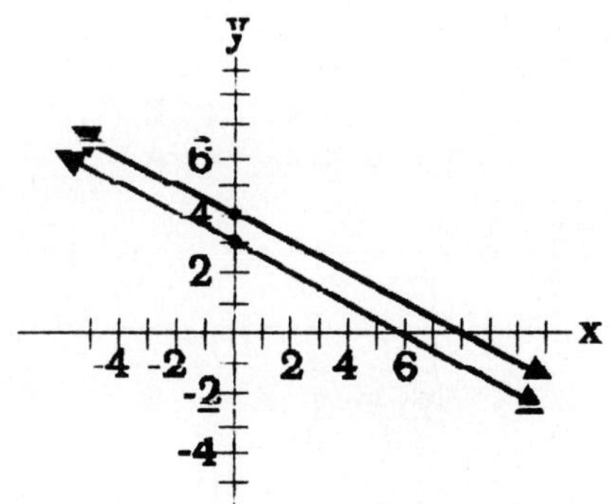

29. $$x + 2y = 8$$
$$-x + x + 2y = -x + 8$$
$$\frac{1}{2}(2y) = (-x + 8)(\frac{1}{2})$$

$$y = \frac{-1}{2}x + 4$$

$m = \frac{-1}{2} \quad b = 4$

$$2x - 3y = 2$$
$$-2x + 2x - 3y = -2x + 2$$
$$-3y = -2x + 2$$

$$\frac{-1}{3}(-3y) = (-2x + 2)(\frac{-1}{3})$$

$$y = \frac{2}{3}x - \frac{2}{3}$$

$m = \frac{2}{3} \quad b = \frac{-2}{3}$

Solution: (4,2)

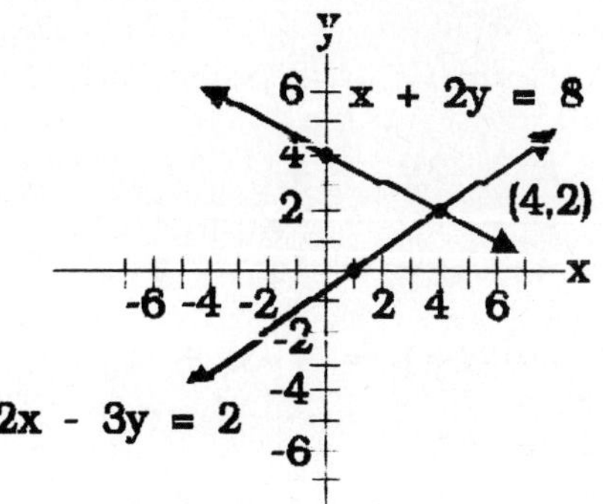

31. $x + y = 5 \qquad 2y = x - 2$

$-x+x+y = -x+5 \qquad \frac{1}{2}(2y) = (x-2)\frac{1}{2}$

$y = -x + 5 \qquad y = \frac{1}{2}x - 1$

$m = -1 \quad b = 5 \quad m = \frac{1}{2} \quad b = -1$

Solution: (4,1)

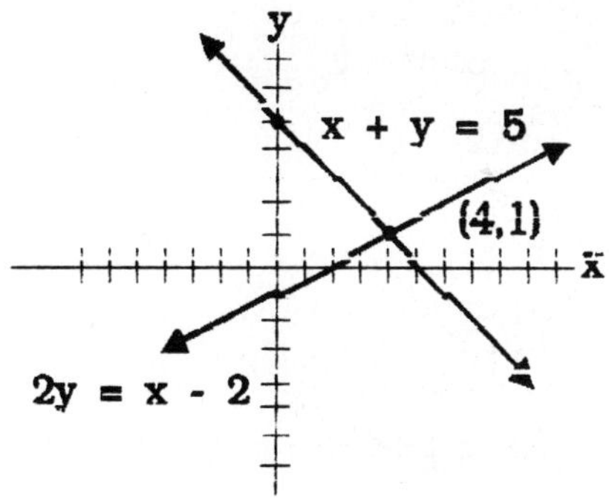

33. $y = 3 \qquad y = 2x - 3$
$m = 0 \quad b = 3 \qquad m = 2 \quad b = -3$
Solution: (3,3)

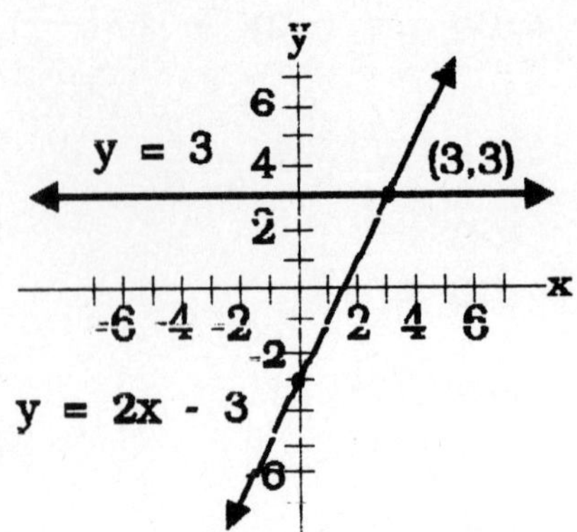

35.
$$x - 2y = 4$$
$$-x + x - 2y = -x + 4$$
$$-2y = -x + 4$$
$$\frac{-1}{2}(-2y) = (-x + 4)(\frac{-1}{2})$$
$$y = \frac{1}{2}x - 2$$
$$m = \frac{1}{2} \quad b = -2$$
$$2x - 4y = 8$$
$$-2x + 2x - 4y = -2x + 8$$
$$-4y = -2x + 8$$
$$(\frac{-1}{4})(-4y) = (-2x + 8)(\frac{-1}{4})$$
$$y = \frac{1}{2}x - 2$$
$$m = \frac{1}{2} \quad b = -2$$
Dependent

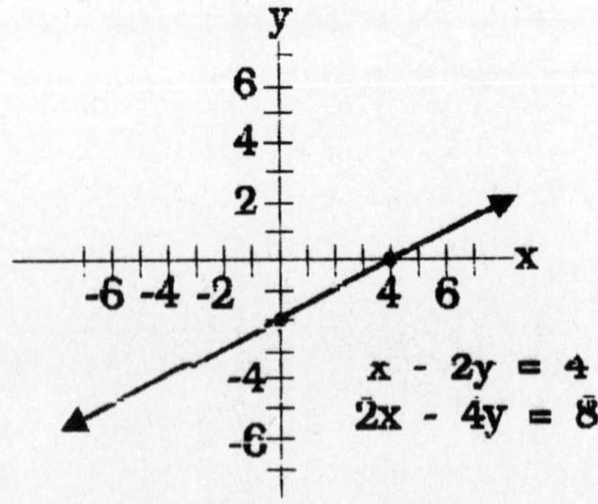

37.
$$2x + y = -2$$
$$-2x + 2x + y = -2x - 2$$
$$y = -2x - 2$$
$$m = -2 \quad b = -2$$
$$6x + 3y = 6$$
$$-6x + 6x + 3y = -6x + 6$$
$$3y = -6x + 6$$
$$\frac{1}{3}(3y) = (-6x + 6)\frac{1}{3}$$
$$y = -2x + 2$$
$$m = -2 \quad b = 2$$
Inconsistent

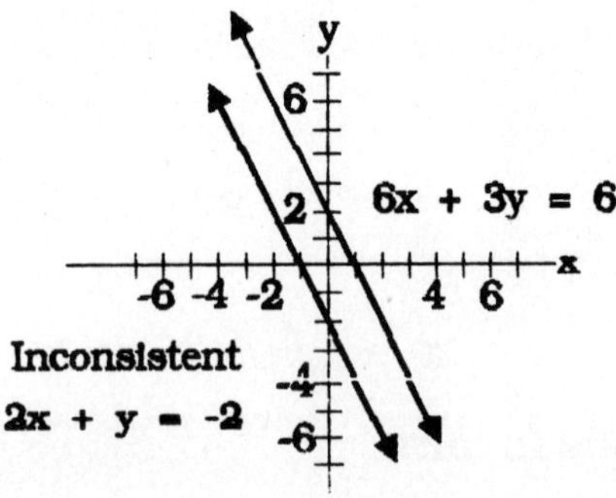

39.
$$4x - 3y = 6$$
$$-4x + 4x - 3y = -4x + 6$$
$$-3y = -4x + 6$$
$$\frac{-1}{3}(-3y) = (-4x + 6)(\frac{-1}{3})$$
$$y = \frac{4}{3}x - 2$$
$$m = \frac{4}{3} \quad b = -2$$
$$2x + 4y = 14$$
$$-2x + 2y + 4y = -2x + 14$$
$$4y = -2x + 14$$
$$\frac{1}{4}(4y) = (-2x + 14)\frac{1}{4}$$
$$y = \frac{-1}{2}x + \frac{7}{2}$$
$$m = \frac{-1}{2} \quad b = \frac{7}{2}$$
Solution: (3,2)

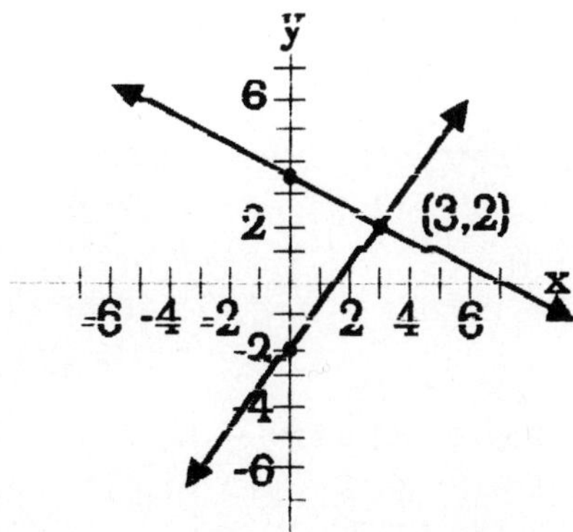

41.

$$2x - 3y = 0$$
$$-2x + 2x - 3y = -2x + 0$$
$$-3y = -2x + 0$$
$$\frac{-1}{3}(-3y) = (-2x + 0)(\frac{-1}{3})$$
$$y = \frac{2}{3}x + 0$$
$$m = \frac{2}{3} \quad b = 0$$
$$x + 2y = 0$$
$$-x + x + 2y = -x + 0$$
$$2y = -x$$
$$\frac{1}{2}(2y) = (-x)\frac{1}{2}$$
$$y = \frac{-1}{2}x$$
$$m = \frac{-1}{2} \quad b = 0$$

Solution: (0,0)

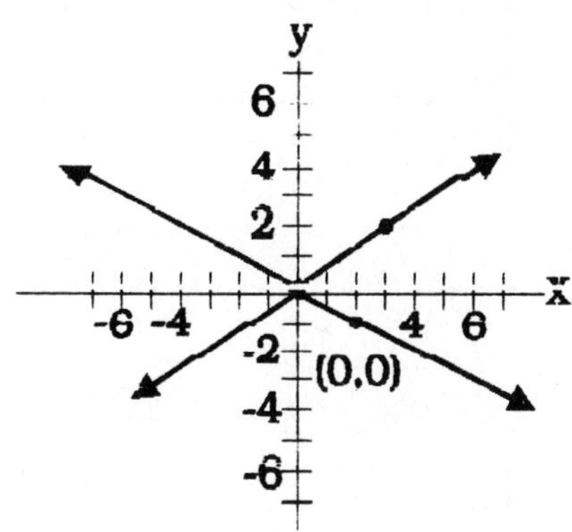

43.

$$6x + 8y = 36$$
$$-6x + 6x + 8y = -6x + 36$$
$$8y = -6x + 36$$
$$\frac{1}{8}(8y) = (-6x + 36)\frac{1}{8}$$
$$y = \frac{-3}{4}x + \frac{9}{2}$$
$$m = \frac{-3}{4} \quad b = \frac{9}{2}$$
$$-3x - 4y = -9$$
$$3x - 3x - 4y = 3x - 9$$
$$-4y = 3x - 9$$
$$\frac{-1}{4}(-4y) = (3x - 9)(\frac{-1}{4})$$
$$y = \frac{-3}{4}x + \frac{9}{4}$$
$$m = \frac{-3}{4} \quad b = \frac{9}{4}$$

Inconsistent

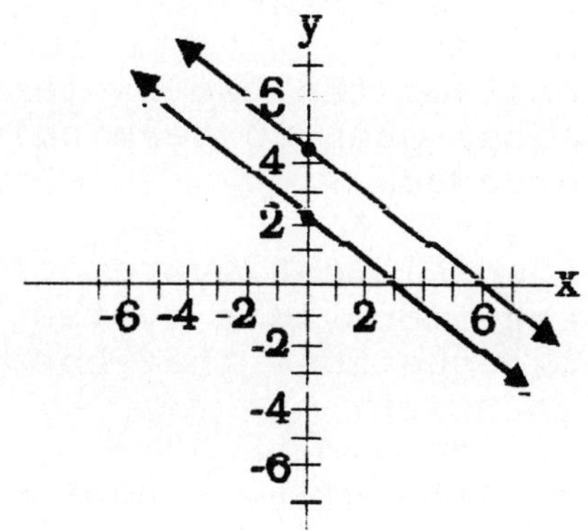

45. Let c = total cost of the system.
n = number of terminals.

Option 1: c = 60,000 + 1500n
Option 2: c = 20,000 + 3500n

n	c
0	60,000
10	75,000
20	90,000
30	105,000

n	c
0	20,000
10	55,000
20	90,000
30	125,000

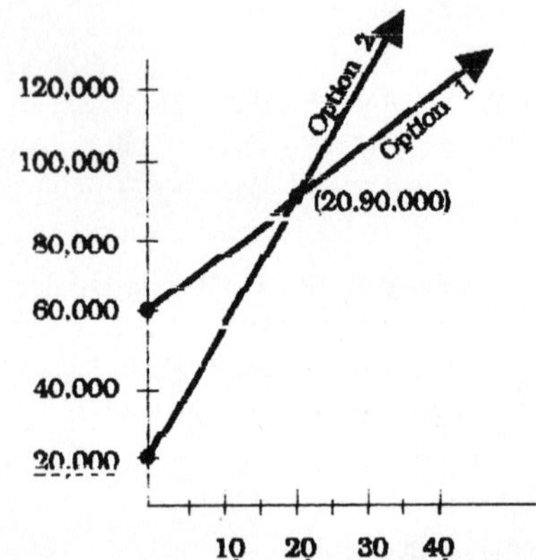

The cost of the two systems are equal when 20 terminals are ordered.

47. a) A consistent system of equations is a system of equations that has a solution.

b) An inconsistent system of equations is a system of equations that does not have a solution.

c) A dependent system of equations is a system of equations that has an infinite number of solutions.

49. When a dependent system of two linear equations is graphed, the result is a single line.

Cumulative Review Exercises

51. $8x^2 + 2x - 15$
$= (2x + 3)(4x - 5)$

53. $\frac{x}{x + 3} - 2 = 0$

$\frac{x}{x + 3} = 2$

$x = 2(x + 3)$
$x = 2x + 6$
$-x = 6$
$x = -6$

Exercise Set 8.3

1. $x + 2y = 4 \quad (1)$
$2x - 3y = 1 \quad (2)$

Solve (1) for x.
$x + 2y - 2y = -2y + 4$
$x = -2y + 4$

Substitute into (2).
$2(-2y + 4) - 3y = 1$
$-4y + 8 - 3y = 1$
$-7y + 8 = 1$
$-7y + 8 - 8 = 1 - 8$
$-7y = -7$

$(\frac{-1}{7})(-7y) = (-7)(\frac{-1}{7})$

$y = 1$

Substitute into (1).
$x + 2(1) = 4$

$$x + 2 - 2 = 4 - 2$$
$$x = 2$$

Solution: (2,1)

Check solution in (2).

$$2x - 3y = 1$$
$$2(2) - 3(1) = 1$$
$$4 - 3 = 1$$
$$1 = 1$$

Solution is correct.

3. $x + y = -2$ (1)
$x - y = (0)$ (2)

Solve (1) for x.

$$x - y + y = -y - 2$$
$$x = -y - 2$$

Substitute into (2).

$$(-y - 2) - y = 0$$
$$-2y - 2 = 0$$
$$-2y - 2 + 2 = 0 + 2$$
$$\frac{-1}{2}(-2y) = 2(\frac{-1}{2})$$
$$y = -1$$

Substitute into (1).

$$x + (-1) = -2$$
$$x - 1 + 1 = -2 + 1$$
$$x = -1$$

Solution: (-1,-1)

Check solution in (2).

$$x - y = 0$$
$$(-1) - (-1) = 0$$
$$-1 + 1 = 0$$
$$0 = 0$$

Solution is correct.

5. $2x + y = 3$ (1)
$2x + y + 5 = 0$ (2)

Solve (1) for y.

$$-2x + 2x + y = -2x + 3$$
$$y = -2x + 3$$

Substitute into (2).

$$2x + (-2x + 3) + 5 = 0$$
$$2x - 2x + 3 + 5 = 0$$
$$3 + 5 = 0$$
$$8 = 0$$

Inconsistent, no solution.

7. $x = 4$ (1)
$x + y + 5 = 0$ (2)

Substitute (1) into (2) and solve for y.

$$(4) + y + 5 = 0$$
$$y + 9 = 0$$
$$y + 9 - 9 = -9$$
$$y = -9$$

Solution: (4,-9)

Check solution in (2).

$$x + y + 5 = 0$$
$$(4) + (-9) + 5 = 0$$
$$-5 + 5 = 0$$
$$0 = 0$$

9. $x - \frac{1}{2} y = 2$ (1)

$y = 2x - 4$ (2)

Substitute (2) into (1) and solve for x.

$$x - \frac{1}{2}(2x - 4) = 2$$
$$x - x + 2 = 2$$
$$0 + 2 = 2$$
$$2 = 2$$

Dependent - infinite number of solutions.

11. $3x + y = -1$ (1)
$y = 3x + 5$ (2)

Substitute (2) into (1) and solve for x.

$$3x + (3x + 5) = -1$$
$$3x + 3x + 5 = -1$$
$$6x + 5 = -1$$
$$6x + 5 - 5 = -1 - 5$$
$$6x = -6$$
$$(\frac{1}{6})(6x) = (-6)(\frac{1}{6})$$

$$x = -1$$

Substitute into (2).

$y = 3(-1) + 5$
$y = -3 + 5$
$y = 2$

Solution: $(-1,2)$

Check solution in (2).

$$\begin{aligned} y &= 3x + 5 \\ (2) &= 3(-1) + 5 \\ 2 &= -3 + 5 \\ 2 &= 2 \end{aligned}$$

The solution is correct.

13. $y = 2x - 13 \quad (1)$
$-4x - 7 = 9y \quad (2)$

Substitute (1) into (2) and solve for x.

$$\begin{aligned} -4x - 7 &= 9(2x - 13) \\ -4x - 7 &= 18x - 117 \\ -4x + 4x - 7 &= 4x + 18x - 117 \\ -7 &= 22x - 117 \\ -7 + 117 &= 22x - 117 + 117 \\ 110 &= 22x \\ \frac{1}{22}(110) &= \frac{1}{22}(22x) \\ 5 &= x \end{aligned}$$

Substitute into (1).

$y = 2(5) - 13$
$y = 10 - 13$
$y = -3$

Solution: $(5,-3)$

Check solution in (2).

$$\begin{aligned} -4x - 7 &= 9y \\ -4(5) - 7 &= 9(-3) \\ -20 - 7 &= -27 \\ -27 &= -27 \end{aligned}$$

Solution is correct.

15. $2x + 3y = 7 \quad (1)$
$6x - 2y = 10 \quad (2)$

Solve (2) for y.

$$\begin{aligned} -6x + 6x - 2y &= 10 - 6x \\ -2y &= 10 - 6x \\ \frac{-1}{2}(-2y) &= (10-6x)(\frac{-1}{2}) \\ y &= -5 + 3x \end{aligned}$$

Substitute into (1) and solve for x.

$$\begin{aligned} 2x + 3(-5 + 3x) &= 7 \\ 2x - 15 + 9x &= 7 \\ 11x - 15 &= 7 \\ 11x - 15 + 15 &= 7 + 15 \\ 11x &= 22 \\ \frac{1}{11}(11x) &= (22)\frac{1}{11} \\ x &= 2 \end{aligned}$$

Substitute into (2).

$$\begin{aligned} 6(2) - 2y &= 10 \\ 12 - 2y &= 10 \\ -12 + 12 - 2y &= 10 - 12 \\ -2y &= -2 \\ (\frac{-1}{2})(-2y) &= (-2)(\frac{-1}{2}) \\ y &= 1 \end{aligned}$$

Solution: $(2,1)$

Check solution in (1).

$$\begin{aligned} 2x + 3y &= 7 \\ 2(2) + 3(1) &= 7 \\ 4 + 3 &= 7 \\ 7 &= 7 \end{aligned}$$

Solution is correct.

17. $3x - y = 14 \quad (1)$
$6x - 2y = 10 \quad (2)$

Solve (1) for y.

$$\begin{aligned} -3x + 3x - y &= -3x + 14 \\ -y &= -3x + 14 \\ (-1)(-y) &= (-3x+14)(-1) \\ y &= 3x - 14 \end{aligned}$$

Substitute into (2) and solve for x.

$6x - 2(3x - 14) = 10$
$6x - 6x + 28 = 10$
$0 + 28 = 10$
$28 = 10$

Inconsistent--no solution.

19. $2x - 7y = 6 \quad (1)$
$5x - 8y = -4 \quad (2)$

Solve (1) for x.

$2x - 7y + 7y = 6 + 7y$
$2x = 6 + 7y$

$$(\frac{1}{2})(2x) = (6 + 7y)(\frac{1}{2})$$

$$x = 3 + \frac{7}{2} y$$

Substitute into (2) and solve for y.

$$5(3 + \frac{7}{2} y) - 8y = -4$$

$$15 + \frac{35}{2} y - 8y = -4$$

$$2(15 + \frac{35}{2} y - 8y) = -4(2)$$

$30 + 35y - 16y = -8$
$19y + 30 = -8$
$19y + 30 - 30 = -8 - 30$
$19y = -38$

$$\frac{1}{19}(19y) = (-38)\frac{1}{19}$$

$y = -2$

Substitute into (1).

$2x - 7(-2) = 6$
$2x + 14 = 6$
$2x + 14 - 14 = 6 - 14$
$2x = -8$

$$(\frac{1}{2})2x = -8(\frac{1}{2})$$

$x = -4$

Solution: $(-4,-2)$

Check solution in (2).

$5x - 8y = -4$
$5(-4) - 8(-2) = -4$
$-20 + 16 = -4$
$-4 = -4$

Solution is correct.

21. $3x + 4y = 10 \quad (1)$
$4x + 5y = 14 \quad (2)$

Solve (1) for x.

$3x + 4y - 4y = 10 - 4y$
$3x = 10 - 4y$

$$(\frac{1}{3})(3x) = (10 - 4y)(\frac{1}{3})$$

$$x = \frac{10 - 4y}{3}$$

Substitute into (2) and solve for y.

$$4(\frac{10 - 4y}{3}) + 5y = 14$$

$$(\frac{40 - 16x}{3}) + 5y = 14$$

$$3(\frac{40 - 16x}{3} + 5y) = 14(3)$$

$40 - 16y + 15y = 42$
$40 - y = 42$
$-40 + 40 - y = 42 - 40$
$-y = 2$
$(-1)(-y) = 2(-1)$
$y = -2$

Substitute into (1) and solve for x.

$3x + 4(-2) = 10$
$3x - 8 = 10$
$3x - 8 + 8 = 10 + 8$
$3x = 18$

$$\frac{1}{3}(3x) = (18)\frac{1}{3}$$

$x = 6$

Solution: (6,-2)

Check solution in (2).

$$4x + 5y = 14$$
$$4(6) + 5(-2) = 14$$
$$24 - 10 = 14$$
$$14 = 14$$

Solution is correct.

23. Solve for the x in the first equation because both 6 and 9 are divisible by 3.

25. You will obtain a true statement such as 2 = 2.

Cumulative Review Exercises

27. $\frac{6}{4} = \frac{8}{x}$

$$6x = 32$$

$$x = \frac{32}{6} = 5\,\frac{1}{3} \text{ in.}$$

29. $\frac{3}{x - 12} + \frac{5}{x - 5}$

$$= \frac{5}{x^2 - 17x + 60}$$

$$\frac{3}{x - 12} + \frac{5}{x - 5}$$

$$= \frac{5}{(x - 12)(x - 5)}$$

$$(x - 12)(x - 5)$$

$$\left[\frac{3}{x - 12} + \frac{5}{x - 5}\right]$$

$$= \left[\frac{5}{(x - 12)(x - 5)}\right]$$

$$(x - 12)(x - 5)$$

$$3(x - 5) + 5(x - 12) = 5$$
$$3x - 15 + 5x - 60 = 5$$
$$8x - 75 = 5$$
$$8x = 80$$
$$x = 10$$

Exercise Set 8.4

1. $x + y = 8 \quad (1)$

$x - y = 4 \quad (2)$

$$2x = 12$$

$$\frac{1}{2}(2x) = (12)\frac{1}{2}$$

$$x = 6$$

Substitute into (1) and solve for y.

$$x + y = 8$$
$$6 + y = 8$$
$$-6 + 6 + y = 8 - 6$$
$$y = 2$$

Solution: (6,2)

Check in both equations

$x + y = 8$	$x - y = 4$
$6 + 2 = 8$	$6 - 2 = 4$
$8 = 8$	$4 = 4$

Solution is correct.

3. $-x + y = 5 \quad (1)$

$x + y = 1 \quad (2)$

$$2y = 6$$

$$\frac{1}{2}(2y) = (6)\frac{1}{2}$$

$$y = 3$$

Substitute into (2) and solve for x.

$$x + y = 1$$
$$x + (3) = 1$$
$$x - 3 + 3 = 1 - 3$$
$$x = -2$$

Solution: $(-2,3)$

Check in both equations.

$$\begin{array}{r|r} -x + y = 5 & x + y = 1 \\ -(-2) + 3 = 5 & -2 + 3 = 1 \\ 5 = 5 & 1 = 1 \end{array}$$

Solution is correct.

5. $x + 2y = 15 \quad (1)$
$x - 2y = -7 \quad (2)$

$$2x = 8$$

$$\frac{1}{2}(2x) = (8)\frac{1}{2}$$

$$x = 4$$

Substitute into (1) and solve for y.

$$\begin{aligned} x + 2y &= 15 \\ (4) + 2y &= 15 \\ -4 + 4 + 2y &= 15 - 4 \\ 2y &= 11 \\ \frac{1}{2}(2y) &= (11)\frac{1}{2} \\ y &= \frac{11}{2} \end{aligned}$$

Solution: $(4,\frac{11}{2})$

Check in both equations.

$$\begin{array}{r|r} x + 2y = 15 & x - 2y = -7 \\ 4 + 2(\frac{11}{2}) = 15 & 4 - 2(\frac{11}{2}) = -7 \\ 4 + 11 = 15 & 4 - 11 = -7 \\ 15 = 15 & -7 = -7 \end{array}$$

Solution is correct.

7. $4x + y = 6 \quad (1)$
$-8x - 2y = 20 \quad (2)$

Multiply (1) by 2:
$2(4x + y) = 6(2)$
$8x + 2y = 12$

$$\begin{aligned} 8x + 2y &= 12 \quad (1) \\ -8x - 2y &= 20 \quad (2) \\ \hline 0 &= 32 \end{aligned}$$

Inconsistent--no solution.

9. $-5x + y = 14 \quad (1)$
$-3x + y = -2 \quad (2)$

Multiply equation (1) by -1:
$-1(-5x + y) = 14(-1)$
$5x - y = -14$

$$\begin{aligned} 5x - y &= -14 \quad (1) \\ -3x + y &= -2 \quad (2) \\ \hline 2x &= -16 \\ \frac{1}{2}(2x) &= (-16)\frac{1}{2} \\ x &= -8 \end{aligned}$$

Substitute into (1) and solve for y.

$$\begin{aligned} -5x + y &= 14 \\ -5(-8) + y &= 14 \\ 40 + y &= 14 \\ 40 - 40 + y &= 14 - 40 \\ y &= -26 \end{aligned}$$

Solution: $(-8,-26)$

Check in both equations

$$\begin{array}{r|r} -5x + y = 14 & -3x + y = -2 \\ -5(-8)+(-26)=14 & -3(-8)+(-26)=-2 \\ 14=14 & -2=-2 \end{array}$$

Solution is correct.

11. $3x + y = 10 \quad (1)$
$3x - 2y = 16 \quad (2)$

Multiply (1) by -1:
$-1(3x + y) = 10(-1)$
$-3x - y = -10$

$$-3x - y = -10 \quad (1)$$
$$3x - 2y = 16 \quad (2)$$

$$-3y = 6$$

$$\frac{-1}{3}(-3y) = 6\left(\frac{-1}{3}\right)$$

$$y = -2$$

Substitute into (1) and solve for x.

$$3x + y = 10$$
$$3x + (-2) = 10$$
$$3x - 2 + 2 = 10 + 2$$
$$3x = 12$$

$$\left(\frac{1}{3}\right)(3x) = (12)\left(\frac{1}{3}\right)$$

$$x = 4$$

Solution: (4,-2)

Check in both equations.

$3x + y = 10$	$3x - 2y = 16$
$3(4)+(-2) = 10$	$3(4)-2(-2) = 16$
$10 = 10$	$16 = 16$

Solution is correct.

13.
$$4x - 3y = 8 \quad (1)$$
$$2x + y = 14 \quad (2)$$
$$-2(2x + y) = -2(14) \quad (2)$$
$$-4x - 2y = -28 \quad (2)$$

$$4x - 3y = 8 \quad (1)$$
$$-4x - 2y = -28 \quad (2)$$

$$-5y = -20$$
$$y = 4$$

Substitute into (1).

$$4x - 3y = 8$$
$$4x - 3(4) = 8$$
$$4x - 12 = 8$$
$$4x = 20$$
$$x = 5$$

Solution: (5,4)

Check in both equations.

$4x - 3y = 8$	$2x + y = 14$
$4(5) - 3(4) = 8$	$2(5) + 4 = 14$
$8 = 8$	$14 = 14$

Solution is correct.

15.
$$5x + 3y = 6 \quad (1)$$
$$2x - 4y = 5 \quad (2)$$
$$-2(5x + 3y) = -2(6) \quad (1)$$
$$5(2x - 4y) = 5(5) \quad (2)$$

$$-10x - 6y = -12 \quad (1)$$
$$10x - 20y = 25 \quad (2)$$

$$-26y = 13$$

$$y = \frac{-1}{2}$$

Substitute into (1).

$$5x + 3\left(\frac{-1}{2}\right) = 6$$

$$2\left(5x - \frac{3}{2}\right) = 2(6)$$

$$10x - 3 = 12$$
$$10x = 15$$

$$x = \frac{3}{2}$$

Solution: $\left(\frac{3}{2}, \frac{-1}{2}\right)$

Check in both equations.

$5x + 3y = 6$	$2x - 4y = 5$
$5\left(\frac{3}{2}\right)+3\left(\frac{-1}{2}\right) = 6$	$2\left(\frac{3}{2}\right)-4\left(\frac{-1}{2}\right) = 5$
$\frac{15}{2} - \frac{3}{2} = 6$	$3 + 2 = 5$
$\frac{12}{2} = 6$	$5 = 5$
$6 = 6$	

Solution is correct.

17.
$$4x - 2y = 6 \quad (1)$$
$$y = 2x - 3 \quad (2)$$
$$2x - y = 3 \quad (2)$$

$$-2(2x - y) = -2(3) \quad (2)$$

$-4x + 2y = -6 \qquad (2)$

$$\begin{array}{rl} -4x + 2y = -6 & (2) \\ 4x - 2y = 6 & (1) \\ \hline 0 = 0 & \end{array}$$

Dependent--Infinite number of solutions.

19. $3x - 2y = -2 \qquad (1)$
$3y = 2x + 4 \qquad (2)$
$2x - 3y = -4 \qquad (2)$

$-2(3x - 2y) = -2(-2) \qquad (1)$
$3(2x - 3y) = 3(-4) \qquad (2)$

$$\begin{array}{rl} -6x + 4y = 4 & (1) \\ 6x - 9y = -12 & (2) \\ \hline -5y = -8 & \end{array}$$

$y = \frac{8}{5}$

Substitute into (1).

$3x - 2(\frac{8}{5}) = -2$

$3x - \frac{16}{5} = -2$

$5(3x - \frac{16}{5}) = 5(-2)$

$15x - 16 = -10$

$15x = 6$

$x = \frac{2}{5}$

Solution: $(\frac{2}{5}, \frac{8}{5})$

Check in both equations.

$3x - 2y = -2$	$3y = 2x + 4$
$3(\frac{2}{5}) - 2(\frac{8}{5}) = -2$	$3(\frac{8}{5}) = 2(\frac{2}{5}) + 4$
$\frac{6}{5} - \frac{16}{5} = -2$	$\frac{24}{5} = \frac{4}{5} + \frac{20}{5}$
$\frac{-10}{5} = -2$	$\frac{24}{5} = \frac{24}{5}$
$-2 = 2$	

Solution is correct.

21. $5x - 4y = 20 \qquad (1)$
$-3x + 2y = -15 \qquad (2)$
$2(-3x + 2y) = 2(-15) \qquad (2)$
$-6x + 4y = -30 \qquad (2)$

$$\begin{array}{rl} 5x - 4y = 20 & (1) \\ -6x + 4y = -30 & (2) \\ \hline -x = -10 & \\ x = 10 & \end{array}$$

Substitute into (1).

$5(10) - 4y = 20$
$50 - 4y = 20$
$-4y = -30$
$y = \frac{15}{2}$

Solution: $(10, \frac{15}{2})$

Check in both equations.

$5x - 4y = 20$	$-3x + 2y = -15$
$5(10) - 4(\frac{15}{2}) = 20$	$-3(10) + 2(\frac{15}{2}) = -15$
$50 - 30 = 20$	$-30 + 15 = -15$
$20 = 20$	$-15 = -15$

Solution is correct.

23. $6x + 2y = 5 \qquad (1)$
$3y = 5x - 8 \qquad (2)$

$5x - 3y = 8 \quad (2)$

$3(6x + 2y) = 3(5) \quad (1)$

$2(5x - 3y) = 2(8) \quad (2)$

$$\begin{array}{rl} 18x + 6y = 15 & (1) \\ 10x - 6y = 16 & (2) \\ \hline 28x \quad = 31 & \end{array}$$

$x = \frac{31}{28}$

Substitute into (1).

$6(\frac{31}{28}) + 2y = 5$

$14(\frac{93}{14} + 2y) = 14(5)$

$93 + 28y = 70$

$28y = -23$

$y = \frac{-23}{28}$

Solution: $(\frac{31}{28}, \frac{-23}{28})$

Check in both equations.

$6x + 2y = 5$	$3y = 5x - 8$
$6(\frac{31}{28})+2(\frac{-23}{28})=5$	$3(\frac{-23}{28})=5(\frac{31}{28})-8$
$\frac{186}{28} - \frac{46}{28} = 5$	$\frac{-69}{28} = \frac{155}{28} - 8$
$\frac{140}{28} = 5$	$\frac{-69}{28} = \frac{155}{28} - \frac{244}{28}$
$5 = 5$	$\frac{-69}{28} = \frac{-69}{28}$

Solution is correct.

25. $4x + 5y = 0 \quad (1)$

$3x = 6y + 4 \quad (2)$

$3x - 6y = 4 \quad (2)$

$-3(4x + 5y) = -3(0) \quad (1)$

$4(3x - 6y) = 4(4) \quad (2)$

$$\begin{array}{rl} -12x - 15y = 0 & (1) \\ 12x - 24y = 16 & (2) \\ \hline -39y = 16 & \end{array}$$

$y = \frac{-16}{39}$

Substitute into (1).

$4x + 5(\frac{-16}{39}) = 0$

$4x - \frac{80}{39} = 0$

$4x = \frac{80}{39}$

$x = \frac{20}{39}$

Solution: $(\frac{20}{39}, \frac{-16}{39})$

Check in both equations.

$4x + 5y = 0$	$3x = 6y + 4$
$4(\frac{20}{39})+5(\frac{-16}{39})=0$	$3(\frac{20}{39})=6(\frac{-16}{39})+4$
$\frac{80}{39} - \frac{80}{39} = 0$	$\frac{60}{39} = \frac{-96}{39}+\frac{156}{39}$
$0 = 0$	$\frac{60}{39} = \frac{60}{39}$

Solution is correct.

27. $x - \frac{1}{2} y = 4 \quad (1)$

$3x + y = 6 \quad (2)$

$2(x - \frac{1}{2} y) = 2(4) \quad (1)$

$$\begin{array}{rl} 2x - y = 8 & (1) \\ 3x + y = 6 & (2) \\ \hline 5x = 14 & \end{array}$$

$x = \frac{14}{5}$

Substitute into (2).

$3(\frac{14}{5}) + y = 6$

$\frac{42}{5} + y = 6$

$y = \frac{-12}{5}$

Solution: $(\frac{14}{5}, \frac{-12}{5})$

Check in both equations

$x - \frac{1}{2}y = 4$	$3x + y = 6$
$\frac{14}{5} - \frac{1}{2}(\frac{-12}{5}) = 4$	$3(\frac{14}{5}) + (\frac{-12}{5}) = 6$
$\frac{14}{5} + \frac{6}{5} = 4$	$\frac{42}{5} - \frac{12}{5} = 6$
$\frac{20}{5} = 4$	$\frac{30}{5} = 6$
$4 = 4$	$6 = 6$

Solution is correct.

29. The variable will disappear and you will obtain a false statement like 0 = 6.

31. a) 1. The equations must be written so that the terms containing the variables are on the left side of the equal sign and the constants are on the right side.

2. Multiply one or both equations so that when the equations are added the resulting sum will contain only one variable.

3. Add the equations to make a single equation in one variable.

4. Solve for the variable.

5. Substitute the variable you found into either original equation and solve for the remaining variable.

b) $3x - 2y = 10$
$3x + 5y = 13$

$5(3x - 2y = 10)$
$2(2x + 5y = 13)$

$15x - 10y = 50$
$4x + 10y = 26$

$19x = 76$
$x = 4$

$3(4) - 2y = 10$
$12 - 2y = 10$
$-2y = -2$
$y = 1$

Solution: (4,1)

33. There are many possible answers. Write the x and y terms with any coefficient then substitute x = 4 and y = -2 to obtain the constant. Repeat the process to get the second equation.

Cumulative Review Exercises

35. $d = rt$
$2420 = 110 \cdot t$
$\frac{2420}{110} = t$
$t = 22$ days

37. $\frac{x}{x^2 - 1} - \frac{3}{x^2 - 16x + 15}$

$$= \frac{x}{(x-1)(x+1)} - \frac{3}{(x-15)(x-1)}$$

$$= \frac{x(x-15)}{(x-1)(x+1)(x-15)} - \frac{3(x+1)}{(x-1)(x+1)(x-15)}$$

$$= \frac{x^2 - 15x - 3x - 3}{(x-1)(x+1)(x-15)}$$

$$= \frac{x^2 - 18x - 3}{(x-1)(x+1)(x-15)}$$

Just for fun

1. $\frac{x+2}{2} - \frac{y+4}{3} = 4 \quad (1)$

$6[\frac{x+2}{2} - \frac{y+4}{3}] = 4(6) \quad (1)$

$3(x+2) - 2(y+4) = 4(6) \quad (1)$
$3x + 6 - 2y - 8 = 24 \quad (1)$
$3x - 2y = 26 \quad (1)$

$\frac{x+y}{2} = \frac{1}{2} + \frac{x-y}{3} \quad (2)$

$6[\frac{x+y}{2}] = [\frac{1}{2} + \frac{x-y}{3}]6 \quad (2)$

$3(x+y) = 3(1) + 2(x-y) \quad (2)$
$3x + 3y = 3 + 2x - 2y \quad (2)$
$x + 5y = 3 \quad (2)$
$(-3)(x+5y) = (3)(-3) \quad (2)$
$-3x - 15y = -9 \quad (2)$

$3x - 2y = 26 \quad (1)$
$-3x - 15y = -9 \quad (2)$

$-17y = 17$
$y = -1$

Substitute into (1).

$\frac{x+2}{2} - \frac{(-1)+4}{3} = 4$

$\frac{x+2}{2} - 1 = 4$

$\frac{x+2}{2} = 5$

$x + 2 = 10$
$x = 8$

Solution: (8,-1)

2. $\frac{5x}{2} + 3y = \frac{9}{2} + y$

$2[\frac{5x}{2} + 3y] = 2[\frac{9}{2} + y]$

$5x + 6y = 9 + 2y$
$5x + 4y = 9$
$5x + 4y = 9$
$-23x - 2y = 48$

$5x + 4y = 9$
$-46x - 4y = 96$

$-41x = 105$

$x = -\frac{105}{41}$

$5(-\frac{105}{41}) + 4y = 9$

$-\frac{525}{41} + 4y = 9$

$41(-\frac{525}{41} + 4y) = 41(9)$

$-525 + 164y = 369$
$164y = 894$

$y = \frac{894}{164} = \frac{447}{82}$

$\frac{1}{4}x - \frac{1}{2}y = 6x + 12$

$4[\frac{1}{4}x - \frac{1}{2}y] = 4[6x + 12]$

$$x - 2y = 24x + 48$$
$$-23x - 2y = 48$$

Exercise Set 8.5

1. Let x = one integer.
   Let y = second integer.

   The sum of two numbers is 37,
   $x + y = 37$ (1).
   One number is one greater than twice the other,
   $x = 2y + 1$ (2).

   Substitute (2) into (1)
   $$x + y = 37$$
   $$(2y + 1) + y = 37$$
   $$2y + 1 + y = 37$$
   $$3y + 1 = 37$$
   $$3y = 37 - 1$$
   $$3y = 36$$
   $$y = \frac{36}{3}$$
   $$y = 12$$

   Substitute 12 into (2).
   $x = 2y + 1$
   $x = 2(12) + 1$
   $x = 25$

   The two numbers are 12 and 25.

3. Let x = an odd integer.
   Let y = the next odd integer.

   The sum of two integers is 76,
   $x + y = 76$ (1).
   The next odd integer,
   $y = x + 2$ (2).

   Substitute (2) into (1).
   $$x + y = 76$$
   $$x + (x + 2) = 76$$
   $$x + x + 2 = 76$$
   $$2x + 2 = 76$$
   $$2x = 76 - 2$$
   $$2x = 74$$
   $$x = \frac{74}{2}$$
   $$x = 37$$

   Substitute 37 into (2).
   $y = x + 2$
   $y = (37) + 2$
   $y = 39$

   The two integers are 37 and 39.

5. Let x = the larger of two integers.
   Let y = the second integer.

   The difference of two integers is 28,
   $x - y = 28$ (1).
   The larger integer is three times the smaller integer,
   $x = 3y$ (2).

   Substitute (2) into (1).
   $$x - y = 28$$
   $$(3y) - y = 28$$
   $$3y - y = 28$$
   $$2y = 28$$
   $$y = \frac{28}{2}$$
   $$y = 14$$

   Substitute 14 into (2).
   $x = 3y$
   $x = 3(14)$
   $x = 42$

   The two integers are 14 and 42.

7. Let x = number of \$2 bills.
   y = number of \$5 bills.

   $x + y = 25$
   $2x + 5y = 101$

   Solve for x.

$x + y = 25$
$x = 25 - y$

Substitute:

$$2x + 5y = 101$$
$$2(25 - y) + 5y = 101$$
$$50 - 2y + 5y = 101$$
$$50 + 3y = 101$$
$$3y = 51$$
$$y = 17$$
$$x = 25 - y = 25 - 17 = 8$$

Paul has 8 \$2 bills and 17 \$5 bills.

9. Let x = speed or rate in still air.
Let y = speed or rate of wind.

A plane can travel 540 mph with the wind,
$x + y = 540$ (1).
A plane can travel 490 mph against the wind,
$x - y = 490$ (2).

$$x + y = 540$$
$$x - y = 490$$

$$2x = 1030$$
$$x = \frac{1030}{2}$$
$$x = 515$$

Substitute 515 into (1).

$$x + y = 540$$
$$515 + y = 540$$
$$y = 540 - 515$$
$$y = 25$$

The speed of the plane in still air is 515 mph.
The speed of the wind is 25 mph.

11.

Speed boats	Distance	Rate	Time
Teresa's	3y	y	3
Jill's	3.2x	x	3.2

Let x = speed of Jill's boat.
y = speed of Teresa's boat.

$$y = x + 4$$
$$3y = 3.2x$$

Substitute:

$$3(x + 4) = 3.2x$$
$$3x + 12 = 3.2x$$

$$12 = 0.2x$$
$$\frac{12}{0.2} = \frac{0.2x}{0.2}$$
$$60 = x$$
$$y = x + 4 = 60 + 4$$
$$= 64$$

Jill's boat travels 60 mph and Teresa's boat travels 64 mph.

13.

Jets	Distance	Rate	Time
United Airlines	3x	x	3
Delta Airlines	3y	y	3

Let x = United's speed.
y = Delta's speed.

$$x = y + 100$$
$$3x + 3y = 2700$$

Substitute:

$$3(y + 100) + 3y = 2700$$
$$3y + 300 + 3y = 2700$$

$$6y + 300 = 2700$$
$$6y = 2400$$
$$y = 400$$
$$x = y + 100$$
$$= 400 + 100$$
$$= 500$$

Speed of United Airlines jet is 500 mph and speed of Delta Airlines jet is 400 mph.

15.

Joggers	Distance	Rate	Time
Micki	5x	5	x
Petra	8y	8	y

Let x = Micki's time.
y = Petra's time.

$$x = y + 0.3$$
$$5x = 8y$$

Substitute:
$$5(y + 0.3) = 8y$$
$$5y + 1.5 = 8y$$
$$1.5 = 3y$$
$$\frac{1.5}{3} = \frac{3y}{3}$$
$$y = 0.5$$

Petra catches up in 0.5 hours.

17. Let x = amount invested at 10%.
y = amount invested at 8%.

The McAdams invested a total of $8000 in two savings accounts,
$$x + y = 8000 \quad (1).$$
Interest of amount invested at 10% plus amount invested at 8% equals $750,
$$0.10x + 0.08y = \$750 \ (2).$$

10(x + y) = 10(8000) gives:
$$10x + 10y = 80{,}000 \quad (1)$$
$-100(0.10x+0.08y)$
= − 100(750) gives:
$$-10x - 8y = -75{,}000 \ (2)$$

$$10x + 10y = 80{,}000 \quad (1)$$
$$-10x - 8y = -75{,}000 \quad (2)$$

$$2y = 5000$$
$$y = \frac{5000}{2}$$
$$y = 2500$$

Substitute 2500 into (1).
$$x + y = 8000$$
$$x + 2500 = 8000$$
$$x = 8000 - 2500$$
$$x = 5500$$

Mr. and Mrs McAdams invested $5,500 at 10% and $2500 at 8%.

19.

	Number of liters	Percent acid	Acid content
25% solution	x	0.25	0.25x
50% solution	y	0.50	0.50y
Mixture	10	0.40	.40(10)

Let x = number of liters of 25% solution.
y = number of liters of 50% solution.

$$x + y = 10$$
$$0.25x + 0.50y = 0.40(10)$$

Solve for x:
$$x + y = 10$$
$$x = 10 - y$$

Substitute:

$$0.25x + 0.50y = 0.40(10)$$
$$0.25(10-y) + 0.50y = 4$$
$$2.5 - 0.25y + 0.50y = 4$$
$$2.5 + 0.25y = 4$$
$$0.25y = 1.5$$
$$\frac{0.25y}{0.25} = \frac{1.5}{0.25}$$
$$y = 6$$
$$x = 10 - y$$
$$= 10 - 6$$
$$= 4$$

Marie needs 4 liters of 25% acid concentration and 6 liters of 50% acid concentration.

21. Let x = fixed charge for typesetting.
Let y = additional charge per booklet.

Total cost for 1000 booklets is $550,
$$x + 1000y = 550 \quad (1)$$
Total cost for 2000 booklets is $800,
$$x + 2000y = 800 \quad (2)$$
$$(-1)[x + 1000y] = (-1)(550)$$
gives:
$$-x - 1000y = -550$$

$$-x - 1000y = -550 \quad (1)$$
$$x + 2000y = 800 \quad (2)$$

$$1000y = 250$$
$$y = 0.25$$

Substitute 0.25 for y in (1).
$$x + 1000(0.25) = 550$$
$$x + 250 = 550$$
$$x = 300$$

The fixed charge for typesetting is $300, and the additional charge per booklet is $0.25.

23. Let x = the number of gallons of 5% butterfat milk.
y = the number of gallons of 0% butterfat milk.
Total gallons of milk in mixture,
$$x + y = 100 \quad (1).$$
How much 5% milk and skim milk should be remixed to make 100 gallons of 3.5% milk,
$$0.05x + 0.00y$$
$$= 0.035(100) \quad (2)$$

$$100(0.05x + 0.00y)$$
$$= 100(3.5)$$
gives:
$$5x - 0y = 350$$
$$5x = 350$$

Solve (2) for x.
$$5x = 350$$
$$x = \frac{350}{5}$$
$$x = 70$$

Substitute 70 into (1).
$$x + y = 100$$
$$70 + y = 100$$
$$y = 100 - 70$$
$$y = 30$$

Jason must mix 70 gallons of 5% butterfat milk with 30 gallons of skim milk.

25. Let x = number of shares of Avon stock.
y = number of shares of Coca-Cola stock.

$$x = 5y$$
$$37x + 75y = 7,800$$
$$37(5y) + 75y = 7800$$
$$185y + 75y = 7800$$
$$260y = 7800$$
$$y = 30$$
$$x = 5y = 5(30)$$
$$= 150$$

Karla bought 150 shares of Avon stock and 30 shares of Coca-Cola stock.

27. Let x = number of ounces of

heavy cream (36% milk fat). Let y = number of ounces of half and half (10.5% milk fat).

Combined quantity of both creams is 16 ounces,

$x + y = 16 \quad (1).$

Milk fat in heavy cream plus milk fat in half and half is equal to 20% milk fat,

$0.36x + 0.105y = 0.2(16) \quad (2).$

$360(x + y) = 360(16)$ gives:

$360x + 360y = 5760 \quad (1).$

$-1000(0.36x + 0.105y) = -1000(3.2)$ gives:

$-360x - 105y = -3200 \quad (2).$

$$\begin{aligned} 360x + 360y &= 5760 \\ -360x - 105y &= -3200 \\ \hline 255y &= 2560 \\ y &= \frac{2560}{255} \\ y &= 10.04 \end{aligned}$$

(rounded)

Substitute 10.04 into (1).

$$\begin{aligned} x + y &= 16 \\ 10.04 + y &= 16 \\ y &= 16 - 10.04 \\ y &= 5.96 \end{aligned}$$

Pierre needs approximately 10.04 ounces of half and half and 5.96 ounces of heavy cream.

Cumulative Review Exercises

29. $3(4x - 3)^2 - 2y^2 - 1$

$= 3[4(3)-3]^2 - 2(-2)^2 - 1$

$= 3[12 - 3]^2 - 2(4) - 1$

$= 3[9]^2 - 8 - 1$

$= 3(81) - 9$

$= 243 - 9 = 234$

31. a) 3rd degree polynomial
    b) 4th degree polynomial
    c) Not a polynomial, because one term has a negative exponent.

Just for fun

	Rate	× Time	= Distance
Older brother	9	x	$9x$
Younger brother	5	y	$5y$

$x = y \quad (1)$

$9x = 5y + \frac{1}{2} \quad (2)$

Substitute (1) into (2).

$$9y = 5y + \frac{1}{2}$$

$$4y = \frac{1}{2}$$

$$y = \frac{1}{8} \text{ hour}$$

Distance $= 9x$

$$= 9\left(\frac{1}{8}\right)$$

$$= \frac{9}{8}$$

$= 1.125$ miles

The school is 1.125 miles from the boy's home.

2. Let x = brass, 70% copper, 30% zinc.
Let y = brass, 40% copper, 60% zinc.
What mixture of each brass will yield 300 grams of brass with 60% copper and

40% zinc?

$0.70x + 0.40y$
$= 0.60(300)$ (1)

Total of 2 mixtures:

$x + y = 300$ (2)

$10(0.70x + 0.40y)$
$= 10[0.60(300)]$ (1)
$-4(x + y) = -4(300)$ (2)

$7x + 4y = 1800$
$-4x - 4y = -1200$

$3x = 600$
$x = 200$

Substitute into (2).

$200 + y = 300$
$y = 100$

You will need 200 grams of brass that are 70% copper, 30% zinc, and 100 grams of brass that are 40% copper, 60% zinc.

Exercise Set 8.6

For all inequality graphs: Solve for y. Graph the line y = mx + b. If y > , shade above the line. If y < , shade below the line. The answer is the intersection of the two shaded areas.

1. $x + y > 2$ (1)
$y > -x + 2$ (1)
Graph: $y = -x + 2$

$x - y < 2$ (2)
$-y < -x + 2$ (2)
$-1(-y) < -1(-x + 2)$ (2)
$y > x - 2$ (2)
Graph: $y = x - 2$

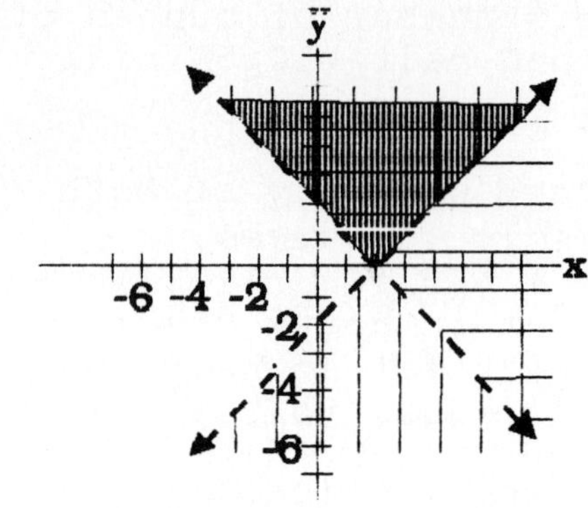

3. $y \leq x$
Graph: $y = x$

$y < -2x + 4$
Graph: $y = -2x + 4$

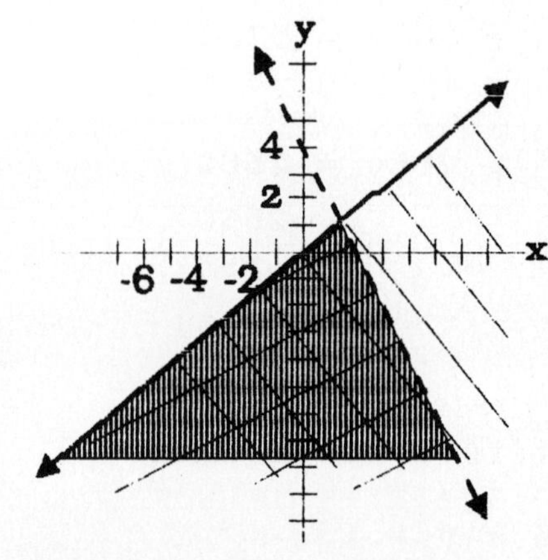

5. $y > x + 1$
Graph: $y = x + 1$

$y \geq 3x + 2$
Graph: $y = 3x + 2$

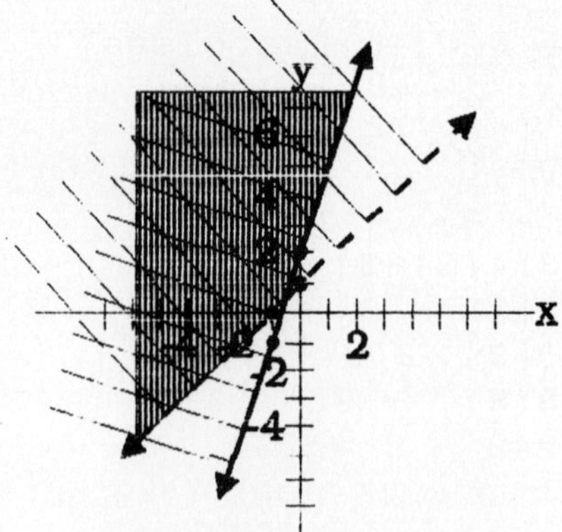

7. $x - 2y < 6$
$-2y < -x + 5$

$y > \frac{1}{2}x - 3$

Graph: $y = \frac{1}{2}x - 3$

$y \le -x + 4$
Graph: $y = -x + 4$

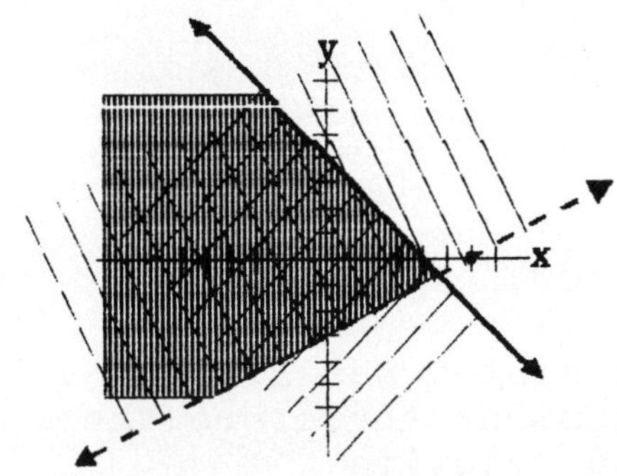

9. $4x + 5y < 20$
$5y < -4x + 20$

$y < \frac{-4}{5}x + 4$

Graph: $y = \frac{-4}{5}x + 4$

$x \ge -3$
Graph: $x = -3$

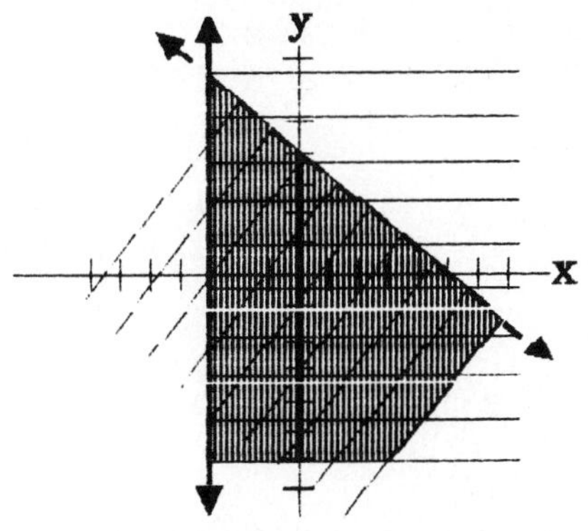

11. $x \le 4$
Graph: $x = 4$

$y \ge -2$

Graph: $y = -2$

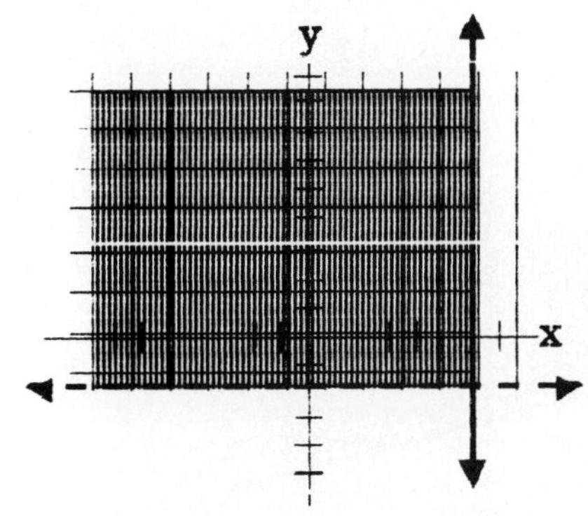

13. $x > -3$
Graph: $x = -3$

$y > 1$
Graph: $y = 1$

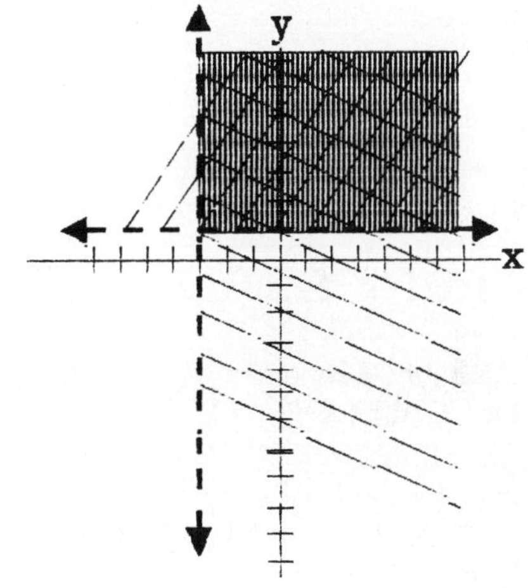

15. $-2x + 3y \ge 6$
$3y \ge 2x + 6$

$y \ge \frac{2}{3}x + 2$

Graph: $y = \frac{2}{3}x + 2$

$x + 4y \ge 4$
$4y \ge -x + 4$

$y \geq \frac{-1}{4} x + 1$

Graph: $y = \frac{-1}{4} x + 1$

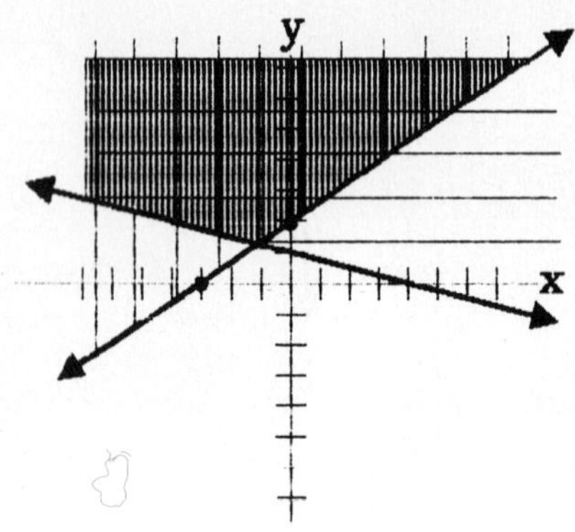

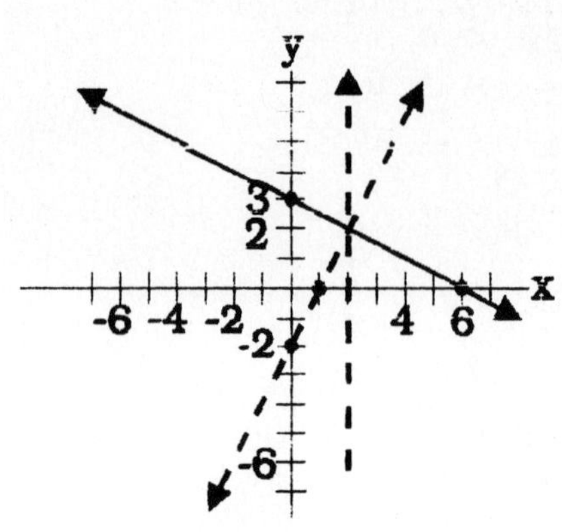

This system has no points in common; therefore there is no solution.

Cumulative Review Exercises

17.
$$\begin{array}{r} x^2 - 2x + 7 \\ 3x - 4 \\ \hline 3x^3 - 6x^2 + 21x \\ - 4x^2 + 8x - 28 \\ \hline 3x^3 - 10x^2 + 29x - 28 \end{array}$$

19. $x^2 - 13x + 42$
$= (x - 6)(x - 7)$

Just for fun

1. $x + 2y \leq 6$
$2y \leq -x + 6$
$$y \leq -\frac{1}{2} x + 3$$

$2x - y < 2$
$-y < -2x + 2$
$y > 2x - 2$

$x > 2$

Review Exercises

1. $y = 3x - 2 \quad 2x + 3y = 5$

a) (0,-2)

$-2 = 3(0)-2 \quad 2(0)+3(-2) = 5$
$-2 = -2 \quad 0+(-6) = 5$
$-6 = 5$

True False

(0,-2) is not a solution.

b) (2,4)

$4 = 3(2)-2 \quad 2(2)+3(4) = 5$
$4 = 6 - 2 \quad 4 + 12 = 5$
$4 = 4 \quad 16 = 5$

True False

(2,4) is not a solution.

c) (1,1)

$1 = 3(1)-2 \quad 2(1)+3(1) = 5$
$1 = 3 - 2 \quad 2 + 3 = 5$
$1 = 1 \quad 5 = 5$

True True

(1,1) is a solution.

3. Consistent--one solution.

5. Dependent, infinite number of solutions.

7. $x + 2y = 8$ $\quad 3x + 6y = 12$

$2y = -x + 8$ $\quad 6y = -3x + 12$

$y = \frac{-1}{2}x + 4$ $\quad y = \frac{-1}{2}x + 2$

No solution.

9. $y = \frac{1}{2}x + 4$ $\quad x - 2y = 8$

$-2y = -x + 8$

$y = \frac{1}{2}x - 4$

Infinite number of solutions.

11. $y = x + 3$ $\quad y = 2x + 5$

Solution: $(-2,1)$

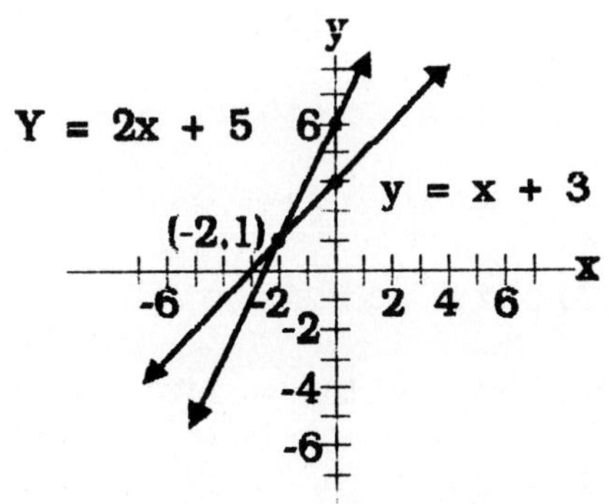

13. $y = 3$ $\quad y = -2x + 5$

Solution: $(1,3)$

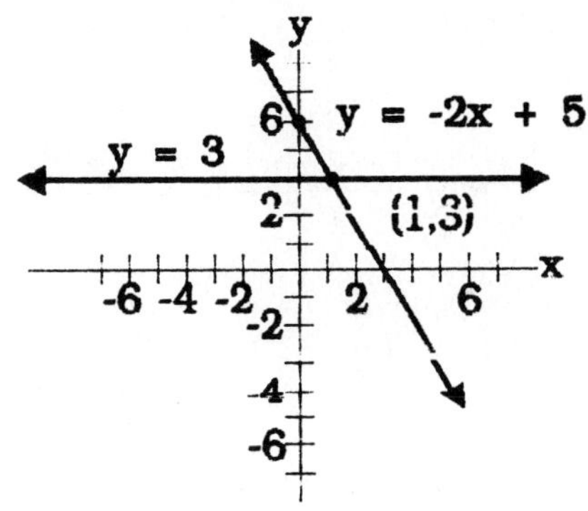

15. $x + 2y = 8$ $\quad 2x - y = -4$

$2y = -x + 8$ $\quad -y = -2x - 4$

$y = \frac{-1}{2}x + 4$ $\quad y = 2x + 4$

Solution: $(0,4)$

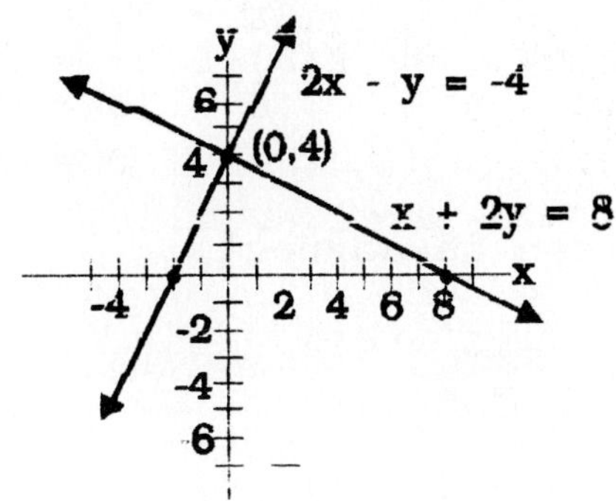

17. $2x + y = 0$ $\quad 4x - 3y = 10$

$y = -2x$ $\quad -3y = -4x + 10$

$y = \frac{4}{3}x - \frac{10}{3}$

Solution: $(1,-2)$

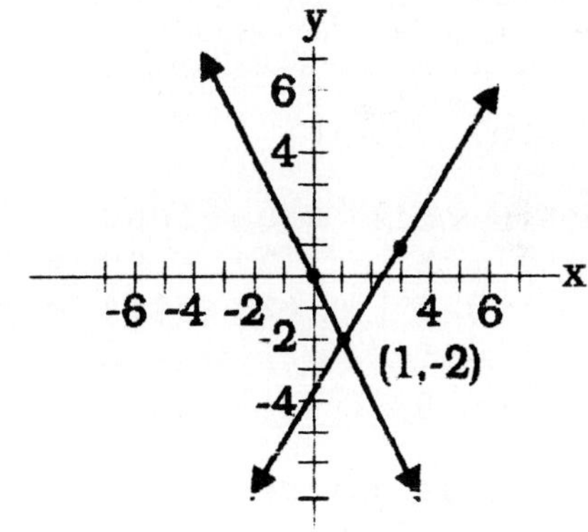

19. $y = 2x - 8$ $\quad (1)$

$2x - 5y = 0$ $\quad (2)$

Substitute (1) into (2).

$2x - 5(2x - 8) = 0$

$2x - 10x + 40 = 0$

$-8x + 40 = 0$

$-8x = -40$

$x = 5$

Substitute into (1).
$y = 2(5) - 8$
$y = 10 - 8$
$y = 2$
Solution: (5,2)

Check in both equations.

$y = 2x - 8$	$2x - 5y = 0$
$2 = 2(5) - 8$	$2(5)-5(2) = 0$
$2 = 10 - 8$	$10 - 10 = 0$
$2 = 2$	$0 = 0$

Solution is correct.

21. $2x + y = 5$ (1)
$3x + 2y = 8$ (2)

$y = -2x + 5$ (1)

Substitute (1) into (2).
$3x + 2(-2x + 5) = 8$
$3x - 4x + 10 = 8$
$-x + 10 = 8$
$-x = -2$
$x = 2$

Substitute into (1).
$2(2) + y = 5$
$4 + y = 5$
$y = 1$

Solution: (2,1)

Check in both equations.

$2x + y = 5$	$3x + 2y = 8$
$2(2) + 1 = 5$	$3(2)+2(1) = 8$
$4 + 1 = 5$	$6 + 2 = 8$
$5 = 5$	$8 = 8$

Solution is correct.

23. $3x + y = 17$ (1)
$2x - 3y = 4$ (2)

$y = -3x + 17$ (1)

Substitute (1) into (2).
$2x - 3(-3x + 17) = 4$
$2x + 9x - 51 = 4$
$11x - 51 = 4$
$11x = 55$
$x = 5$

Substitute into (1).
$3(5) + y = 17$
$15 + y = 17$
$y = 2$

Solution: (5,2)

Check in both equations.

$3x + y = 17$	$2x - 3y = 4$
$3(5) + 2 = 17$	$2(5) - 3(2) = 4$
$15 + 2 = 17$	$10 - 6 = 4$
$17 = 17$	$4 = 4$

Solution is correct.

25. $4x - 2y = 10$ (1)
$y = 2x + 3$ (2)

Substitute (2) into (1).
$4x - 2(2x + 3) = 10$
$4x - 4x - 6 = 10$
$-6 = 10$

Inconsistent--no solution.

27. $2x - 3y = 8$ (1)
$6x + 5y = 10$ (2)

$2x = 3y + 8$ (1)

$x = \frac{3}{2}y + 4$ (1)

Substitute (1) into (2).

$6(\frac{3}{2}y + 4) + 5y = 10$

$9y + 24 + 5y = 10$
$24 + 14y = 10$
$14y = -14$
$y = -1$

Substitute into (1).
$2x - 3(-1) = 8$
$2x + 3 = 8$
$2x = 5$
$x = \frac{5}{2}$

Solution: $(\frac{5}{2},-1)$

Check in both equations.

$2x - 3y = 8$	$6x + 5y = 10$
$2(\frac{5}{2})-3(-1) = 8$	$6(\frac{5}{2})+5(-1) = 10$
$5 + 3 = 8$	$15 - 5 = 10$
$8 = 8$	$10 = 10$

Solution is correct.

29. $x + y = 6$ (1)
$x - y = 10$ (2)

$2x = 16$
$x = 8$

Substitute into (1).

$8 + y = 6$
$y = -2$

Solution: $(8,-2)$

Check in both equations.

$x + y = 6$	$x - y = 10$
$8 + (-2) = 6$	$8 - (-2) = 10$
$6 = 6$	$10 = 10$

Solution is correct.

31. $2x + 3y = 4$ (1)
$x + 2y = -6$ (2)

$-2(x + 2y) = -2(-6)$ (2)
$-2x - 4y = 12$ (2)

$2x + 3y = 4$ (1)
$-2x - 4y = 12$ (2)

$-y = 16$
$y = -16$

Substitute into (1).
$2x + 3(-16) = 4$
$2x - 48 = 4$
$2x = 52$
$x = 26$

Solution: $(26,-16)$

Check in both equations.

$2x + 3y = 4$	$x + 2y = -6$
$2(26)+3(-16) = 4$	$26+2(-16) = -6$
$52 - 48 = 4$	$26 - 32 = -6$
$4 = 4$	$-6 = -6$

Solution is correct.

33. $4x - 3y = 8$ (1)
$2x + 5y = 8$ (2)

$-2(2x + 5y) = -2(8)$ (2)
$-4x - 10y = -16$ (2)

$4x - 3y = 8$ (1)
$-4x - 10y = -16$ (2)

$-13y = -8$

$y = \frac{8}{13}$

Substitute into (1).

$4x - 3(\frac{8}{13}) = 8$

$4x - \frac{24}{13} = \frac{104}{13}$

$4x = \frac{128}{13}$

$\frac{1}{4}(4x) = \frac{1}{4}(\frac{128}{13})$

$x = \frac{32}{13}$

Solution: $(\frac{32}{13},\frac{8}{13})$

Check in both equations.

$4x - 3y = 8$	$2x + 5y = 8$
$4(\frac{32}{13})-3(\frac{8}{13})=\frac{104}{13}$	$2(\frac{32}{13})+5(\frac{5}{13})=\frac{104}{13}$
$\frac{128}{13} - \frac{24}{13}=\frac{104}{13}$	$\frac{64}{13} + \frac{40}{13}=\frac{104}{13}$
$\frac{104}{13}=\frac{104}{13}$	$\frac{104}{13}=\frac{104}{13}$

Solution is correct.

35. $2x + y = 9$ (1)
$-4x - 2y = 4$ (2)

$2(2x + y) = 2(9)$ (1)
$4x + 2y = 18$ (1)

$4x + 2y = 18$ (1)
$-4x - 2y = 4$ (2)

$0 = 22$

Inconsistent--no solution.

37. $3x + 4y = 10$ (1)
$-6x - 8y = -20$ (2)

$2(3x + 4y) = 2(10)$ (1)
$6x + 8y = 20$ (1)

$6x + 8y = 20$ (1)
$-6x - 8y = -20$ (2)

$0 = 0$

Dependent--infinite solutions.

39. x = the larger integer.
y = the smaller integer.

$x + y = 48$ (1)
$x = 2y - 3$ (2)

Substitute (2) into (1).
$(2y - 3) + y = 48$
$3y - 3 = 48$
$3y = 51$
$y = 17$

Substitute into (2).
$x = 2(17) - 3$
$x = 34 - 3$
$x = 31$

The two numbers are 17 and 31.

41. Let x = number of pounds of Green Turf's grass seed.
y = number of pounds of Agway's grass seed.

$x + y = 40$
$0.60x + 0.45y = 20.25$

Solve for x:
$x + y = 40$
$x = 40 - y$

Substitute:
$0.60x + 0.45y = 20.25$
$0.60(40 - y) + 0.45y = 20.25$
$24 - 0.60y + 0.45y = 20.25$
$24 - 0.15y = 20.25$
$-0.15y = -3.75$
$$\frac{-0.15y}{-0.15} = \frac{-3.75}{-0.15}$$
$y = 25$
$x = 40 - y = 40 - 25 = 15$

15 pounds of Green Turf and 25 pounds of Agway were used.

43. x = speed in still air.
y = speed of wind.

$x + y = 600$ (1)
$x - y = 530$ (2)

$2x = 1130$
$x = 565$

Substitute into (1).
$565 + y = 600$
$y = 35$

Speed of the plane in still air is 565 mph. The speed of the wind is 35 mph.

45. $2x - 3y \le 6$ (1)
$x + 4y > 4$ (2)

$-3y \le -2x + 6$ (1)

$y \ge \frac{2}{3}x - 2$ (1)

$4y \ge -x + 4$ (2)

$y \geq \frac{-1}{4}x + 1 \qquad (2)$

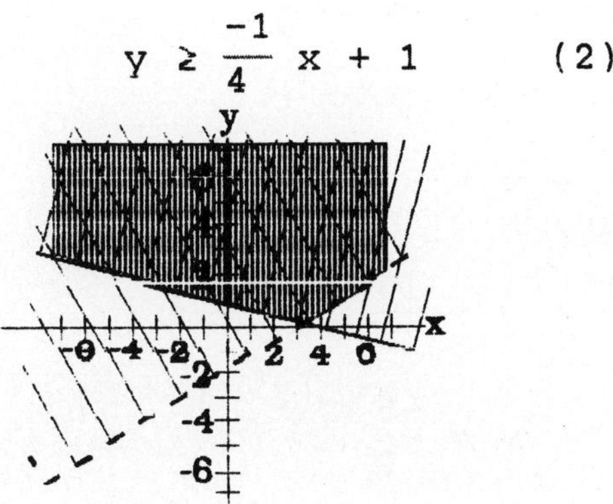

47. $x < 2 \qquad (1)$
    $y \geq -1 \qquad (2)$

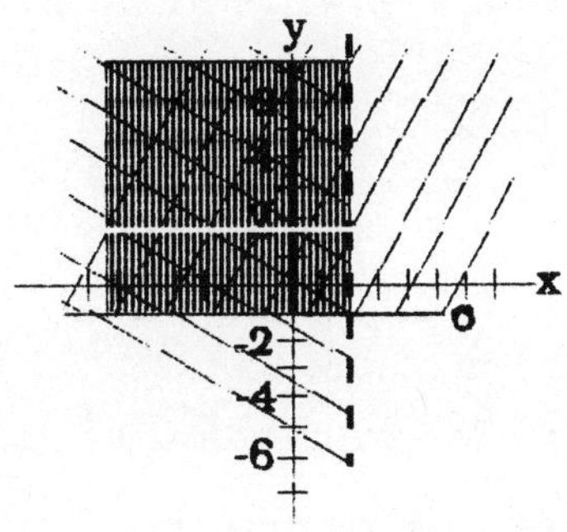

Practice Test

1. $x + 2y = -6 \qquad 3x + 2y = -12$

a) $(0,-6)$

$0+2(-6) = -6 \qquad 3(0)+2(-6) = -12$
$-12 = -6 \qquad -12 = -12$
False True
$(0,-6)$ is not a solution.

b) $(-3,\frac{-3}{2})$

$-3+2(\frac{-3}{2}) = -6 \qquad 3(-3)+2(\frac{-3}{2}) = -12$

$-3 + (-3) = -6 \qquad -9 + (-3) = -12$
$-6 = -6 \qquad -12 = -12$
True True

$(-3,\frac{-3}{2})$ is a solution.

c) $(2,-4)$

$(2)+2(-4) = -6 \qquad 3(2)+2(-4) = -12$
$2 - 8 = -6 \qquad 6 - 8 = -12$
$-6 = -6 \qquad -2 = -12$
True False

$(2,-4)$ is not a solution.

2. Inconsistent, no solution.

3. Consistent, one solution.

4. Dependent, infinite number of solutions.

5. $3y = 6x - 9 \qquad 2x - y = 6$

$y = \frac{6}{3}x - \frac{9}{3} \qquad y = 2x-6$

$y = 2x - 3 \qquad y = 2x-6$
No solution.

6. $3x + 2y = 10 \qquad 3x - 2y = 10$
$2y = -3x + 10 \qquad -2y = -3x + 10$

$y = \frac{-3}{2}x + \frac{10}{2} \qquad y = \frac{-3}{-2}x + \frac{10}{-2}$

$y = \frac{-3}{2}x + 5 \qquad y = \frac{3}{2}x - 5$

One solution.

7. $y = 3x - 2$
$y = -2x + 8$

The solution is $(2,4)$.

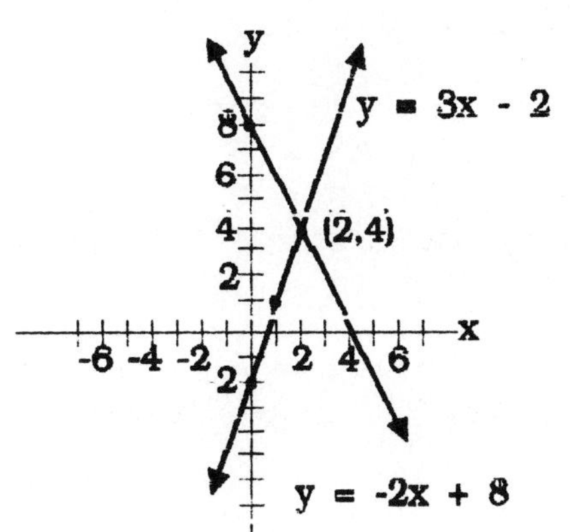

8. $3x - 2y = -3$
$3x + y = 6$

Solution:
$3x - 2y = -3$ Ordered Pair
Let $x = 0$,

then $y = \frac{3}{2}$ $(0,\frac{3}{2})$

Let $y = 0$,

then $x = -1$ $(-1,0)$

Solution:
$3x + y = 6$ Ordered Pair
Let $x = 0$,
then $y = 6$ $(0,6)$
Let $y = 0$,
then $x = 2$ $(2,0)$

The solution is (1,3).

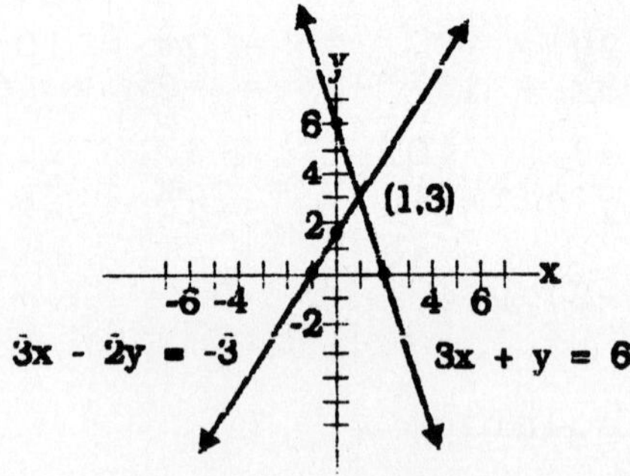

9. $3x + y = 8$ (1)
$x - y = 6$ (2)

$y = -3x + 8$ (1)

Substitute (1) into (2).

$$x - (-3x + 8) = 6$$
$$x + 3x - 8 = 6$$
$$4x - 8 = 6$$
$$4x = 14$$
$$x = \frac{14}{4}$$
$$x = \frac{7}{2}$$

Substitute into (1).

$$3(\frac{7}{2}) + y = 8$$
$$\frac{21}{2} + y = \frac{16}{2}$$
$$y = \frac{-5}{2}$$

Solution: $(\frac{7}{2},\frac{-5}{2})$

Check in both equations.

$3(\frac{7}{2}) + (\frac{-5}{2}) = 8$	$\frac{7}{2} - (\frac{-5}{2}) = 6$
$\frac{21}{2} - \frac{5}{2} = \frac{16}{2}$	$\frac{7}{2} + \frac{5}{2} = 6$
$\frac{16}{2} = \frac{16}{2}$	$\frac{12}{2} = 6$
$8 = 8$	$6 = 6$

The solution is correct.

10. $4x - 3y = 9$ (1)
$2x + 4y = 10$ (2)

$2x = -4y + 10$ (2)

$x = \frac{-4}{2} y + \frac{10}{2}$ (2)

$x = -2y + 5$ (2)

Substitute (2) into (1).

$$4(-2y + 5) - 3y = 9$$
$$-8y + 20 - 3y = 9$$
$$-11y + 20 = 9$$
$$-11y = -11$$
$$y = 1$$

Substitute into (2).

$$2x + 4(1) = 10$$
$$2x + 4 = 10$$
$$2x = 6$$

$x = 3$

The solution is (3,1).

Check in both equations.

$$\begin{array}{r|r} 4(3)-3(1) = 9 & 2(3)+4(1) = 10 \\ 12 - 3 = 9 & 6 + 4 = 10 \\ 9 = 9 & 10 = 10 \end{array}$$

Solution is correct.

11. $2x + y = 5 \quad (1)$
$x + 3y = -10 \quad (2)$

$-2(x + 3y) = -2(-10) \quad (2)$
$-2x - 6y = 20 \quad (2)$

$$\begin{array}{rl} 2x + y = 5 & (1) \\ -2x - 6y = 20 & (2) \\ \hline -5y = 25 & \\ y = -5 & \end{array}$$

Substitute into (1).

$2x + (-5) = 5$
$2x - 5 = 5$
$2x = 10$
$x = 5$

Solution: (5,-5)

Check in both equations.

$$\begin{array}{r|r} 2(5)+(-5) = 5 & 5+3(-5) = -10 \\ 10 - 5 = 5 & 5 - 15 = -10 \\ 5 = 5 & -10 = -10 \end{array}$$

Solution is correct.

12. $3x + 2y = 12 \quad (1)$
$-2x + 5y = 8 \quad (2)$

$2(3x + 2y) = 2(12) \quad (1)$
$6x + 4y = 24 \quad (1)$

$3(-2x + 5y) = 3(8) \quad (2)$
$-6x + 15y = 24 \quad (2)$

$$\begin{array}{rl} 6x + 4y = 24 & (1) \\ -6x + 15y = 24 & (2) \\ \hline 19y = 48 & \end{array}$$

$y = \frac{48}{19}$

Substitute into (1).

$3x + 2[\frac{48}{19}] = 12$

$3x + \frac{96}{19} = \frac{228}{19}$

$3x = \frac{132}{19}$

$x = \frac{44}{19}$

Solution: $[\frac{44}{19}, \frac{48}{19}]$

Check in both equations.

$3[\frac{44}{19}] + 2[\frac{48}{19}] = 12$

$\frac{132}{19} + \frac{96}{19} = 12$

$\frac{228}{19} = 12$

$12 = 12$

$-2[\frac{44}{19}] + 5[\frac{48}{19}] = 8$

$\frac{-88}{19} + \frac{240}{19} = 8$

$\frac{152}{19} = 8$

$8 = 8$

Solution is correct.

13. m = miles driven.

Budget Car: $B = \$40 + 0.08m$
Hertz: $H = \$45 + 0.03m$

The cost of driving the

Budget Car will equal the cost of driving the Hertz car when,

$$40 + 0.08m = 45 + 0.03m$$
$$100(40+0.08m) = 100(45+0.03m)$$
$$4000 + 8m = 4500 + 3m$$
$$4000 + 5m = 4500$$
$$5m = 500$$
$$m = 100$$

The cost of driving the Budget Car will equal the cost of driving the Hertz car when 100 miles have been driven.

14. x = pounds of cashews at \$6.00 per pound.
y = pounds of peanuts at \$4.50 per pound.
20 = pounds of mix at \$5.00 per pound.

Combined weight of cashews and peanuts,
$x + y = 20$ (1).
Total cost of cashews and peanuts,
$6x + 4.5y = 5(20)$ (2).

Solve (1) for x:
$x + y = 20$ (1)
$x = -y + 20$ (1)

Substitute (1) into (2).

$$6x + 4.5y = 100$$
$$6(-y + 20) + 4.5y = 100$$
$$-6y + 120 + 4.5y = 100$$
$$-1.5y + 120 = 100$$
$$-1.5y = -20$$
$$y = \frac{20}{1.5}$$
$$y = 13\ \frac{1}{3}$$

Substitute into (1).

$$x + \frac{20}{1.5} = 20$$
$$x = \frac{20(1.5)}{1.5} - \frac{20}{1.5}$$
$$x = \frac{30}{1.5} - \frac{20}{1.5}$$
$$x = \frac{10}{1.5}$$
$$x = 6\ \frac{2}{3}$$

The mixture contains 13 1/3 pounds of peanuts and 6 2/3 pounds of cashews.

15. $2x + 4y < 8$ (1)
$x - 3y \geq 6$ (2)

$4y < -2x + 8$ (1)

$y < \frac{-1}{2}x + 2$ (1)

$-3y \geq -x + 6$ (2)

$y \leq \frac{1}{3}x - 2$ (2)

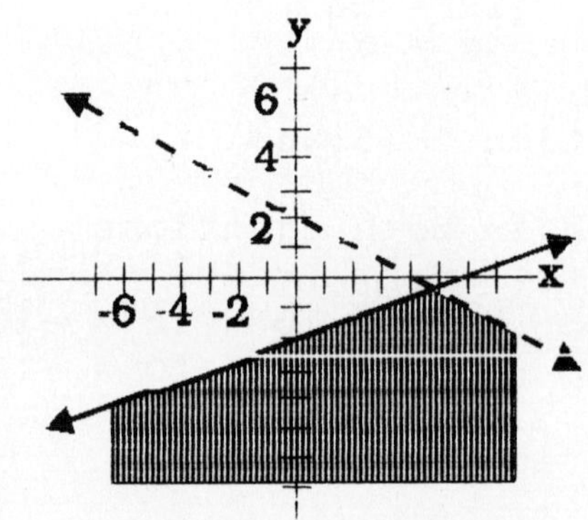

Cumulative Review Test

1. $\dfrac{|-4| + |-16| \div 2^2}{3 - [2 - (4 \div 2)]}$

$$= \frac{4 + 16 \div 4}{3 - [2 - (2)]}$$

$$= \frac{4 + 2}{3 - 0} = \frac{8}{3}$$

2. $4(x - 2) + 6(x - 3)$

$= 2 - 4x$

$4x - 8 + 6x - 18 = 2 - 4x$

$10x - 26 = 2 - 4x$

$14x - 26 = 2$

$14x = 28$

$x = 2$

3. $3x^2 - 13x + 12 = 0$

$(3x - 4)(x - 3) = 0$

$3x - 4 = 0 \qquad x - 3 = 0$

$3x = 4 \qquad x = 3$

$x = \frac{4}{3}$

4. $\frac{1}{3}(x + 2) + \frac{1}{4} = 8$

$\frac{1}{3}x + \frac{2}{3} + \frac{1}{4} = 8$

$12(\frac{1}{3}x + \frac{2}{3} + \frac{1}{4}) = 12(8)$

$4x + 8 + 3 = 96$

$4x + 11 = 96$

$4x = 85$

$x = \frac{85}{4}$

5. $\frac{1}{x - 3} + \frac{1}{x + 3} = \frac{1}{x^2 - 9}$

$(x-3)(x+3)[\frac{1}{x-3} + \frac{1}{x+3}]$

$= [\frac{1}{(x-3)(x+3)}]\cdot(x+3)(x-3)$

$x + 3 + x - 3 = 1$

$2x = 1$

$x = \frac{1}{2}$

6. $\frac{6}{10} = \frac{4}{x}$

$6x = 4(10)$

$6x = 40$

$x = \frac{40}{6} = \frac{20}{3} = 6\frac{2}{3}$ inches

7. $(x^5y^3)^4(2x^3y^5)$

$= (x^{20}y^{12})(2x^3y^5) = (2x^{23}y^{17})$

8. $6x^2 - 11x + 4$

$= (3x - 4)(2x - 1)$

9. $\frac{4}{x^2 - 9} - \frac{3}{x^2 - 9x + 18}$

$= \frac{4}{(x - 3)(x + 3)}$

$- \frac{3}{(x - 3)(x - 6)}$

$= \frac{4(x - 6)}{(x - 3)(x + 3)(x - 6)}$

$- \frac{3(x + 3)}{(x - 3)(x + 3)(x - 6)}$

$= \frac{4x - 24 - 3x - 9}{(x - 3)(x + 3)(x - 6)}$

$= \frac{x - 33}{(x - 3)(x + 3)(x - 6)}$

10. $\frac{x^2 - 7x + 12}{2x^2 - 11x + 12} \div \frac{x^2 - 9}{x - 16}$

$= \frac{x^2 - 7x + 12}{2x^2 - 11x + 12} \cdot \frac{x^2 - 16}{x^2 - 9}$

$= \frac{(x - 3)(x - 4)}{(2x - 3)(x - 4)}$

$\cdot \frac{(x - 4)(x + 4)}{(x - 3)(x + 3)}$

$= \frac{(x - 4)(x + 4)}{(2x - 3)(x + 3)}$

11. $2x - 3y = 6$

x	y
0	-2
3	0
6	2

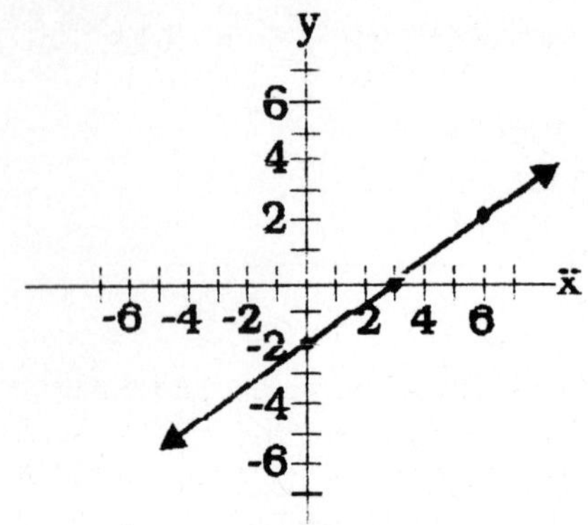

12. $3x + 2y = 9$

x	y
0	$\frac{9}{2}$
3	0

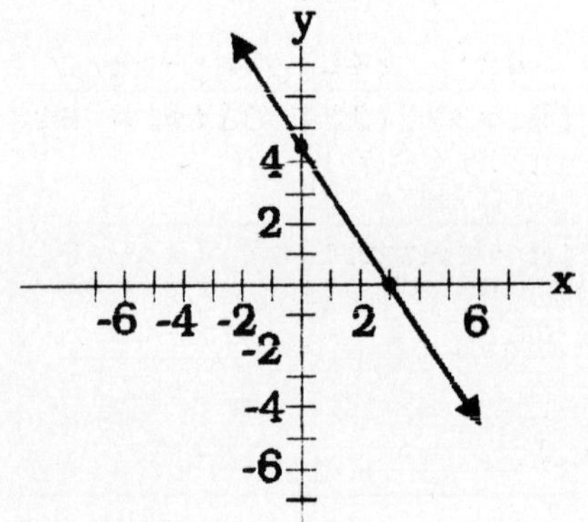

13. $2x - y < 6$
$-y < -2x + 6$
$y > 2x - 6$

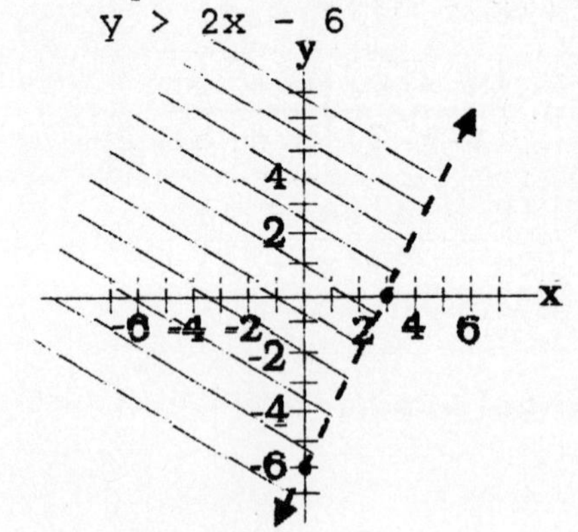

14. $3x = 2y + 8$
$3x - 8 = 2y$
$\frac{3}{2}x - 4 = y$
$-4y = -6x + 12$
$y = \frac{3}{2}x - 3$

The slopes of the lines are the same; therefore the lines are parallel. There is no solution.

15. $x + 2y = 2 \quad 2x - 3y = -3$

x	y
0	1
2	0
4	-1

x	y
0	1
$-\frac{3}{2}$	0
-3	-1

Solution: (0,1)

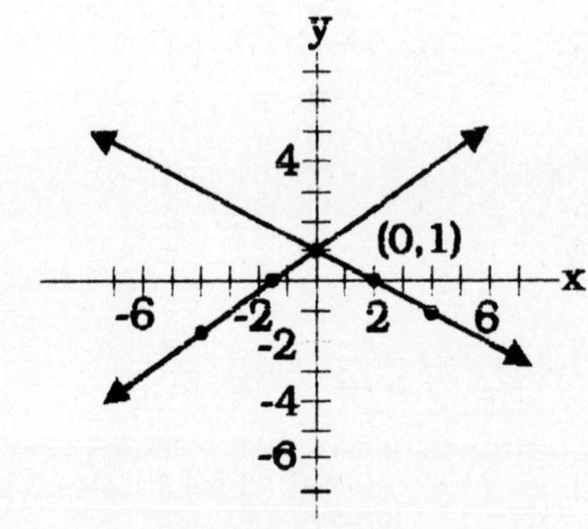

16. $2x + 3y = 4 \quad (1)$
$x - 4y = 6 \quad (2)$

$-2(x - 4y) = -2(6) \quad (2)$
$-2x + 8y = -12 \quad (2)$

$$\begin{array}{rl} 2x + 3y = 4 & (1) \\ -2x + 8y = -12 & (2) \\ \hline 11y = -8 & \end{array}$$

$$y = -\frac{8}{11}$$

Substitute:

$$2x + 3y = 4$$

$$2x + 3(-\frac{8}{11}) = 4$$

$$2x - \frac{24}{11} = 4$$

$$11(2x - \frac{24}{11}) = 11(4)$$

$$22x - 24 = 44$$

$$22x = 68$$

$$x = \frac{68}{22} = \frac{34}{11}$$

Solution: $(\frac{34}{11}, -\frac{8}{11})$

17. $\frac{40 \text{ units}}{15 \text{ minutes}} = \frac{160 \text{ units}}{x \text{ minutes}}$

$40x = 2400$

$x = 60$ minutes

18. a) Let x = amount of sales.

$$20{,}000 + 0.10x = 10{,}000 + 0.12x$$

$$20{,}000 - 0.02x = 10{,}000$$

$$-0.02x = -10{,}000$$

$$\frac{-0.02x}{-0.02} = \frac{-10{,}000}{-0.02}$$

$$x = \$500{,}000$$

b) $20{,}000 + 0.10x$

$= 20{,}000 + 0.10\ (200{,}000)$

$= 20{,}000 + 20{,}000$

PCR: $= \$40{,}000$

$10{,}000 + 0.12x$

$= 10{,}000 + 0.12\ (200{,}000)$

$= 10{,}000 + 24{,}000$

ARA: $= \$34{,}000$

Higher income: PCR

Publishing Company

19. Let x = number of liters of 20% hydrochloric acid.
y = number of liters of 35% hydrochloric acid.

$$x + y = 10$$

$$0.20x + 0.25y = 0.25(10)$$

Solve for x:

$x + y = 10$

$x = 10 - y$

Substitute:

$$0.20x + 0.35y = 0.25(10)$$

$$0.20(10 - y) + 0.35y = 2.5$$

$$2 - 0.20y + 0.35y = 2.5$$

$$2 + 0.15y = 2.5$$

$$0.15y = 0.5$$

$$\frac{0.15y}{0.15} = \frac{0.5}{0.15}$$

$$y = 3\frac{1}{3}$$

$$x = 10 - y = 10 - 3\frac{1}{3} = 6\frac{2}{3}$$

Mix 6 2/3 liters of 20% acid with 3 1/3 liters of 35% acid.

20. Let x = amount of time to clean the pizza shop together.

$$\frac{x}{50} + \frac{x}{60} = 1$$

$$300[\frac{x}{50} + \frac{x}{60}] = 1(300)$$

$$6x + 5x = 300$$

$$11x = 300$$

$x = \frac{300}{11} = 27\frac{3}{11}$ min. or

or 27.3 min.

CHAPTER NINE

Exercise Set 9.1

1. $\sqrt{1} = 1$

3. $\sqrt{0} = 0$

5. $-\sqrt{81} = -9$

7. $\sqrt{121} = 11$

9. $-\sqrt{16} = -4$

11. $\sqrt{144} = 12$

13. $\sqrt{169} = 13$

15. $-\sqrt{1} = -1$

17. $\sqrt{81} = 9$

19. $-\sqrt{121} = -11$

21. $\sqrt{\frac{1}{4}} = \frac{\sqrt{1}}{\sqrt{4}} = \frac{1}{2}$

23. $\sqrt{\frac{9}{16}} = \frac{\sqrt{9}}{\sqrt{16}} = \frac{3}{4}$

25. $-\sqrt{\frac{4}{25}} = \frac{-\sqrt{4}}{\sqrt{25}} = \frac{-2}{5}$

27. $\sqrt{8} = 2.828$

29. $\sqrt{15} = 3.873$

31. $\sqrt{80} = 8.944$

33. $\sqrt{81} = 9.000$

35. $\sqrt{97} = 9.849$

37. $\sqrt{3} = 1.732$

39. True

41. True

43. False

45. True

47. False

49. True

51. $\sqrt{7} = (7)^{1/2}$

53. $\sqrt{17} = (17)^{1/2}$

55. $\sqrt{5x} = (5x)^{1/2}$

57. $\sqrt{12x^2} = (12x^2)^{1/2}$

59. $\sqrt{19xy^2} = (19xy^2)^{1/2}$

61. $\sqrt{40x^3} = (40x^3)^{1/2}$

63. When we see a variable in a square-root, we assume the variable is nonnegative because the square root of a negative number is not a real number.

65. The square root of a negative number is not a real number, because no real number when squared will be a negative number.

Cumulative Review Exercises

67. $\frac{4x}{x^2 + 6x + 9} - \frac{2x}{x + 3}$

$= \frac{x + 1}{x + 3}$

$(x + 3)^2[\frac{4x}{(x + 3)^2} - \frac{2x}{x + 3}]$

$$= \left[\frac{x + 1}{x + 3}\right](x + 3)^2$$

$$4x - 2x(x + 3) = (x + 1)(x + 3)$$

$$4x - 2x^2 - 6x = x^2 + 4x + 3$$

$$-2x^2 - 2x = x^2 + 4x + 3$$

$$0 = 3x^2 + 6x + 3$$

$$0 = 3(x^2 + 2x + 1)$$
$$0 = 3(x + 1)(x + 1)$$
$$x + 1 = 0$$
$$x = -1$$

69. $(x_1, y_1) = (-5, 3)$
$(x_2, y_2) = (6, 7)$

$$m = \frac{y_2 - y_1}{x_2 - x_1}$$

$$m = \frac{7 - 3}{6 - (-5)} = \frac{4}{6 + 5} = \frac{4}{11}$$

Slope is $\frac{4}{11}$

Exercise Set 9.2

1. $\sqrt{16} = 4$

3. $\sqrt{8} = \sqrt{4} \cdot \sqrt{2} = 2\sqrt{2}$

5. $\sqrt{96} = \sqrt{16} \cdot \sqrt{6} = 4\sqrt{6}$

7. $\sqrt{32} = \sqrt{16} \cdot \sqrt{2} = 4\sqrt{2}$

9. $\sqrt{160} = \sqrt{16} \cdot \sqrt{10} = 4\sqrt{10}$

11. $\sqrt{48} = \sqrt{16} \cdot \sqrt{3} = 4\sqrt{3}$

13. $\sqrt{108} = \sqrt{36} \cdot \sqrt{3} = 6\sqrt{3}$

15. $\sqrt{156} = \sqrt{4}\ \sqrt{39} = 2\sqrt{39}$

17. $\sqrt{256} = \sqrt{16}\ \sqrt{16} = 4 \cdot 4 = 16$

19. $\sqrt{900} = \sqrt{30} \cdot \sqrt{30} = 30$

21. $\sqrt{y^6} = \sqrt{y^2} \cdot \sqrt{y^2} \cdot \sqrt{y^2}$
$= y \cdot y \cdot y = y^3$

23. $\sqrt{x^2y^4} = \sqrt{x^2} \cdot \sqrt{y^4} = xy^2$

25. $\sqrt{x^9y^{12}} = \sqrt{x^8 \cdot x \cdot y^{12}}$
$= \sqrt{x^8} \cdot \sqrt{x} \cdot \sqrt{y^{12}}$
$= x^4y^6\sqrt{x}$

27. $\sqrt{a^2b^4c} = \sqrt{a^2}\ \sqrt{b^4}\ \sqrt{c} = ab^2\sqrt{c}$

29. $\sqrt{3x^3} = \sqrt{3}\ \sqrt{x^2}\ \sqrt{x} = x\sqrt{3}\ \sqrt{x}$
$= x\sqrt{3x}$

31. $\sqrt{50x^2y^3}$
$= \sqrt{25}\ \sqrt{2} \cdot \sqrt{x^2} \cdot \sqrt{y^2} \cdot \sqrt{y}$
$= 5\sqrt{2} \cdot x \cdot y \cdot \sqrt{y}$
$= 5xy\ \sqrt{2}\ \sqrt{y}$
$= 5xy\ \sqrt{2y}$

33. $\sqrt{200y^5z^{12}}$
$= \sqrt{100} \cdot \sqrt{2} \cdot \sqrt{y^4}$
$\cdot \sqrt{y} \cdot \sqrt{z^{12}}$
$= 10\sqrt{2} \cdot y^2 \cdot \sqrt{y} \cdot z^6$
$= 10y^2z^6 \cdot \sqrt{2} \cdot \sqrt{y}$
$= 10y^2z^6 \cdot \sqrt{2y}$

35. $\sqrt{243q^2b^3c}$
$= \sqrt{81}\ \sqrt{3}\ \sqrt{q^2}\ \sqrt{b^2}\ \sqrt{b}\ \sqrt{c}$
$= 9\sqrt{3}\ qb\ \sqrt{b}\ \sqrt{c}$
$= 9qb\ \sqrt{3bc}$

37. $\sqrt{128x^3yz^5}$
$= \sqrt{64 \cdot 2 \cdot x^2 \cdot x \cdot y \cdot z^4 \cdot z}$

$= \sqrt{64} \cdot \sqrt{x^2} \cdot \sqrt{z^4} \cdot \sqrt{2xyz}$

$= 8xz^2\sqrt{2xyz}$

39. $\sqrt{250x^4yz} = \sqrt{25}\ \sqrt{10}\ \sqrt{x^4}\ \sqrt{yz}$

$= 5\sqrt{10} \cdot x^2 \cdot \sqrt{yz}$

$= 5x^2\ \sqrt{10yz}$

41. $\sqrt{8} \cdot \sqrt{3} = \sqrt{24} = \sqrt{4}\ \sqrt{6}$

$= 2\sqrt{6}$

43. $\sqrt{18} \cdot \sqrt{3} = \sqrt{18 \cdot 3} = \sqrt{54}$

$= \sqrt{9} \cdot \sqrt{6} = 3\sqrt{6}$

45. $\sqrt{75}\ \sqrt{6} = \sqrt{450} = \sqrt{225} \cdot \sqrt{2}$

$= 15\sqrt{2}$

47. $\sqrt{3x} \cdot \sqrt{5x} = \sqrt{15} \cdot \sqrt{x^2}$

$= \sqrt{15} \cdot x$

$= x\sqrt{15}$

49. $\sqrt{5x^2}\ \sqrt{8x^3} = \sqrt{40x^5}$

$= \sqrt{4}\ \sqrt{10}\ \sqrt{x^4}\ \sqrt{x}$

$= 2\sqrt{10}\ x^2\ \sqrt{x}$

$= 2x^2\ \sqrt{10x}$

51. $\sqrt{12x^2y}\ \sqrt{6xy^3} = \sqrt{72x^3y^4}$

$= \sqrt{36}\ \sqrt{2}\ \sqrt{x^2}\ \sqrt{x}\ \sqrt{y^4}$

$= 6\sqrt{2}\ x\sqrt{x}\ y^2$

$= 6xy^2\ \sqrt{2x}$

53. $\sqrt{18xy^4}\ \sqrt{3x^2y} = \sqrt{54x^3y^5}$

$= \sqrt{9}\ \sqrt{6}\ \sqrt{x^2}\ \sqrt{x}\ \sqrt{y^4}\ \sqrt{y}$

$= 3\sqrt{6}\ x\sqrt{x} \cdot y^2\sqrt{y}$

$= 3xy^2\ \sqrt{6xy}$

55. $\sqrt{15xy^6}\ \sqrt{6xyz} = \sqrt{90x^2y^7z}$

$= \sqrt{9}\ \sqrt{10}\ \sqrt{x^2}\ \sqrt{y^6}\ \sqrt{y}\ \sqrt{z}$

$= 3\sqrt{10} \cdot xy^3\ \sqrt{yz}$

$= 3xy^3\ \sqrt{10yz}$

57. $\sqrt{9x^4y^6} \cdot \sqrt{4x^2y^4} = \sqrt{36x^6y^{10}}$

$= \sqrt{36}\ \sqrt{x^6}\ \sqrt{y^{10}}$

$= 6x^3y^5$

59. $(\sqrt{4x})^2 = (\sqrt{4x})(\sqrt{4x})$

$= \sqrt{16x^2}$

$= \sqrt{16}\ \sqrt{x^2}$

$= 4x$

61. $(\sqrt{13x^4y^6})^2$

$= (\sqrt{13x^4y^6})(\sqrt{13x^4y^6})$

$= \sqrt{169x^8y^{12}}$

$= \sqrt{169}\ \sqrt{x^8}\ \sqrt{y^{12}}$

$= \sqrt{13x^4y^6}$

63. $\sqrt{16x^{\square}y^6} = 4x^2y^3$

$\sqrt{16x^4y^6} = 4x^2y^3$

Because $(x^2)^2 = x^4$

65. $\sqrt{4x^{\square}y^{\square}} = 2x^3y^2\ \sqrt{y}$

$\sqrt{4x^6y^5} = 2x^3y^2\ \sqrt{y}$

Because $2^2 = 4$ and

$(y^2\ \sqrt{y})^2 = y^5$

67. $\sqrt{2x^{\square}y^5} \cdot \sqrt{\square x^3y^{\square}} = 4x^7y^6\ \sqrt{x}$

$\sqrt{2x^{12}y^5} \cdot \sqrt{8x^3y^7} = 4x^7y^6\ \sqrt{x}$

Because $4^2 = 16$,

$(x^{7}\sqrt{x})^2 = x^{15}$, and

$(y^6)^2 = y^{12}$

69. $\dfrac{3x^2 - 16x - 12}{3x^2 - 10x - 8} \div \dfrac{x^2 - 7x + 6}{3x^2 - 11x - 4}$

$= \dfrac{3x^2 - 16x - 12}{3x^2 - 10x - 8} \cdot \dfrac{3x^2 - 11x - 4}{x^2 - 7x + 6}$

$= \dfrac{(3x + 2)(x - 6)}{(3x + 2)(x - 4)} \cdot \dfrac{(3x + 1)(x - 4)}{(x - 6)(x - 1)}$

$= \dfrac{(3x + 1)}{(x - 1)}$

71. $3x - 4y = 6$

x	y
0	$-\frac{3}{2}$
2	0
-2	-3

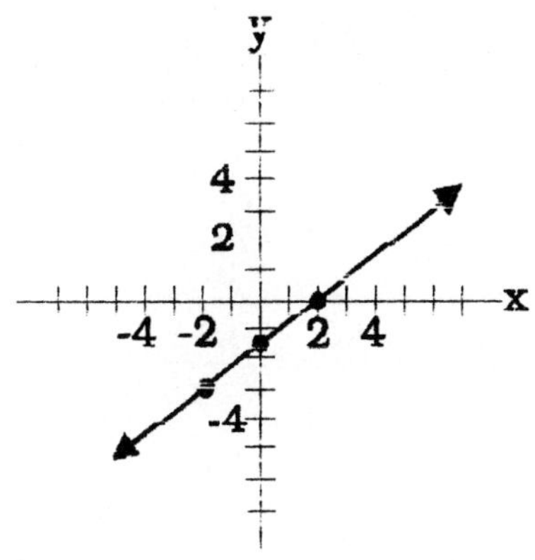

Exercise Set 9.3

1. $\sqrt{\dfrac{12}{3}} = \sqrt{4} = 2$

3. $\sqrt{\dfrac{27}{3}} = \sqrt{9} = 3$

5. $\dfrac{\sqrt{18}}{\sqrt{2}} = \sqrt{\dfrac{18}{2}} = \sqrt{9} = 3$

7. $\sqrt{\dfrac{1}{25}} = \dfrac{\sqrt{1}}{\sqrt{25}} = \dfrac{1}{5}$

9. $\sqrt{\dfrac{9}{49}} = \dfrac{\sqrt{9}}{\sqrt{49}} = \dfrac{3}{7}$

11. $\dfrac{\sqrt{10}}{\sqrt{490}} = \sqrt{\dfrac{10}{490}} = \sqrt{\dfrac{1}{49}} = \dfrac{\sqrt{1}}{\sqrt{49}}$

$= \dfrac{1}{7}$

13. $\sqrt{\dfrac{40x^3}{2x}} = \sqrt{20x^2} = \sqrt{4}\ \sqrt{5}\ \sqrt{x^2}$

$= 2\ \sqrt{5x^2}$

$= 2x\ \sqrt{5}$

15. $\sqrt{\dfrac{9xy^4}{3y^3}} = \sqrt{3xy}$

17. $\sqrt{\dfrac{25x^6y}{45x^6y^3}} = \sqrt{\dfrac{5}{9y^2}} = \dfrac{\sqrt{5}}{\sqrt{9y^2}}$

$= \dfrac{\sqrt{5}}{3y}$

19. $\sqrt{\dfrac{72xy}{72x^3y^5}} = \sqrt{\dfrac{1}{x^2y^4}} = \dfrac{\sqrt{1}}{\sqrt{x^2y^4}}$

$$= \frac{1}{xy^2}$$

21. $\frac{\sqrt{32x^5}}{\sqrt{8x}} = \sqrt{\frac{32x^5}{8x}} = \sqrt{4x^4} = 2x^2$

23. $\frac{\sqrt{16x^4y}}{\sqrt{25x^6y^3}} = \sqrt{\frac{16x^4y}{25x^6y^3}}$

$$= \sqrt{\frac{16}{25x^2y^2}}$$

$$= \frac{\sqrt{16}}{\sqrt{25x^2y^2}}$$

$$= \frac{4}{5xy}$$

25. $\frac{\sqrt{45xy^6}}{\sqrt{9xy^4z^2}} = \sqrt{\frac{45xy^6}{9xy^4z^2}}$

$$= \sqrt{\frac{5y^2}{z^2}}$$

$$= \frac{\sqrt{5y^2}}{\sqrt{z^2}}$$

$$= \frac{y\sqrt{5}}{z}$$

27. $\frac{3}{\sqrt{2}} = \frac{3\sqrt{2}}{\sqrt{2}\ \sqrt{2}} = \frac{3\sqrt{2}}{2}$

29. $\frac{4}{\sqrt{8}} = \frac{4}{\sqrt{8}} \cdot \frac{\sqrt{8}}{\sqrt{8}} = \frac{4\sqrt{8}}{8}$

$$= \frac{4\sqrt{4 \cdot 2}}{8}$$

$$= \frac{4\sqrt{4} \cdot \sqrt{2}}{8}$$

$$= \frac{4 \cdot 2 \cdot \sqrt{2}}{8}$$

$$= \frac{8\sqrt{2}}{8}$$

$$= \sqrt{2}$$

31. $\frac{5}{\sqrt{10}} = \frac{5\sqrt{10}}{\sqrt{10}\ \sqrt{10}} = \frac{5\sqrt{10}}{10}$

$$= \frac{\sqrt{10}}{2}$$

33. $\sqrt{\frac{2}{5}} = \frac{\sqrt{2}}{\sqrt{5}} = \frac{\sqrt{2}}{\sqrt{5}} \cdot \frac{\sqrt{5}}{\sqrt{5}}$

$$= \frac{\sqrt{10}}{5}$$

35. $\sqrt{\frac{3}{15}} = \sqrt{\frac{1}{5}} = \frac{\sqrt{1}}{\sqrt{5}} = \frac{1}{\sqrt{5}}$

$$= \frac{1}{\sqrt{5}} \cdot \frac{\sqrt{5}}{\sqrt{5}} = \frac{\sqrt{5}}{5}$$

37. $\sqrt{\frac{x^2}{2}} = \frac{\sqrt{x^2}}{\sqrt{2}} = \frac{x}{\sqrt{2}}$

$$= \frac{x}{\sqrt{2}} \cdot \frac{\sqrt{2}}{\sqrt{2}} = \frac{x\sqrt{2}}{2}$$

39. $\sqrt{\frac{x^2}{8}} = \frac{\sqrt{x^2}}{\sqrt{8}} = \frac{x}{\sqrt{8}}$

$$= \frac{x}{\sqrt{8}} \cdot \frac{\sqrt{8}}{\sqrt{8}} = \frac{x\sqrt{8}}{8}$$

$$= \frac{x \cdot \sqrt{4}\ \sqrt{2}}{8}$$

$$= \frac{2x\sqrt{2}}{8}$$

$= \dfrac{x\sqrt{2}}{4}$

41. $\sqrt{\dfrac{x^4}{5}} = \dfrac{\sqrt{x^4}}{\sqrt{5}} = \dfrac{x^2}{\sqrt{5}}$

$= \dfrac{x^2}{\sqrt{5}} \cdot \dfrac{\sqrt{5}}{\sqrt{5}}$

$= \dfrac{x^2\sqrt{5}}{5}$

43. $\sqrt{\dfrac{x^6}{15y}} = \dfrac{\sqrt{x^6}}{\sqrt{15y}} = \dfrac{x^3}{\sqrt{15y}}$

$= \dfrac{x^3}{\sqrt{15y}} \cdot \dfrac{\sqrt{15y}}{\sqrt{15y}}$

$= \dfrac{x^3\sqrt{15y}}{15y}$

45. $\sqrt{\dfrac{8x^4y^2}{32x^2y^3}} = \sqrt{\dfrac{x^2}{4y}}$

$= \dfrac{\sqrt{x^2}}{\sqrt{4}\sqrt{y}} = \dfrac{x}{2\sqrt{y}}$

$= \dfrac{x}{2\sqrt{y}} \cdot \dfrac{\sqrt{y}}{\sqrt{y}} = \dfrac{x\sqrt{y}}{2y}$

47. $\sqrt{\dfrac{18yz}{75x^4y^5z^3}} = \sqrt{\dfrac{6}{25x^4y^4z^2}}$

$= \dfrac{\sqrt{6}}{\sqrt{25x^4y^4z^2}} = \dfrac{\sqrt{6}}{5x^2y^2z}$

49. $\dfrac{\sqrt{90x^4y}}{\sqrt{2x^5y^5}} = \sqrt{\dfrac{90x^4y}{2x^5y^5}}$

$= \sqrt{\dfrac{45}{xy^4}} = \dfrac{\sqrt{9}\sqrt{5}}{\sqrt{x}\sqrt{y^4}}$

$= \dfrac{3\sqrt{5}}{y^2\sqrt{x}}$

$= \dfrac{3\sqrt{5}}{y^2\sqrt{x}} \cdot \dfrac{\sqrt{x}}{\sqrt{x}}$

$= \dfrac{3\sqrt{5x}}{y^2x}$

$= \dfrac{3\sqrt{5x}}{xy^2}$

51. Conditions necessary for a square root to be simplfied:
1. No perfect square factors in any radicand.
2. No radicand contains a fraction.
3. No square roots in any denominator.

53. $3x^2 - 12x - 96$

$= 3(x^2 - 4x - 32)$
$= 3(x + 4)(x - 8)$

55. $x + \dfrac{24}{x} = 10$

$x(x + \dfrac{24}{x}) = 10(x)$

$x^2 + 24 = 10x$
$x^2 - 10x + 24 = 0$
$(x - 6)(x - 4) = 0$
$x - 6 = 0 \quad x - 4 = 0$
$x = 6 \quad x = 4$

Exercise Set 9.4

1. $4\sqrt{3} - 2\sqrt{3} = 2\sqrt{3}$

3. $6\sqrt{7} - 8\sqrt{7} = -2\sqrt{7}$

5. $2\sqrt{3} - 2\sqrt{3} - 4\sqrt{3} + 5$

$= 5 - 4\sqrt{3}$

7. $4\sqrt{x} + \sqrt{x} = 5\sqrt{x}$

9. $-\sqrt{x} + 6\sqrt{x} - 2\sqrt{x}$

$= -1\sqrt{x} + 6\sqrt{x} - 2\sqrt{x} = 3\sqrt{x}$

11. $3\sqrt{y} - \sqrt{y} + 3$

$= 3\sqrt{y} - 1\sqrt{y} + 3$

$= 2\sqrt{y} + 3$

$= 3 + 2\sqrt{y}$

13. $\sqrt{x} + \sqrt{y} + x + 3\sqrt{y}$

$= x + \sqrt{x} + 4\sqrt{y}$

15. $3 + 4\sqrt{x} - 6\sqrt{x} = 3 - 2\sqrt{x}$

17. $\sqrt{8} - \sqrt{12}$

$= \sqrt{4} \cdot \sqrt{2} - \sqrt{4} \cdot \sqrt{3}$

$= 2\sqrt{2} - 2\sqrt{3}$

19. $\sqrt{200} - \sqrt{72}$

$= \sqrt{100} \cdot \sqrt{2} - \sqrt{36} \cdot \sqrt{2}$

$= 10\sqrt{2} - 6\sqrt{2}$

$= 4\sqrt{2}$

21. $\sqrt{125} + \sqrt{20}$

$= \sqrt{25} \cdot \sqrt{5} + \sqrt{4} \cdot \sqrt{5}$

$= 5\sqrt{5} + 2\sqrt{5}$

$= 7\sqrt{5}$

23. $4\sqrt{50} - \sqrt{72} + \sqrt{8}$

$= 4 \cdot \sqrt{25}\,\sqrt{2}$

$- \sqrt{36}\,\sqrt{2} + \sqrt{4}\,\sqrt{2}$

$= 4 \cdot 5\sqrt{2} - 6\sqrt{2} + 2\sqrt{2}$

$= 20\sqrt{2} - 6\sqrt{2} + 2\sqrt{2}$

$= 16\sqrt{2}$

25. $-6\sqrt{75} + 4\sqrt{125}$

$= -6 \cdot \sqrt{25}\,\sqrt{3} + 4 \cdot \sqrt{25}\,\sqrt{5}$

$= (-6)(5)\sqrt{3} + (4)(5)\sqrt{5}$

$= -30\sqrt{3} + 20\sqrt{5}$

27. $5\sqrt{250} - 9\sqrt{80}$

$= 5 \cdot \sqrt{25} \cdot \sqrt{10} - 9\sqrt{16}\,\sqrt{5}$

$= 5 \cdot 5 \cdot \sqrt{10} - 9 \cdot 4\sqrt{5}$

$= 25\sqrt{10} - 36\sqrt{5}$

29. $8\sqrt{64} - \sqrt{96} = 8 \cdot 8 - \sqrt{16}\,\sqrt{6}$

$= 64 - 4\sqrt{6}$

31. $(3 + \sqrt{2})(3 - \sqrt{2})$

$= \overset{F}{3 \cdot 3} + \overset{O}{3(-\sqrt{2})} + \overset{I}{\sqrt{2}(3)}$

$+ \overset{L}{\sqrt{2}(-\sqrt{2})}$

$= 9 - 3\sqrt{2} + 3\sqrt{2} - 2$
$= 7$

33. $(6 - \sqrt{5})(6 + \sqrt{5})$

$= \overset{F}{6(6)} + \overset{O}{6\sqrt{5}} + \overset{I}{(-\sqrt{5})6}$

$+ \overset{L}{(-\sqrt{5})(\sqrt{5})}$

$= 36 + 6\sqrt{5} - 6\sqrt{5} - 5$
$= 31$

35. $(\sqrt{x} + 3)(\sqrt{x} - 3)$

$$= \overset{F}{(\sqrt{x})(\sqrt{x})} + \overset{O}{(\sqrt{x})(-3)} + \overset{I}{3(\sqrt{x})} + \overset{L}{3(-3)}$$

$$= x - 3\sqrt{x} + 3\sqrt{x} - 9$$

$$= x - 9$$

37. $(\sqrt{6} + x)(\sqrt{6} - x)$

$$= \overset{F}{\sqrt{6}(\sqrt{6})} - \overset{O}{\sqrt{6} \cdot x} + \overset{I}{x\sqrt{6}} + \overset{L}{(x)(-x)}$$

$$= 6 - \sqrt{6} \cdot x + \sqrt{6} \cdot x - x^2$$

$$= 6 - x^2$$

39. $(\sqrt{x} + y)(\sqrt{x} - y)$

$$= \overset{F}{\sqrt{x}\sqrt{x}} + \overset{O}{\sqrt{x}(-y)} + \overset{I}{y\sqrt{x}} + \overset{L}{y(-y)}$$

$$= x + x\sqrt{y} - x\sqrt{y} - y^2$$

$$= x - y^2$$

41. $(2\sqrt{x} + 3\sqrt{y})(2\sqrt{x} - 3\sqrt{y})$

$$= \overset{F}{2\sqrt{x}(2\sqrt{x})} + \overset{O}{2\sqrt{x}(-3\sqrt{y})} + \overset{I}{3\sqrt{y}(2\sqrt{x})} + \overset{L}{3\sqrt{y}(-3\sqrt{y})}$$

$$= 4x - 6\sqrt{xy} + 6\sqrt{xy} - 9y$$

$$= 4x - 9y$$

43. $\dfrac{4}{2 + \sqrt{3}} = \dfrac{4(2 - \sqrt{3})}{(2 + \sqrt{3})(2 - \sqrt{3})}$

$$= \frac{4(2 - \sqrt{3})}{1}$$

$$= 4(2 - \sqrt{3})$$

45. $\dfrac{3}{\sqrt{2} + 5} = \dfrac{3(\sqrt{2} - 5)}{(\sqrt{2} + 5)(\sqrt{2} - 5)}$

$$= \frac{3(\sqrt{2} - 5)}{-23}$$

$$= \frac{-3(\sqrt{2} - 5)}{23}$$

$$= \frac{-3\sqrt{2} + 15}{23}$$

47. $\dfrac{2}{\sqrt{2} + \sqrt{3}} = \dfrac{2(\sqrt{2} - \sqrt{3})}{(\sqrt{2} + \sqrt{3})(\sqrt{2} - \sqrt{3})}$

$$= \frac{2(\sqrt{2} - \sqrt{3})}{-1}$$

$$= -2(\sqrt{2} - \sqrt{3})$$

49. $\dfrac{8}{\sqrt{5} - \sqrt{8}}$

$$= \frac{8}{(\sqrt{5} - \sqrt{8})} \cdot \frac{(\sqrt{5} + \sqrt{8})}{(\sqrt{5} + \sqrt{8})}$$

$$= \frac{8(\sqrt{5} + \sqrt{8})}{-3}$$

$$= \frac{-8(\sqrt{5} + \sqrt{8})}{3}$$

$$= \frac{-8(\sqrt{5} + 2\sqrt{2})}{3}$$

51. $\dfrac{2}{6 + \sqrt{x}}$

$$= \frac{2}{(6 + \sqrt{x})} \cdot \frac{(6 - \sqrt{x})}{(6 - \sqrt{x})}$$

$$= \frac{2(6 - \sqrt{x})}{36 - x}$$

53. $\dfrac{6}{4 - \sqrt{y}}$

$$= \frac{6}{(4 - \sqrt{y})} \cdot \frac{(4 + \sqrt{y})}{(4 + \sqrt{y})}$$

$$= \frac{6(4 + \sqrt{y})}{16 - y}$$

55. $\frac{4}{\sqrt{x} - y}$

$= \frac{4}{(\sqrt{x} - y)} \cdot \frac{(\sqrt{x} + y)}{(\sqrt{x} + y)}$

$= \frac{4(\sqrt{x} + y)}{x - y^2}$

57. $\frac{x}{\sqrt{x} + \sqrt{y}}$

$= \frac{x(\sqrt{x} - \sqrt{y})}{(\sqrt{x} + \sqrt{y})(\sqrt{x} - \sqrt{y})}$

$= \frac{x(\sqrt{x} - \sqrt{y})}{(x - y)}$

59. $\frac{\sqrt{x}}{\sqrt{5} + \sqrt{x}}$

$= \frac{\sqrt{x}}{(\sqrt{5} + \sqrt{x})} \cdot \frac{(\sqrt{5} - \sqrt{x})}{(\sqrt{5} - \sqrt{x})}$

$= \frac{\sqrt{x}(\sqrt{5} - \sqrt{x})}{5 - x}$

$= \frac{\sqrt{5x} - x}{5 - x}$

61. a) To rationalize the the denominator of $a/b+\sqrt{c}$, multiply the numerator and denominator by $(b - \sqrt{c})$.

b) $\frac{a}{b + \sqrt{c}} \cdot \frac{b - \sqrt{c}}{b - \sqrt{c}}$

$= \frac{a(b - \sqrt{c})}{b^2 - c}$

Cumulative Review Exercises

63. $2x^2 - x - 36 = 0$

$(2x - 9)(x + 4) = 0$

$2x - 9 = 0 \quad x + 4 = 0$

$2x = 9 \quad x = -4$

$x = \frac{9}{2}$

65. Let x = time for Mrs. Moreno to stack the wood herself.

$\frac{12}{20} + \frac{12}{x} = 1$

$20x[\frac{12}{20} + \frac{12}{x}] = 1(20x)$

$12x + 240 = 20x$

$240 = 8x$

$30 = x$

$x = 30$ minutes

Exercise Set 9.5

1. $\sqrt{x} = 8$

$(\sqrt{x})^2 = (8)^2$

$x = 64$

Ck: $\sqrt{x} = 8$

$\sqrt{64} = 8$

$8 = 8$

True

3. $\sqrt{x} = -3$

$(\sqrt{x})^2 = (-3)^2$

$x = 9$

Ck: $\sqrt{x} = -3$

$\sqrt{9} = -3$

$3 = -3$

False

Since $3 = -3$ is false, the number 9 is an extraneous root and is not a solution.

5. $\sqrt{x+5} = 3$

$(\sqrt{x+5})^2 = (3)^2$

$x + 5 = 9$

$x+5 - 5 = 9 - 5$

Ck: $\sqrt{x+5} = 3$

$\sqrt{(4)+5} = 3$

$\sqrt{9} = 3$

$3 = 3$

$x = 4$ True

7. $\sqrt{2x+4} = -6$

$(\sqrt{2x+4})^2 = (-6)^2$

$2x + 4 = 36$

$2x = 32$

$x = 16$

Ck: $\sqrt{2x+4} = -6$

$\sqrt{2(16)+4} = -6$

$\sqrt{32+4} = -6$

$\sqrt{36} = -6$

$6 = -6$

False

Since $6 = -6$ is false, the number 16 is an extraneous root and is not a solution.

9. $\sqrt{x} + 3 = 5$

$\sqrt{x} +3-3 = 5-3$

$\sqrt{x} = 2$

$(\sqrt{x})^2 = 2^2$

$x = 4$

Ck: $\sqrt{x} + 3 = 5$

$\sqrt{(4)} + 3 = 5$

$2 + 3 = 5$

$5 = 5$

True

11. $6 = 4 + \sqrt{x}$

$6-4 = 4 -4+ \sqrt{x}$

$2 = \sqrt{x}$

$2^2 = (\sqrt{x})^2$

$4 = x$

Ck: $6 = 4 + \sqrt{x}$

$6 = 4 + \sqrt{(4)}$

$6 = 4 + 2$

$6 = 6$

True

13. $4 + \sqrt{x} = 2$

$\sqrt{x} = 2-4$

$\sqrt{x} = -2$

$(\sqrt{x})^2 = (-2)^2$

$x = 4$

Ck: $4 + \sqrt{x} = 2$

$4 + \sqrt{4} = 2$

$4 + 2 = 2$

$6 = 2$

False

Since $6 = 2$ is false, the number 4 is an extraneous root and is not a solution.

15. $\sqrt{2x - 5} = x - 4$

$(\sqrt{2x - 5})^2 = (x - 4)^2$

$2x - 5 = x^2 - 8x + 16$

$2x - 2x - 5 + 5$

$= x^2 - 8x - 2x + 16 + 5$

$0 = x^2 - 10x + 21$

$0 = (x-7)(x-3)$

$x - 7 = 0 \quad x - 3 = 0$

$x-7+7 = 0+7 \quad x-3+3 = 0+3$

$x = 7 \quad x = 3$

Ck: $x = 7$

$\sqrt{2x-5} = x - 4$

$\sqrt{2(7)-5} = 7 - 4$

$\sqrt{14-5} = 3$

$\sqrt{9} = 3$

$3 = 3$

True

The solution is 7.

Ck: $x = 3$

$\sqrt{2x-5} = x - 4$

$\sqrt{2(3)-5} = (3)-4$

$\sqrt{1} = -1$

$1 = -1$

False

Since $1 = -1$ is false, the number 3 is an extraneous root and not a solution.

17. $\sqrt{2x - 6} = \sqrt{5x - 27}$

$(\sqrt{2x - 6})^2 = (\sqrt{5x - 27})^2$

$2x - 6 = 5x - 27$

$2x - 2x - 6 = 5x - 2x - 27$

$-6 = 3x - 27$

$-6 + 27 = 3x - 27 + 27$

$21 = 3x$

$\frac{21}{3} = \frac{3x}{3}$

$7 = x$

Ck: $\sqrt{2x - 6} = \sqrt{5x - 27}$

$\sqrt{2(7) - 6} = \sqrt{5(7) - 27}$

$\sqrt{14 - 6} = \sqrt{35 - 27}$

$\sqrt{8} = \sqrt{8}$

True

19. $\sqrt{3x + 3} = \sqrt{5x - 1}$

$$(\sqrt{3x + 3})^2 = (\sqrt{5x - 1})^2$$
$$3x + 3 = 5x - 1$$
$$3x - 3x + 3 = 5x - 3x - 1$$
$$3 = 2x - 1$$
$$3 + 1 = 2x - 1 + 1$$
$$4 = 2x$$
$$\frac{4}{2} = \frac{2x}{2}$$
$$2 = x$$

Ck: $\sqrt{3x + 3} = \sqrt{5x - 1}$

$$\sqrt{3(2) + 3} = \sqrt{5(2) - 1}$$
$$\sqrt{6 + 3} = \sqrt{10 - 1}$$
$$\sqrt{9} = \sqrt{9}$$
$$3 = 3$$

True

21. $\sqrt{3x + 9} = 2\sqrt{x}$

$$(\sqrt{3x + 9})^2 = (2\sqrt{x})^2$$
$$3x + 9 = 4x$$
$$3x - 3x + 9 = 4x - 3x$$
$$9 = x$$

Ck: $\sqrt{3x + 9} = 2\sqrt{x}$

$$\sqrt{3(9) + 9} = 2\sqrt{9}$$
$$\sqrt{27 + 9} = 2\sqrt{9}$$
$$\sqrt{36} = 2 \cdot 3$$
$$6 = 6$$

True

23. $\sqrt{4x - 5} = \sqrt{x + 9}$

$$(\sqrt{4x - 5})^2 = (\sqrt{x + 9})^2$$
$$4x - 5 = x + 9$$
$$4x - x - 5 = x - x + 9$$
$$3x - 5 = 9$$
$$3x - 5 + 5 = 9 + 5$$
$$3x = 14$$
$$\frac{3x}{3} = \frac{14}{3}$$
$$x = \frac{14}{3}$$

Ck: $\sqrt{4x - 5} = \sqrt{x + 9}$

$$\sqrt{4\left(\frac{14}{3}\right) - 5} = \sqrt{\frac{14}{3} + 9}$$
$$\sqrt{\frac{56}{3} - 4} = \sqrt{\frac{14}{3} + 9}$$
$$\sqrt{\frac{56}{3} - \frac{15}{3}} = \sqrt{\frac{14}{3} + \frac{27}{3}}$$
$$\sqrt{\frac{41}{3}} = \sqrt{\frac{41}{3}}$$

True

25. $3\sqrt{x} = \sqrt{x + 8}$

$$(3\sqrt{x})^2 = (\sqrt{x + 8})^2$$
$$9x = x + 8$$
$$8x = 8$$
$$x = 1$$

Ck: $3\sqrt{x} = \sqrt{x + 8}$

$$3\sqrt{1} = \sqrt{1 + 8}$$
$$3\sqrt{1} = \sqrt{9}$$
$$3(1) = 3$$
$$3 = 3$$

True

27. $4\sqrt{x} = x + 3$

$$(4\sqrt{x})^2 = (x + 3)^2$$
$$16x = x^2 + 6x + 9$$
$$16x - 16x = x^2 + 6x - 16x + 9$$
$$0 = x^2 - 10x + 9$$
$$0 = (x - 9)(x - 1)$$

$x - 9 = 0$ $\quad$ $x - 1 = 0$

$x - 9 + 9 = 0 + 9$ $\quad$ $x - 1 + 1 = 0 + 1$

$x = 9$ $\quad$ $x = 1$

Ck: $x = 9$	Ck: $x = 1$
$4\sqrt{x} = x + 3$	$4\sqrt{x} = x + 3$
$4\sqrt{(9)} = 9+3$	$4\sqrt{(1)} = (1)+3$
$4(3) = 12$	$4(1) = 4$
$12 = 12$	$4 = 4$
True	True
9 is a solution	1 is a solution

29. $\sqrt{2x - 3} = 2\sqrt{3x - 2}$

$$(\sqrt{2x - 3})^2 = (2\sqrt{3x - 2})^2$$
$$2x - 3 = 4(3x - 2)$$
$$2x - 3 = 12x - 8$$
$$2x - 2x - 3 = 12x - 2x - 8$$
$$-3 = 10x - 8$$
$$-3 + 8 = 10x - 8 + 8$$
$$5 = 10x$$
$$\frac{5}{10} = \frac{10x}{10}$$
$$\frac{1}{2} = x$$

Ck: $\sqrt{2x - 3} = 2\sqrt{3x - 2}$

$$\sqrt{2(\tfrac{1}{2}) - 3} = 2\sqrt{\frac{3(1)}{2} - 2}$$
$$\sqrt{1 - 3} = 2\sqrt{\frac{3}{2} - \frac{4}{2}}$$
$$\sqrt{-1} = 2\sqrt{\frac{-1}{2}}$$
$$\sqrt{-2} = 2\,\frac{\sqrt{-1}}{\sqrt{2}}$$
$$\sqrt{-2} = 2 \cdot \frac{\sqrt{-1}}{\sqrt{2}} \cdot \frac{\sqrt{2}}{\sqrt{2}}$$
$$\sqrt{-2} = \frac{2 \cdot \sqrt{-1}}{2}$$
$$\sqrt{-2} = \sqrt{-1}$$

False

Since $\sqrt{-2} = \sqrt{-1}$ is false, the number 1/2 is an extraneous root and not a solution.

31. $\sqrt{x^2 + 5} = x + 5$

$$(\sqrt{x + 5})^2 = (x + 5)^2$$
$$x^2 + 5 = (x + 5)(x + 5)$$
$$x^2 + 5 = x^2 + 10x + 25$$
$$x^2 + 5 - x^2 = 10x + 25$$
$$5 = 10x + 25$$
$$-20 = 10x$$
$$-2 = x$$

Ck: $\sqrt{x^2 + 5} = x + 5$

$$\sqrt{(-2)^2 + 5} = (-2) + 5$$
$$\sqrt{4 + 5} = -2 + 5$$
$$\sqrt{9} = 3$$
$$3 = 3$$

True

33. $x - 4\sqrt{x} + 3 = 0$

$$4 - 4\sqrt{x} + 4\sqrt{x} + 3 = 0 + 4\sqrt{x}$$
$$(x + 3) = 4\sqrt{x}$$
$$(x + 3)^2 = (4\sqrt{x})^2$$
$$x^2 + 6x + 9 = 16x$$
$$x^2 + 6x - 16x + 9 = 16x - 16x$$
$$x^2 - 10x + 9 = 0$$

$x - 9 = 0$	$x - 1 = 0$
$x-9+9 = 0+9$	$x-1+1 = 0 + 1$
$x = 9$	$x = 1$

Ck: $x = 9$

$$x - 4\sqrt{x} + 3 = 0$$
$$9 - 4\sqrt{9} + 3 = 0$$
$$9 - 4 \cdot 3 + 3 = 0$$
$$9 - 12 + 3 = 0$$
$$0 = 0$$

True

Ck: $x = 1$

$$x - 4\sqrt{x} + 3 = 0$$

$$1 - 4\sqrt{1} + 3 = 0$$
$$1 - 4(1) + 3 = 0$$
$$1 - 4 + 3 = 0$$
$$0 = 0$$

True

Both 9 and 1 are solutions.

35. There may be extraneous roots.

Cumulative Review Exercises

37. $3x - 2y = 6$ $y = 2x - 4$

x	y
0	-3
2	0
4	3

x	y
0	-4
2	0
3	2

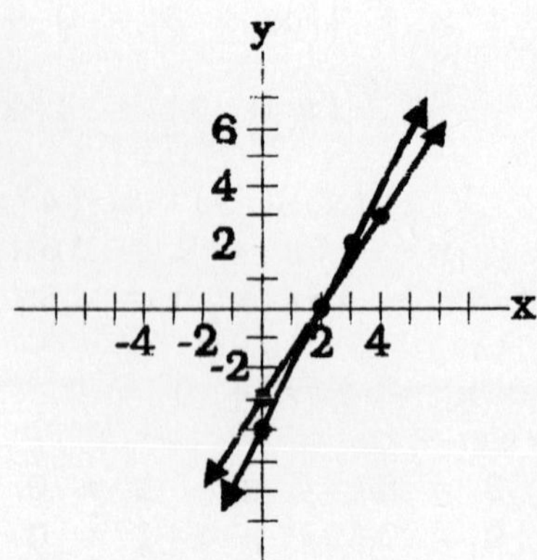

Solution: (2,0)

39. $3x - 2y = 6$ (1)
$y = 2x - 4$ (2)

$3x - 2y + 6$ (1)
$-2x + y = -4$ (2)

$3x - 2y = 6$ (1)
$2(-2x+y) = 2(-4)$ (2)

$3x - 2y = 6$ (1)
$-4x + 2y = -8$ (2)

$$-x = -2$$
$$x = 2$$
$$y = 2x - 4 = 2(2) - 4$$
$$= 4 - 4 = (0)$$

Solution: (2,0)

Just for fun

1. $$\sqrt{x} + 2 = \sqrt{x + 16}$$
$$(\sqrt{x} + 2)^2 = (\sqrt{x + 16})^2$$
$$x + 4\sqrt{x} + 4 = x + 16$$
$$x - x + 4\sqrt{x} + 4 = x - x + 16$$
$$4\sqrt{x} + 4 - 4 = 16 - 4$$
$$4\sqrt{x} = 12$$
$$\frac{4\sqrt{x}}{4} = \frac{12}{4}$$
$$\sqrt{x} = 3$$
$$(\sqrt{x})^2 = (3)^2$$
$$x = 9$$

Ck: $$\sqrt{x} + 2 = \sqrt{x + 16}$$
$$\sqrt{9} + 2 = \sqrt{9 + 16}$$
$$3 + 2 = \sqrt{25}$$
$$5 = 5$$

True

2. $$\sqrt{x + 1} = 2 - \sqrt{x}$$
$$(\sqrt{x + 1})^2 = (2 - \sqrt{x})^2$$
$$x + 1 = 4 - 4\sqrt{x} + x$$
$$x - x + 1 = 4 - 4\sqrt{x} + x - x$$

$$1 = 4 - 4\sqrt{x}$$

$$1 - 4 = 4 - 4 - 4\sqrt{x}$$

$$-3 = -4\sqrt{x}$$

$$\frac{-3}{-4} = \frac{-4\sqrt{x}}{-4}$$

$$\frac{3}{4} = \sqrt{x}$$

$$\left(\frac{3}{4}\right)^2 = (\sqrt{x})^2$$

$$\frac{9}{16} = x$$

Ck: $x = \frac{9}{16}$

$$\sqrt{x + 1} = 2 - \sqrt{x}$$

$$\sqrt{\frac{9}{16} + 1} = 2 - \sqrt{\frac{9}{16}}$$

$$\sqrt{\frac{9}{16} + \frac{16}{16}} = 2 - \frac{\sqrt{9}}{\sqrt{16}}$$

$$\sqrt{\frac{25}{16}} = 2 - \frac{3}{4}$$

$$\frac{\sqrt{25}}{\sqrt{16}} = 2 - \frac{3}{4}$$

$$\frac{5}{4} = 2 - \frac{3}{4}$$

$$\frac{5}{4} = \frac{8}{4} - \frac{3}{4}$$

$$\frac{5}{4} = \frac{5}{4}$$

True

3. $\sqrt{x + 7} = 5 - \sqrt{x - 8}$

$$(\sqrt{x + 7})^2 = (5 - \sqrt{x - 8})^2$$

$$x + 7 = 25 - 10\sqrt{x - 8} + (x - 8)$$

$$x + 7 = 25 - 10\sqrt{x - 8} + x - 8$$

$$x + 7 = 17 - 10\sqrt{x - 8} + x$$

$$x - x + 7 = 17 - 10\sqrt{x - 8} + x - x$$

$$7 = 17 - 10\sqrt{x - 8}$$

$$7 - 17 = 17 - 17 - 10\sqrt{x - 8}$$

$$-10 = -10\sqrt{x - 8}$$

$$\frac{-10}{-10} = \frac{-10\sqrt{x - 8}}{-10}$$

$$1 = \sqrt{x - 8}$$

$$1^2 = (\sqrt{x - 8})^2$$

$$1 = x - 8$$

$$1 + 8 = x - 8 + 8$$

$$9 = x$$

Ck: $\sqrt{x + 7} = 5 - \sqrt{x - 8}$

$$\sqrt{9 + 7} = 5 - \sqrt{9 - 8}$$

$$\sqrt{16} = 5 - \sqrt{1}$$

$$4 = 5 - 1$$

$$4 = 4$$

True

Exercise Set 9.6

1. $a^2 + b^2 = c^2$

$$(x)^2 + (5)^2 = (12)^2$$

$$x^2 + 25 = 144$$

$$x^2 = 144 - 25$$

$$x^2 = 119$$

$$\sqrt{x^2} = \sqrt{119}$$

$$x = \sqrt{119}$$

$$x = 10.91$$

3. $a^2 + b^2 = c^2$

$$10^2 + 8^2 = x^2$$
$$100 + 64 = x^2$$
$$164 = x^2$$
$$\sqrt{164} = \sqrt{x^2}$$
$$\sqrt{164} = x$$
$$12.81 = x$$

5. $$a^2 + b^2 = c^2$$
$$x^2 + 15^2 = 20^2$$
$$x^2 + 225 = 400$$
$$x^2 + 225 - 225 = 400 - 225$$
$$x^2 = 175$$
$$\sqrt{x^2} = \sqrt{175}$$
$$x^2 = \sqrt{175}$$
$$x = 13.23$$

7. $$a^2 + b^2 = c^2$$
$$(\sqrt{5})^2 + (6)^2 = x^2$$
$$5 + 36 = x^2$$
$$41 = x^2$$
$$\sqrt{41} = \sqrt{x^2}$$
$$\sqrt{41} = x$$
$$6.40 = x$$

9. $$a^2 + b^2 = c^2$$
$$(\sqrt{5})^2 + 12^2 = x^2$$
$$5 + 144 = x^2$$
$$149 = x^2$$
$$\sqrt{149} = \sqrt{x^2}$$
$$\sqrt{149} = x$$
$$12.21 = x$$

11. $$a^2 + b^2 = c^2$$
$$4^2 + x^2 = 12^2$$
$$16 + x^2 = 144$$
$$16 - 16 + x^2 = 144 - 16$$
$$x^2 = 128$$
$$\sqrt{x^2} = \sqrt{128}$$
$$x = \sqrt{128}$$
$$x = 11.31$$

13. $$c^2 = a^2 + b^2$$
$$(x)^2 = (120)^2 + (53.3)^2$$
$$x^2 = 14,400 + 2840.89$$
$$x^2 = 17,240.89$$
$$\sqrt{x^2} = \sqrt{17,240.89}$$
$$x = \sqrt{17,240.89}$$
$$x = 131.30 \text{ yards}$$

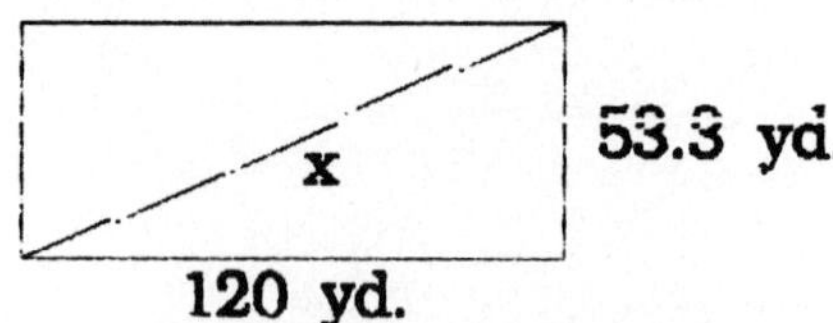

15. $$c^2 = a^2 + b^2$$
$$c^2 = (1.5)^2 + (4)^2$$
$$c^2 = 2.25 + 16$$
$$c^2 = 18.25$$
$$\sqrt{c^2} = \sqrt{18.25}$$
$$c = \sqrt{18.25}$$
$$c = 4.27 \text{ meters}$$

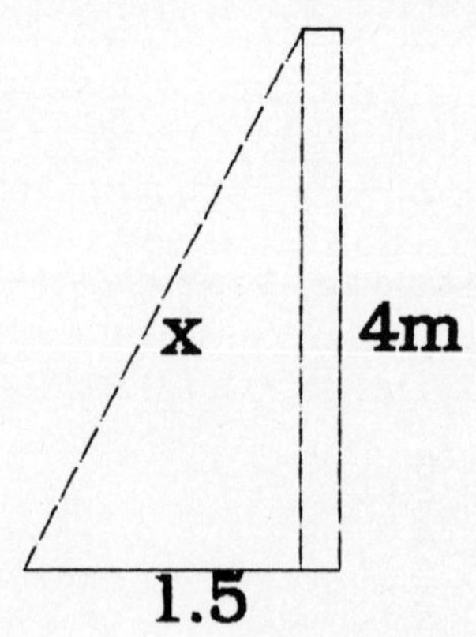

17. $x_1 \quad y_1 \qquad x_2 \quad y_2$

$(-4\ ,\ 3) \qquad (-1\ ,\ 4)$

$d = \sqrt{(x_2-x_1)^2+(y_2-y_1)^2}$

$d = \sqrt{[(-1-(-4))]^2+(4-3)^2}$

$d = \sqrt{(-1+4)^2+(4-3)^2}$

$d = \sqrt{(3)^2+(1)^2}$

$d = \sqrt{9 + 1}$

$d = \sqrt{10}$
$d = 3.16$

19. $x_1 \quad y_1 \qquad x_2 \quad y_2$

$(-8, 4) \qquad (4, -8)$

$a = \sqrt{(x_2-x_1)^2+(y_2-y_1)^2}$

$a = \sqrt{[4-(-8)]^2+(-8-4)^2}$

$a = \sqrt{(4 + 8)^2+(-12)^2}$

$a = \sqrt{(12)^2+(-12)^2}$

$a = \sqrt{144 + 144}$

$a = \sqrt{288}$
$a = 16.97$

21. $A = s^2 \qquad A = 144$
$144 = s^2$

$\sqrt{144} = \sqrt{s^2}$
$12 = s$
$s = 12$ inches

23. $A = \pi r^2 \qquad \pi = 3.14$
Aa = 80 square feet

$80 = \pi r^2 \qquad 80 = 3.14(r^2)$

$\frac{80}{3.14} = r^2$

$25.48 = r^2$

$\sqrt{25.48} = \sqrt{r^2}$

$r = \sqrt{25.48}$
$r = 5.05$ feet

25. $c^2 = a^2 + b^2$

$(x)^2 = 9^2 + 12^2$

$x^2 = 81 + 144$

$x^2 = 225$

$\sqrt{x^2} = \sqrt{225}$
$x = 15$ inches

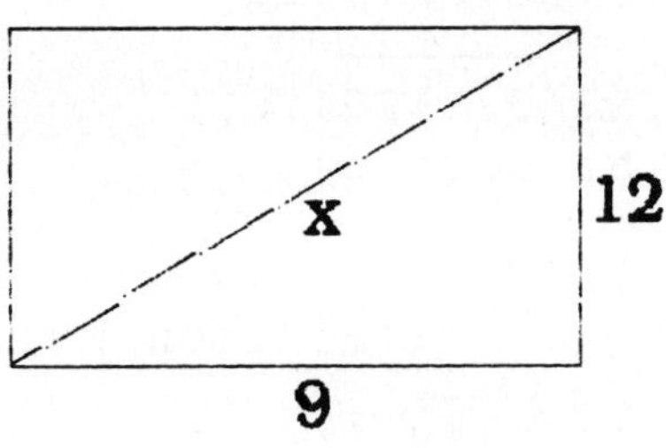

27. $V = \sqrt{2gh}$
g = 32 per second
h = 80 feet

$V = \sqrt{2(32)(80)}$

$V = \sqrt{5120}$
v = 71.55 feet/sec.

29. $V = \sqrt{2gh}$
g = 32 feet per second
h = 1250 feet

$V = \sqrt{2(32)(1250)}$

$V = \sqrt{80,000}$
V = 282.84 feet/sec.

31. $T = 2\pi \sqrt{\frac{L}{32}} \qquad \pi = 3.14$
L = 40 feet

$T = 2\pi \sqrt{\frac{40}{32}}$

$T = 2(3.14)\sqrt{1.25}$

$T = 6.28\sqrt{1.25}$
$T = 7.02$ seconds

33. $T = 2\pi\sqrt{\frac{L}{32}}$ $\pi = 3.14$
$L = 10$ feet

$T = 2(3.14)\sqrt{\frac{10}{32}}$

$T = 2(3.14)\sqrt{0.3125}$

$T = 6.28\sqrt{0.3125}$
$T = 3.51$ seconds

35. $N = 0.2(\sqrt{R})^3$ $R = 1418$ million km

$N = 0.2(\sqrt{1418})^3$
$N = 10{,}679.34$ days

37. $V_e = \sqrt{2gR}$
$g = 9.75$ meters per second
$R = 6{,}370{,}000$ meters

$V_e = \sqrt{2(9.75)(6.370{,}000)}$

$V_e = \sqrt{19.5(6{,}370{,}000)}$

$V_e = \sqrt{124{,}215{,}000}$

$V_e = 11{,}145.18$ meters/sec.

39. $(4x^{-4}y^3)^{-1} = 4^{-1}x^4y^{-3}$

$= \frac{x^4}{4y^3}$

41. $\frac{5x^4 - 9x^3 + 6x^2 - 4x - 3}{3x^2}$

$= \frac{5x^4}{3x^2} - \frac{9x^3}{3x^2} + \frac{6x^2}{3x^2} - \frac{4x}{3x^2}$

$- \frac{3}{3x^2}$

$= \frac{5x^2}{3} - 3x + 2 - \frac{4}{3x} - \frac{1}{x^2}$

Just for fun

1.

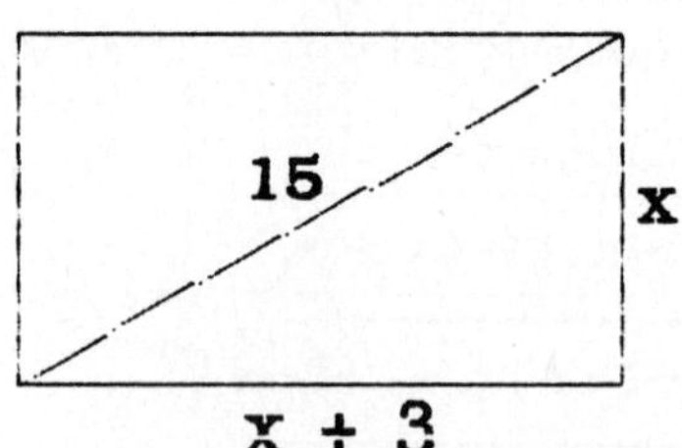

$15^2 = x^2 + (x + 3)^2$

$225 = x^2 + x^2 + 6x + 9$

$2x^2 + 6x - 216 = 0$
$2(x^2 + 3x - 108) = 0$
$2(x - 9)(x + 12) = 0$

$x - 9 = 0$ $x + 12 = 0$
$x = 9$ $x = -12$
Width = 9 feet
Length = 9 + 3 = 12 feet

2. $v = \sqrt{2gh}$

$v = \sqrt{2(\frac{32}{6})100}$

$v = \sqrt{1066.67}$
$v = 32.66$ ft/sec.

3. $T = 2\pi\sqrt{\frac{L}{32}}$

$2 = 2\pi\sqrt{\frac{L}{32}}$

$$1 = 3.14\sqrt{\frac{L}{32}}$$

$$1 = \frac{9.86}{32}L$$

$$1 = 0.308L$$

$$L = 3.25 \text{ feet}$$

4. $d = \sqrt{a^2+b^2+c^2}$

$d = \sqrt{(37)^2+(15)^2+(9)^2}$

$d = \sqrt{1369+225+81}$

$d = \sqrt{1675}$

$d = 40.93$ inches

Exercise Set 9.7

1. $\sqrt[3]{8} = 2$

3. $\sqrt[3]{-8} = -2$

5. $\sqrt[4]{16} = 2$

7. $\sqrt[4]{81} = 3$

9. $\sqrt[3]{-1} = -1$

11. $\sqrt[3]{64} = 4$

13. $\sqrt[3]{54} = \sqrt[3]{27} \cdot \sqrt[3]{2} = 3\sqrt[3]{2}$

15. $\sqrt[3]{16} = \sqrt[3]{8} \cdot \sqrt[3]{2} = 2\sqrt[3]{2}$

17. $\sqrt[3]{81} = \sqrt[3]{27}\,\sqrt[3]{3} = 3\sqrt[3]{3}$

19. $\sqrt[4]{32} = \sqrt[4]{16}\,\sqrt[4]{2} = 2\sqrt[4]{2}$

21. $\sqrt[3]{40} = \sqrt[3]{8}\,\sqrt[3]{5} = 2\sqrt[3]{5}$

23. $\sqrt[3]{x^3} = x$

25. $\sqrt[4]{y^{12}} = y^{12/4} = y^3$

27. $\sqrt[3]{x^{12}} = x^{12/3} = x^4$

29. $\sqrt[4]{y^4} = y^{4/4} = y^1 = y$

31. $\sqrt[3]{x^{15}} = x^{15/3} = x^5$

33. $8^{4/3} = (\sqrt[3]{8})^4 = 2^4 = 16$

35. $16^{3/4} = (\sqrt[4]{16})^3 = 2^3 = 8$

37. $1^{5/3} = (\sqrt[3]{1})^5 = 1^5 = 1$

39. $9^{3/2} = (\sqrt[2]{9})^3 = 3^3 = 27$

41. $16^{3/4} = (\sqrt[4]{16})^3 = 2^3 = 8$

43. $125^{4/3} = (\sqrt[3]{125})^4 = 5^4 = 625$

45. $27^{-2/3} = \frac{1}{27^{2/3}} = \frac{1}{(\sqrt[3]{27})^2}$

$= \frac{1}{3^2} = \frac{1}{9}$

47. $8^{-5/3} = \frac{1}{8^{5/3}} = \frac{1}{(\sqrt[3]{8})^5}$

$= \frac{1}{2^5} = \frac{1}{32}$

49. $\sqrt[3]{x^7} = x^{7/3}$

51. $\sqrt[3]{x^4} = x^{4/3}$

53. $\sqrt[4]{y^{15}} = y^{15/4}$

55. $\sqrt[4]{y^{21}} = y^{21/4}$

57. $\sqrt[3]{x} \cdot \sqrt[3]{x} = x^{1/3} \cdot x^{1/3}$

$= x^{1/3+1/3} = x^{2/3}$

59. $\sqrt[4]{x^2} \cdot \sqrt[4]{x^2} = x^{2/4} \cdot x^{2/4}$

$= x^{4/4} = x$

61. $(\sqrt[3]{x^2})^6 = (x^{2/3})^6 = x^{6/1 \cdot 2/3}$

$= x^{12/3} = x^4$

63. $(\sqrt[4]{x^2})^4 = (x^{2/4})^4 = x^{2/4 \cdot 4/1} = x^2$

65. $(\sqrt[3]{x})^2 = \sqrt[3]{x^2}$

$(\sqrt[3]{8})^2 = \sqrt[3]{8^2}$

$(2)^2 = \sqrt[3]{64}$

$4 = 4$

67. a) To change an expression in radical form to exponential form, the expression under the radical sign is the base to a fractional exponent whose numerator is the power of the expression and whose denominator is the index number of the radical.

b) $\sqrt[4]{x^9} = x^{9/4}$

Cumulative Review Exercises

69. $3x^2 - 28x + 32$

$= (3x - 4)(x - 8)$

71. $\sqrt{\dfrac{64x^3y^7}{2x^4}} = \sqrt{\dfrac{32y^7}{x}}$

$= \dfrac{\sqrt{32y^7}}{\sqrt{x}} \cdot \dfrac{\sqrt{x}}{\sqrt{x}} = \dfrac{\sqrt{32xy^7}}{x}$

$= \dfrac{\sqrt{16} \cdot \sqrt{2} \cdot \sqrt{y^6} \cdot \sqrt{xy}}{x}$

$= \dfrac{4y^3\sqrt{2xy}}{x}$

Just for fun

1. $\sqrt[3]{xy} \cdot \sqrt[3]{x^2y^2}$

$= (xy)^{1/3} \cdot (x^2y^2)^{1/3}$

$= x^{1/3}\, y^{1/3}\, x^{2/3}\, y^{2/3}$

$= x^{1/3+2/3} \cdot y^{1/3+2/3}$

$= x^{2/3}\, y^{3/3}$

$= xy$

2. $\sqrt[4]{3x^2y} \cdot \sqrt[4]{27x^6y^3}$

$= (3x^2y)^{1/4}\,(27x^6y^3)^{1/4}$

$= (3x^2y)^{1/4}\,(3^3x^6y^3)^{1/4}$

$= 3^{1/4}\, x^{2/4}\, y^{1/4}\, 3^{3/4}\, x^{6/4}\, y^{3/4}$

$= 3^{1/4+3/4}\, x^{2/4+6/4}\, y^{1/4+3/4}$

$= 3^{4/4}\, x^{8/4}\, y^{4/4}$

$= 3x^2y$

3. $\sqrt[4]{32} - \sqrt[4]{2} = \sqrt[4]{16} \cdot \sqrt[4]{2} - \sqrt[4]{2}$

$= 2\sqrt[4]{2} - 1\sqrt[4]{2}$

$= 1\sqrt[4]{2}$

$= \sqrt[4]{2}$

4. $\sqrt[3]{3x^3y} + \sqrt[3]{24x^3y}$

$= \sqrt[3]{3}\, \sqrt[3]{x^3}\, \sqrt[3]{y}$

$+ \sqrt[3]{8}\, \sqrt[3]{3}\, \sqrt[3]{x^3}\, \sqrt[3]{y}$

$= \sqrt[3]{3} \cdot x \sqrt[3]{y}$

$+ 2\sqrt[3]{3} \cdot x \sqrt[3]{y}$

$= x\sqrt[3]{3y} + 2x\sqrt[3]{3y}$

$= 3x\sqrt[3]{3y}$

5. $\dfrac{1}{\sqrt[3]{2}} = \dfrac{1}{2^{1/3}}$

$= \dfrac{1}{2^{1/3}} \cdot \dfrac{2^{2/3}}{2^{2/3}}$

$$= \frac{2^{2/3}}{2^{1/3+2/3}}$$

$$= \frac{2^{2/3}}{2^{3/3}}$$

$$= \frac{2^{2/3}}{2^1}$$

$$= \frac{\sqrt[3]{2^2}}{2}$$

$$= \frac{\sqrt[3]{4}}{2}$$

6. $\frac{1}{\sqrt[3]{x}} = \frac{1}{x^{1/3}}$

$$= \frac{1}{x^{1/3}} \cdot \frac{x^{2/3}}{x^{2/3}} = \frac{x^{2/3}}{x^{1/3+2/3}}$$

$$= \frac{x^{2/3}}{x^{3/3}} = \frac{x^{2/3}}{x}$$

$$= \frac{\sqrt[3]{x^2}}{x}$$

Chapter 9 Review

1. $\sqrt{25} = 5$

3. $-\sqrt{81} = -9$

5. $\sqrt{26x} = (26x)^{1/2}$

7. $\sqrt{32} = \sqrt{16} \cdot \sqrt{2} = 4\sqrt{2}$

9. $\sqrt{45x^5y^4} = \sqrt{9x^4y^4} \cdot \sqrt{5x}$

$$= 3x^2y^2\sqrt{5x}$$

11. $\sqrt{15x^5yz^3} = \sqrt{x^4z^2} \cdot \sqrt{15xyz}$

$$= x^2z\sqrt{15xyz}$$

13. $\sqrt{8} \cdot \sqrt{12} = \sqrt{96} = \sqrt{16} \cdot \sqrt{6}$

$$= 4\sqrt{6}$$

15. $\sqrt{18x} \cdot \sqrt{2xy}$

$$= \sqrt{36x^2} \cdot \sqrt{y} = 6x\sqrt{y}$$

17. $\sqrt{20xy^4} \cdot \sqrt{5xy^3} = \sqrt{100x^2y^7}$

$$= \sqrt{100x^2y^6}\sqrt{y}$$

$$= 10xy^3\sqrt{y}$$

19. $\frac{\sqrt{32}}{\sqrt{2}} = \sqrt{\frac{32}{2}} = \sqrt{16} = 4$

21. $\sqrt{\frac{7}{28}} = \sqrt{\frac{1}{4}} = \frac{\sqrt{1}}{\sqrt{4}} = \frac{1}{2}$

23. $\sqrt{\frac{5x}{12}} = \frac{\sqrt{5x}}{\sqrt{4 \cdot 3}} \cdot \frac{\sqrt{3}}{\sqrt{3}}$

$$= \frac{\sqrt{15x}}{6}$$

25. $\sqrt{\frac{x^2}{2}} = \frac{\sqrt{x^2}}{\sqrt{2}}$

$$= \frac{x}{\sqrt{2}} \cdot \frac{\sqrt{2}}{\sqrt{2}} = \frac{x\sqrt{2}}{2}$$

27. $\sqrt{\frac{60xy^5}{4x^5y^3}} = \sqrt{\frac{15y^2}{x^4}} = \frac{\sqrt{15y^2}}{\sqrt{x^4}}$

$$= \frac{y\sqrt{15}}{x^2}$$

29. $\frac{\sqrt{90}}{\sqrt{8x^3y^2}} = \sqrt{\frac{90}{8x^3y^2}}$

$$= \frac{\sqrt{45}}{\sqrt{4x^3y^2}} = \frac{\sqrt{9}\ \sqrt{5}}{\sqrt{4x^3y^2}} \cdot \frac{\sqrt{x}}{\sqrt{x}}$$

$$= \frac{3\sqrt{5x}}{2x^2y}$$

31. $\dfrac{3}{1+\sqrt{2}} = \dfrac{3}{(1+\sqrt{2})}\cdot\dfrac{(1-\sqrt{2})}{(1-\sqrt{2})}$

$= \dfrac{3(1-\sqrt{2})}{1-2}$

$= \dfrac{-3(1-\sqrt{2})}{1}$

$= -3 + 3\sqrt{2}$

33. $\dfrac{\sqrt{3}}{2+\sqrt{x}} = \dfrac{\sqrt{3}(2-\sqrt{x})}{(2+\sqrt{x})(2-\sqrt{x})}$

$= \dfrac{2\sqrt{3}-\sqrt{3x}}{4-x}$

35. $\dfrac{\sqrt{5}}{\sqrt{x}+\sqrt{3}} = \dfrac{\sqrt{5}(\sqrt{x}-\sqrt{3})}{(\sqrt{x}+\sqrt{3})(\sqrt{x}-\sqrt{3})}$

$= \dfrac{\sqrt{5x}-\sqrt{15}}{x-3}$

37. $6\sqrt{2} - 8\sqrt{2} + \sqrt{2} = -\sqrt{2}$

39. $\sqrt{x} + 3\sqrt{x} - 4\sqrt{x} = 0$

41. $7\sqrt{40} - 2\sqrt{10}$

$= 7\sqrt{4}\ \sqrt{10} - 2\sqrt{10}$

$= 7\cdot 2\sqrt{10} - 2\sqrt{10}$

$= 14\sqrt{10} - 2\sqrt{10}$

$= 12\sqrt{10}$

43. $3\sqrt{18} + 5\sqrt{50} - 2\sqrt{32}$

$= 3\sqrt{9}\ \sqrt{2} + 5\sqrt{25}\ \sqrt{2}$

$- 2\sqrt{16}\ \sqrt{2}$

$= 3\cdot 3\sqrt{2} + 5\cdot 5\sqrt{2} - 2\cdot 4\sqrt{2}$

$= 9\sqrt{2} + 25\sqrt{2} - 8\sqrt{2}$

$= 26\sqrt{2}$

45. $\sqrt{x} = 9$ Ck: $\sqrt{81} = 9$

$(\sqrt{x})^2 = 9^2$ $\quad 9 = 9$

$x = 81$ $\quad$ True

47. $\sqrt{x-3} = 6$ Ck: $\sqrt{39-3} = 6$

$(\sqrt{x-3})^2 = 6^2$ $\quad \sqrt{36} = 6$

$x - 3 = 36$ $\quad 6 = 6$

$\quad +3 \quad +3$ $\quad$ True

$x = 39$

49. $\sqrt{2x+4} = \sqrt{3x-5}$

$(\sqrt{2x+4})^2 = (\sqrt{3x-5})^2$

$2x + 4 = 3x - 5$

$-2x \quad\quad -2x$

$4 = x - 5$

$5 + 4 = x - 5 + 5$

$9 = x$

Ck: $\sqrt{2(9)+4} = \sqrt{3\cdot 9 - 5}$

$\sqrt{18+4} = \sqrt{27-5}$

$\sqrt{22} = \sqrt{22}$

True

51. $\sqrt{x^2+4} = x + 2$

$(\sqrt{x^2+4})^2 = (x+2)^2$

$x^2 + 4 = x^2 + 4x + 4$

$-x^2 \quad = \quad -x^2$

$4 = 4x + 4$

$-4 = \quad -4$

$\dfrac{0}{4} = \dfrac{4x}{4}$

$0 = x$

Ck: $\sqrt{0^2+4} = 0 + 2$

$\sqrt{4} = 2$

$$2 = 2$$
True

53. $$3\sqrt{2x + 3} = 9$$
$$(3\sqrt{2x + 3})^2 = 9^2$$
$$9(2x + 3) = 81$$
$$18x + 27 = 81$$
$$-27 \quad -27$$
$$\frac{18x}{18} = \frac{54}{18}$$
$$x = 3$$

Ck: $$3\sqrt{2(3) + 3} = 9$$
$$3\sqrt{6 + 3} = 9$$
$$3\sqrt{9} = 9$$
$$3 \cdot 3 = 9$$
$$9 = 9$$
True

55. $$c^2 = a^2 + x^2$$
$c = 13 \quad a = 9$
$$13^2 = 9^2 + x^2$$
$$169 = 81 + x^2$$
$$-81 \quad -81$$
$$88 = x^2$$
$$\sqrt{88} = \sqrt{x^2}$$
$$\sqrt{4}\ \sqrt{22} = x$$
$$2\sqrt{22} = x$$

57. $$x^2 = a^2 + b^2$$
$a = 5 \quad b = 6$
$$x^2 = 5^2 + 6^2$$
$$x^2 = 25 + 36$$
$$x = 61$$
$$\sqrt{x^2} = \sqrt{61}$$
$$x = \sqrt{61}$$

59. $$c^2 = a^2 + b^2$$
$a = 6$ inches $\quad b = 15$ inches
$$c^2 = 6^2 + 15^2$$
$$c^2 = 36 + 225$$
$$c^2 = 261 \text{ inches}$$
$$c = \sqrt{261} \approx 16.15 \text{ inches}$$

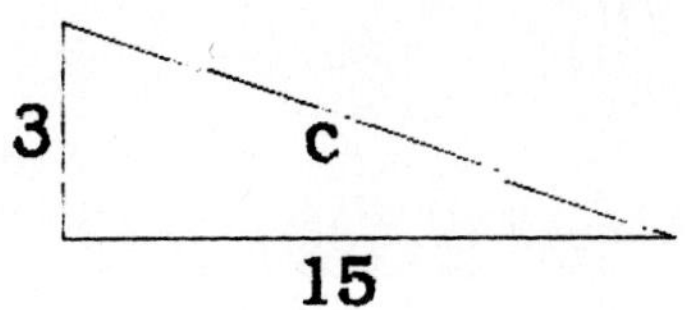

61. $$d = \sqrt{(x_2-x_1)^2 + (y_2-y_1)^2}$$
$$(x_1, y_1) = (6, 5)$$
$$d = \sqrt{(-6-6)^2 + (8 - 5)^2}$$
$$(x_2, y_2 = (-6, 8)$$
$$d = \sqrt{(-12)^2 + 3^2}$$
$$d = \sqrt{144 + 9}$$
$$d = \sqrt{153}$$
$$d = 12.37$$

63. $\sqrt[3]{8} = \sqrt[3]{2^3} = 2$

65. $\sqrt[4]{16} = \sqrt[4]{2^4} = 2$

67. $\sqrt[3]{24} = \sqrt[3]{8} \cdot \sqrt[3]{3} = 2\ \sqrt[3]{3}$

69. $\sqrt[3]{48} = \sqrt[3]{8} \cdot \sqrt[3]{6} = 2\ \sqrt[3]{6}$

71. $\sqrt[4]{96} = \sqrt[4]{16} \cdot \sqrt[4]{6} = 2\ \sqrt[4]{6}$

73. $\sqrt[3]{x^{12}} = (x^{1/3})^{12} = x^4$

75. $\sqrt[4]{y^{20}} = (y^{1/4})^{20} = y^5$

77. $16^{1/2} = \sqrt{16} = 4$

79. $64^{2/3} = (\sqrt[3]{64})^2 = 4^2 = 16$

81. $25^{3/2} = (\sqrt{25})^{3} = 5^{3} = 125$

83. $\sqrt[3]{x^{10}} = (x^{1/3})^{10} = x^{10/3}$

85. $\sqrt{x^{5}} = (x^{1/2})^{5} = x^{5/2}$

87. $\sqrt[4]{x^{7}} = (x^{1/4})^{7} = x^{7/4}$

89. $\sqrt[3]{x} \cdot \sqrt[3]{x} = x^{1/3} \cdot x^{1/3}$

$= x^{1/3+1/3}$

$= x^{2/3}$

$= \sqrt[3]{x^{2}}$

91. $\sqrt[4]{x^{2}} \cdot \sqrt[4]{x^{6}}$

$= x^{2/4} \cdot x^{6/4} = x^{2/4+6/4}$

$= x^{8/4}$

$= x^{2}$

93. $(\sqrt[3]{x^{2}})^{3} = (x^{2/3})^{3} = (x^{2/3})^{3/1}$

$= x^{2/3 \cdot 3/1}$

$= x^{2}$

95. $(\sqrt[4]{x^{3}})^{8} = (x^{3/4})^{8} = x^{3/4 \cdot 8/1}$

$= x^{24/4}$

$= x^{6}$

Practice Test

1. $\sqrt{3xy} = (3xy)^{1/2}$

2. $\sqrt{(x + 3)^{2}} = x + 3$

3. $\sqrt{96} = \sqrt{16}\sqrt{6} = 4\sqrt{6}$

4. $\sqrt{12x^{2}} = \sqrt{4x^{2}}\sqrt{3} = 2x\sqrt{3}$

5. $\sqrt{32x^{4}y^{5}} = \sqrt{16x^{4}y^{4}} \cdot \sqrt{2y}$

$= 4x^{2}y^{2}\sqrt{2y}$

6. $\sqrt{8x^{2}y}\sqrt{6xy} = \sqrt{48x^{3}y^{2}}$

$= \sqrt{16x^{2}y^{2}}\sqrt{3x}$

$= 4xy\sqrt{3x}$

7. $\sqrt{15xy^{2}} \cdot \sqrt{5x^{3}y^{3}} = \sqrt{75x^{4}y^{5}}$

$= \sqrt{25x^{4}y^{4}}\sqrt{3y}$

$= 5x^{2}y^{2}\sqrt{3y}$

8. $\sqrt{\frac{5}{125}} = \sqrt{\frac{1}{25}} = \frac{\sqrt{1}}{\sqrt{25}} = \frac{1}{5}$

9. $\frac{\sqrt{3xy^{2}}}{\sqrt{48x^{3}}} = \sqrt{\frac{3xy^{2}}{48x^{3}}} = \sqrt{\frac{y^{2}}{16x^{2}}} = \frac{y}{4x}$

10. $\frac{1}{\sqrt{2}} = \frac{1 \cdot \sqrt{2}}{\sqrt{2}\sqrt{2}} = \frac{\sqrt{2}}{\sqrt{4}} = \frac{\sqrt{2}}{2}$

11. $\sqrt{\frac{4x}{5}} = \frac{\sqrt{4x}}{\sqrt{5}} \cdot \frac{\sqrt{5}}{\sqrt{5}} = \frac{2\sqrt{5x}}{5}$

12. $\sqrt{\frac{40x^{2}y^{5}}{6x^{3}y^{7}}} = \sqrt{\frac{20}{3xy^{2}}}$

$= \frac{\sqrt{4}\sqrt{5}}{\sqrt{3xy^{2}}} \cdot \frac{\sqrt{3x}}{\sqrt{3x}}$

$= \frac{2\sqrt{15x}}{3xy}$

13. $\frac{3}{(2 + \sqrt{5})} \cdot \frac{(2 - \sqrt{5})}{(2 - \sqrt{5})}$

$= \frac{3(2 - \sqrt{5})}{4 - 5}$

$= \frac{3(2 - \sqrt{5})}{-1}$

$= -3(2 - \sqrt{5})$

14. $\frac{6}{\sqrt{x} - 3}$

$= \frac{6}{(\sqrt{x} - 3)} \cdot \frac{(\sqrt{x} + 3)}{(\sqrt{x} + 3)}$

$= \frac{6(\sqrt{x} + 3)}{x - 9}$

15. $\sqrt{48} + \sqrt{75} + 2\sqrt{3}$

$= \sqrt{16}\ \sqrt{3} + \sqrt{25}\ \sqrt{3} + 2\sqrt{3}$

$= 4\sqrt{3} + 5\ \sqrt{3} + 2\sqrt{3}$

$= 11\ \sqrt{3}$

16. $4\sqrt{x} - 6\sqrt{x} - \sqrt{x} = -3\sqrt{x}$

17. $\sqrt{x + 5} = 9$ Ck: $\sqrt{76 + 5} = 9$

$(\sqrt{x + 5})^2 = 9^2$ $\qquad \sqrt{81} = 9$

$x + 5 = 81$ $\qquad 9 = 9$

$-5 \quad -5$ $\qquad$ True

$x = 76$

18. $2\sqrt{x - 4} + 4 = x$

$2\sqrt{x - 4} + 4 - 4 = x - 4$

$2\sqrt{x - 4} = x - 4$

$(2\sqrt{x - 4})^2 = (x - 4)^2$

$4(x - 4) = x^2 - 8x + 16$

$4x - 16 = x^2 - 8x + 16$

$4x - 16 - 4x + 16$

$= x^2 - 8x + 16 - 4x + 16$

$0 = x^2 - 12x + 32$

$0 = (x - 8)(x - 4)$

$x - 8 = 0$ $\qquad x - 4 = 0$

$x-8+8 = 0+8$ $\qquad x-4+4 = 0+4$

$x = 8$ $\qquad x = 4$

Ck: $x = 8$ $\qquad$ Ck: $x = 4$

$2\sqrt{8-4} + 4 = 8$ $\qquad 2\sqrt{4-4} + 4 = 4$

$2\sqrt{4} + 4 = 8$ $\qquad 2\sqrt{0} + 4 = 4$

$2(2) + 4 = 8$ $\qquad 4 = 4$

$8 = 8$ $\qquad$ True

19. $a = 5 \quad b = 9$

$x^2 = a^2 + b^2$

$x^2 = 5^2 + 9^2$

$x^2 = 25 + 81$

$x^2 = 106$

$x = \sqrt{106} = 10.30$

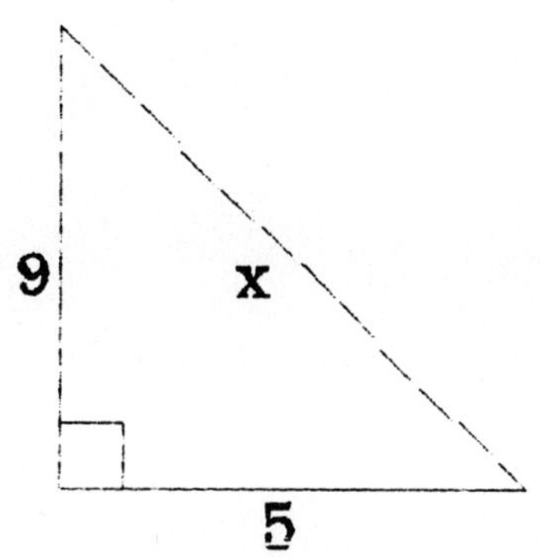

20. $d = \sqrt{(x_2-x_1)^2 + (y_2-y_1)^2}$

$x_1, y_1 = (-2, -4) \qquad x_2, y_2 = (5, 1)$

$d = \sqrt{[5-(-2)]^2 + [1-(-4)]^2}$

$d = \sqrt{(5+2)^2 + (1+4)^2}$

$d = \sqrt{7^2 + 5^2}$

$d = \sqrt{49 + 25}$

$d = \sqrt{74}$

$d = 8.60$

21. $27^{-4/3} = \frac{1}{27^{4/3}} = \frac{1}{(\sqrt[3]{27})^4}$

$$= \frac{1}{3^4} = \frac{1}{81}$$

22. $\sqrt[3]{x^4} \cdot \sqrt[3]{x^{11}} = \sqrt[3]{x^{15}}$

$$= (x^{1/3})^{15} = x^5$$

CHAPTER TEN

Exercise Set 10.1

1. $x^2 = 16$

$\sqrt{x^2} = \pm\sqrt{16}$

$x = \pm 4$

3. $x^2 = 100$

$\sqrt{x^2} = \pm\sqrt{100}$

$x = \pm 10$

5. $y^2 = 36$

$\sqrt{y^2} = \pm\sqrt{36}$

$y = \pm 6$

7. $x^2 = 10$

$\sqrt{x^2} = \pm\sqrt{10}$

$x = \pm\sqrt{10}$

9. $x^2 = 8$

$\sqrt{x^2} = \pm\sqrt{8}$

$x = \pm\sqrt{4 \cdot 2}$

$x = \pm 2\sqrt{2}$

11. $3x^2 = 12$

$\frac{3x^2}{3} = \frac{12}{3}$

$x^2 = 4$

$\sqrt{x^2} = \pm\sqrt{4}$

$x = \pm 2$

13. $2w^2 = 34$

$\frac{2w^2}{2} = \frac{34}{2}$

$w^2 = 17$

$\sqrt{w^2} = \pm\sqrt{17}$

$w = \pm\sqrt{17}$

15. $2x^2 + 1 = 19$

$\quad\quad -1 \quad -1$

$2x^2 = 18$

$\frac{2x^2}{2} = \frac{18}{2}$

$x^2 = 9$

$\sqrt{x^2} = \pm\sqrt{9}$

$x = \pm 3$

17. $4w^2 - 3 = 12$

$\quad\quad +3 \quad +3$

$4w^2 = 15$

$\frac{4w^2}{4} = \frac{15}{4}$

$w^2 = \frac{15}{4}$

$\sqrt{w^2} = \pm\sqrt{\frac{15}{4}}$

$w = \pm\frac{\sqrt{15}}{2}$

19. $5x^2 - 9 = 30$

$\quad\quad +9 \quad +9$

$5x^2 = 39$

$\frac{5x^2}{5} = \frac{39}{5}$

$x^2 = \frac{39}{5}$

$\sqrt{x^2} = \pm\sqrt{\frac{39}{5}}$

$x = \pm\sqrt{\frac{39 \cdot 5}{5 \cdot 5}}$

$$x = \pm \frac{\sqrt{195}}{5}$$

21. $(x + 1)^2 = 4$

$$x + 1 = \pm \sqrt{4}$$
$$x + 1 = \pm 2$$

$$\begin{array}{rcl} x + 1 &=& 2 \\ -1 && -1 \\ \hline x &=& 1 \end{array} \qquad \begin{array}{rcl} x + 1 &=& -2 \\ -1 && -1 \\ \hline x &=& -3 \end{array}$$

23. $(x - 3)^2 = 16$

$$x - 3 = \pm \sqrt{16}$$
$$x - 3 = \pm 4$$

$$\begin{array}{rcl} x - 3 &=& 4 \\ +3 && +3 \\ \hline x &=& 7 \end{array} \qquad \begin{array}{rcl} x - 3 &=& -4 \\ +3 && +3 \\ \hline x &=& -1 \end{array}$$

25. $(x + 4)^2 = 36$

$$x + 4 = \pm \sqrt{36}$$
$$x + 4 = \pm 6$$

$$\begin{array}{rcl} x + 4 &=& 6 \\ -4 && -4 \\ \hline x &=& 2 \end{array} \qquad \begin{array}{rcl} x + 4 &=& -6 \\ -4 && -4 \\ \hline x &=& -10 \end{array}$$

27. $(x - 1)^2 = 12$

$$x - 1 = \pm \sqrt{12}$$
$$x - 1 = \pm \sqrt{4 \cdot 3}$$
$$x - 1 = \pm 2\sqrt{3}$$

$$\begin{array}{rcl} x - 1 &=& 2\sqrt{3} \\ +1 && +1 \\ \hline x &=& 1 + 2\sqrt{3} \end{array} \qquad \begin{array}{rcl} x - 1 &=& -2\sqrt{3} \\ +1 && +1 \\ \hline x &=& 1 - 2\sqrt{3} \end{array}$$

29. $(x + 6)^2 = 20$

$$x + 6 = \pm \sqrt{20}$$
$$x + 6 = \pm \sqrt{4 \cdot 5}$$
$$x + 6 = \pm 2\sqrt{5}$$

$$\begin{array}{rcl} x + 6 &=& 2\sqrt{5} \\ -6 && -6 \\ \hline x &=& -6 + 2\sqrt{5} \end{array} \qquad \begin{array}{rcl} x + 6 &=& -2\sqrt{5} \\ -6 && -6 \\ \hline x &=& -6 - 2\sqrt{5} \end{array}$$

31. $(x + 2)^2 = 25$

$$x + 2 = \pm \sqrt{25}$$
$$x + 2 = \pm 5$$

$$\begin{array}{rcl} x + 2 &=& 5 \\ -2 && -2 \\ \hline x &=& 3 \end{array} \qquad \begin{array}{rcl} x + 2 &=& -5 \\ -2 && -2 \\ \hline x &=& -7 \end{array}$$

33. $(x - 9)^2 = 100$

$$x - 9 = \pm \sqrt{100}$$
$$x - 9 = \pm 10$$

$$\begin{array}{rcl} x - 9 &=& 10 \\ +9 && +9 \\ \hline x &=& 19 \end{array} \qquad \begin{array}{rcl} x - 9 &=& -10 \\ +9 && +9 \\ \hline x &=& -1 \end{array}$$

35. $(2x + 3)^2 = 18$

$$2x + 3 = \pm \sqrt{18}$$
$$2x + 3 = \pm \sqrt{9 \cdot 2}$$
$$2x + 3 = \pm 3\sqrt{2}$$

$$\begin{array}{rcl} 2x + 3 &=& 3\sqrt{2} \\ -3 && -3 \\ \hline \dfrac{2x}{2} &=& \dfrac{-3 + 3\sqrt{2}}{2} \end{array}$$

$$x = \frac{-3 + 3\sqrt{2}}{2}$$

$$2x + 3 = -3\sqrt{2}$$
$$\quad -3 \quad -3$$

$$\frac{2x}{2} = \frac{-3 - 3\sqrt{2}}{2}$$

$$x = \frac{-3 - 3\sqrt{2}}{2}$$

37. $(4x + 1)^2 = 20$

$$4x + 1 = \pm \sqrt{20}$$

$$4x + 1 = \pm \sqrt{4 \cdot 5}$$

$$4x + 1 = \pm 2\sqrt{5}$$

$$4x + 1 = 2\sqrt{5}$$
$$\quad -1 \quad -1$$

$$\frac{4x}{4} = \frac{-1 + 2\sqrt{5}}{4}$$

$$x = \frac{-1 + 2\sqrt{5}}{4}$$

$$4x + 1 = -2\sqrt{5}$$
$$\quad -1 \quad -1$$

$$\frac{4x}{4} = \frac{-1 - 2\sqrt{5}}{4}$$

$$x = \frac{-1 - 2\sqrt{5}}{4}$$

39. $(2x - 6)^2 = 18$

$$2x - 6 = \pm \sqrt{18}$$

$$2x - 6 = \pm \sqrt{9 \cdot 2}$$

$$2x - 6 = \pm 3\sqrt{2}$$

$$2x - 6 = 3\sqrt{2}$$
$$\quad +6 \quad +6$$

$$\frac{2x}{2} = \frac{6 + 3\sqrt{2}}{2}$$

$$x = \frac{6 + 3\sqrt{2}}{2}$$

$$2x - 6 = -3\sqrt{2}$$
$$\quad +6 \quad +6$$

$$\frac{2x}{2} = \frac{6 - 3\sqrt{2}}{2}$$

$$x = \frac{6 - 3\sqrt{2}}{2}$$

41. x = width
$2x$ = length

$$x(2x) = 80$$

$$\frac{2x^2}{2} = \frac{80}{2}$$

$$x^2 = 40$$

$$x = \pm \sqrt{40}$$

$$x = \pm \sqrt{4 \cdot 10}$$

$x = \pm 2\sqrt{10}$ discard negative values

width $x = 2\sqrt{10} \approx 6.32$ feet

length $2x = 4\sqrt{10} \approx 12.65$ feet

43. $x^2 = 36$ is one possible answer.

45. $x^2 - \square = 27$

$x^2 - 9 = 27$ has solutions of 6 and -6. The solutions needed an equation equivalent to $x^2 = 36$.

Cumulative Review Exercises

47. $4x^2 - 10x - 24$

$$= 2(2x^2 - 5x - 12)$$

$= 2(2x + 3)(x - 4)$

49. Let x = amount invested at 6%.
y = amount invested at 8%.

$$x + y = 10{,}000$$
$$0.06x + 0.08y = 760$$

Solve for x:

$$x + y = 10{,}000$$
$$x = 10{,}000 - y$$

Substitute:

$$0.06x + 0.08y = 760$$
$$0.06(10{,}000 - y) + 0.08y = 760$$
$$600 - 0.06y + 0.08y = 760$$
$$600 + 0.02y = 760$$
$$0.02y = 160$$
$$\frac{0.02y}{0.02} = \frac{160}{0.02}$$

$y = 8000$
$x = 10{,}000 - y = 10{,}000 - 8{,}000 = 2000$
Collette invested $8000 at 8% and $2000 at 6%.

Exercise Set 10.2

1. $x^2 + 2x - 3 = 0$

$$x^2 + 2x = 3$$
$$x^2 + 2x + 1 = 3 + 1$$
$$(x + 1)^2 = 4$$
$$x + 1 = \pm\sqrt{4}$$
$$x + 1 = \pm 2$$

$$x + 1 = 2 \qquad x + 1 = -2$$
$$-1 \quad -1 \qquad -1 \quad -1$$
$$x = 1 \qquad x = -3$$

$$\frac{1}{2}b = \frac{1}{2}(2) = 1$$

$$\left(\frac{1}{2}b\right)^2 = 1^2 = 1$$

3. $x^2 - 4x - 5 = 0$

$$x^2 - 4x = 5$$
$$x^2 - 4x + 4 = 5 + 4$$
$$(x - 2)^2 = 9$$
$$x - 2 = \pm\sqrt{9}$$
$$x - 2 = \pm 3$$

$$x - 2 = 3 \qquad x - 2 = -3$$
$$+2 \quad +2 \qquad +2 \quad +2$$
$$x = 5 \qquad x = -1$$

$$\frac{1}{2}b = \frac{1}{2}(-4) = -2$$

$$\left(\frac{1}{2}b\right)^2 = (-2)^2 = 4$$

5. $x^2 + 3x + 2 = 0$

$$x^2 + 3x = -2$$
$$x^2 + 3x + \frac{9}{4} = -2 + \frac{9}{4}$$
$$\left(x + \frac{3}{2}\right)^2 = \frac{-8 + 9}{4}$$
$$x + \frac{3}{2} = \pm\sqrt{\frac{1}{4}}$$
$$x + \frac{3}{2} = \pm\frac{1}{2}$$

$$x + \frac{3}{2} = \frac{1}{2} \qquad x + \frac{3}{2} = \frac{-1}{2}$$
$$\frac{-3}{2} \quad \frac{-3}{2} \qquad \frac{-3}{2} \quad \frac{-3}{2}$$
$$x = -1 \qquad x = -2$$

$$\frac{1}{2}b = \frac{1}{2}(3) = \frac{3}{2}$$

$$(\frac{1}{2}b)^2 = (\frac{3}{2})^2 = \frac{9}{4}$$

7. $x^2 - 8x + 15 = 0$

$$x^2 - 8x = -15$$
$$x^2 - 8x + 16 = -15 + 16$$
$$(x - 4)^2 = 1$$
$$x - 4 = \pm\sqrt{1}$$
$$x - 4 = \pm 1$$

$$\begin{array}{ll} x - 4 = 1 & x - 4 = -1 \\ \underline{\quad +4 \quad +4} & \underline{\quad +4 \quad +4} \\ x = 5 & x = 3 \end{array}$$

$$\frac{1}{2}b = \frac{1}{2}(-8) = -4$$

$$(\frac{1}{2}b)^2 = (-4)^2 = 16$$

9. $x^2 = -6x - 9$

$$x^2 + 6x = -9$$
$$x^2 + 6x + 9 = -9 + 9$$
$$(x + 3)^2 = 0$$
$$x + 3 = 0$$
$$\frac{-3}{x} = \frac{-3}{-3}$$

$$\frac{1}{2}b = \frac{1}{2}(6) = 3$$

$$(\frac{1}{2}b)^2 = 3^2 = 9$$

11. $x^2 = -5x - 6$

$$x^2 + 5x = -6$$
$$x^2 + 5x + \frac{25}{4} = -6 + \frac{25}{4}$$
$$(x + \frac{5}{2})^2 = \frac{-24 + 25}{4}$$
$$x + \frac{5}{2} = \pm\sqrt{\frac{1}{4}}$$
$$x + \frac{5}{2} = \pm\frac{1}{2}$$

$$\begin{array}{ll} x + \frac{5}{2} = \frac{1}{2} & x + \frac{5}{2} = \frac{-1}{2} \\ \underline{\quad \frac{-5}{2} \quad \frac{-5}{2}} & \underline{\quad \frac{-5}{2} \quad \frac{-5}{2}} \\ x = \frac{-4}{2} & x = \frac{-6}{2} \\ x = -2 & x = -3 \end{array}$$

$$\frac{1}{2}b = \frac{1}{2}(5) = \frac{5}{2}$$

$$(\frac{1}{2}b)^2 = (\frac{5}{2})^2 = \frac{25}{4}$$

13. $x^2 + 9x + 18 = 0$

$$x^2 + 9x = -18$$
$$x^2 + 9x + \frac{81}{4} = -18 + \frac{81}{4}$$
$$(x + \frac{9}{2})^2 = \frac{-72 + 81}{4}$$
$$x + \frac{9}{2} = \pm\sqrt{\frac{9}{4}}$$
$$x + \frac{9}{2} = \pm\frac{3}{2}$$

$$\begin{array}{ll} x + \frac{9}{2} = \frac{3}{2} & x + \frac{9}{2} = \frac{-3}{2} \\ \underline{\quad \frac{-9}{2} \quad \frac{-9}{2}} & \underline{\quad \frac{-9}{2} \quad \frac{-9}{2}} \\ x = \frac{-6}{2} & x = \frac{-12}{2} \\ x = -3 & x = -6 \end{array}$$

$$\frac{1}{2}b = \frac{1}{2}(9) = \frac{9}{2}$$

$(\frac{1}{2}b)^2 = (\frac{9}{2})^2 = \frac{81}{4}$

15.
$$x^2 = 15x - 56$$
$$x^2 - 15x = -56$$
$$x^2 - 15x + \frac{225}{4} = -56 + \frac{225}{4}$$
$$(x - \frac{15}{2})^2 = \frac{-224 + 225}{4}$$
$$x - \frac{15}{2} = \pm\sqrt{\frac{1}{4}}$$
$$x - \frac{15}{2} = \pm\frac{1}{2}$$

$x - \frac{15}{2} = \frac{1}{2}$ $\quad$ $x - \frac{15}{2} = \frac{-1}{2}$

$+\frac{15}{2}$ $+\frac{15}{2}$ $\quad$ $+\frac{15}{2}$ $+\frac{15}{2}$

$x = \frac{16}{2}$ $\quad$ $x = \frac{14}{2}$

$x = 8$ $\quad$ $x = 7$

$\frac{1}{2}b = \frac{1}{2}(-15) = \frac{-15}{2}$

$(\frac{1}{2}b)^2 = (\frac{-15}{2})^2 = \frac{225}{4}$

17.
$$-4x = -x^2 + 12$$
$$x^2 - 4x = 12$$
$$x^2 - 4x + 4 = 12 + 4$$
$$(x - 2)^2 = 16$$
$$x - 2 = \pm\sqrt{16}$$
$$x - 2 = \pm 4$$

$x - 2 = 4$ $\quad$ $x - 2 = -4$

$+2$ $+2$ $\quad$ $+2$ $+2$

$x = 6$ $\quad$ $x = -2$

$\frac{1}{2}b = \frac{1}{2}(-4) = -2$

$(\frac{1}{2}b)^2 = (-2)^2 = 4$

19.
$$x^2 + 2x - 6 = 0$$
$$x^2 + 2x = 6$$
$$x^2 + 2x + 1 = 6 + 1$$
$$(x + 1)^2 = 7$$
$$x + 1 = \pm\sqrt{7}$$

-1 $\quad$ -1

$$x = -1 \pm\sqrt{7}$$

$\frac{1}{2}b = \frac{1}{2}(2) = 1$

$(\frac{1}{2}b)^2 = 1^2 = 1$

21.
$$6x + 6 = -x^2$$
$$x^2 + 6x = -6$$
$$x^2 + 6x + 9 = -6 + 9$$
$$(x + 3)^2 = 3$$
$$x + 3 = \pm\sqrt{3}$$

-3 $\quad$ -3

$$x = -3 \pm\sqrt{3}$$

$\frac{1}{2}b = \frac{1}{2}(6) = 3$

$(\frac{1}{2}b)^2 = 3^2 = 9$

23.
$$-x^2 + 5x = -8$$
$$-1(x^2 + 5x) = -1(-8)$$
$$x^2 - 5x = 8$$
$$x^2 - 5x + \frac{25}{4} = 8 + \frac{25}{4}$$
$$(x - \frac{5}{2})^2 = \frac{32 + 25}{4}$$

$$x - \frac{5}{2} = \pm\sqrt{\frac{57}{4}}$$

$$x - \frac{5}{2} = \pm\frac{\sqrt{57}}{2}$$

$$+\frac{5}{2} \qquad +\frac{5}{2}$$

$$x = \frac{5 \pm \sqrt{57}}{2}$$

$$\frac{1}{2}b = \frac{1}{2}(-5) = \frac{-5}{2}$$

$$(\frac{1}{2}b)^2 = (\frac{-5}{2})^2 = \frac{25}{4}$$

25. $2x^2 + 4x - 6 = 0$

$$\frac{1}{2}(2x^2 + 4x) = \frac{1}{2}(6)$$

$$x^2 + 2x = 3$$

$$x^2 + 2x + 1 = 3 + 1$$

$$(x + 1)^2 = 4$$

$$x + 1 = \pm\sqrt{4}$$

$$x + 1 = \pm 2$$

$$x + 1 = 2 \qquad x + 1 = -2$$

$$-1 \quad -1 \qquad -1 \quad -1$$

$$x = 1 \qquad x = -3$$

$$\frac{1}{2}b = \frac{1}{2}(2) = 1$$

$$(\frac{1}{2}b)^2 = 1^2 = 1$$

27. $2x^2 + 18x + 4 = 0$

$$\frac{1}{2}(2x^2 + 18x) = \frac{1}{2}(-4)$$

$$x^2 + 9x = -2$$

$$x + 9x + \frac{81}{4} = -2 + \frac{81}{4}$$

$$(x + \frac{9}{2})^2 = \frac{-8 + 81}{4}$$

$$x + \frac{9}{2} = \pm\sqrt{\frac{73}{4}}$$

$$x + \frac{9}{2} = \pm\frac{\sqrt{73}}{2}$$

$$\frac{-9}{2} \qquad \frac{-9}{2}$$

$$x = \frac{-9 \pm \sqrt{73}}{2}$$

$$\frac{1}{2}b = \frac{1}{2}(9) = \frac{9}{2}$$

$$(\frac{1}{2}b)^2 = (\frac{9}{2})^2 = \frac{81}{4}$$

29. $3x^2 + 33x + 72 = 0$

$$\frac{1}{3}(3x^2 + 33x) = \frac{1}{3}(-72)$$

$$x^2 + 11x = -24$$

$$x^2 + 11x + \frac{121}{4} = -24 + \frac{121}{4}$$

$$(x + \frac{11}{2})^2 = \frac{-96 + 121}{4}$$

$$x + \frac{11}{2} = \pm\sqrt{\frac{25}{4}}$$

$$x = \frac{11}{2} = \pm\frac{5}{2}$$

$$x + \frac{11}{2} = \frac{5}{2} \qquad x + \frac{11}{2} = \frac{-5}{2}$$

$$\frac{-11}{2} \quad \frac{-11}{2} \qquad \frac{-11}{2} \quad \frac{-11}{2}$$

$$x = \frac{-6}{2} \qquad x = \frac{-16}{2}$$

$$x = -3 \qquad x = -8$$

$$\frac{1}{2}b = \frac{1}{2}(11) = \frac{11}{2}$$

$$(\frac{1}{2}b)^2 = (\frac{11}{2})^2 = \frac{121}{4}$$

31. $2x^2 + 10x - 3 = 0$

$$\frac{1}{2}(2x^2 + 10x) = \frac{1}{2}(3)$$

$$x^2 + 5x = \frac{3}{2}$$

$$x^2 + 5x + \frac{25}{4} = \frac{3}{2} + \frac{25}{4}$$

$$(x + \frac{5}{2})^2 = \frac{6 + 25}{4}$$

$$x + \frac{5}{2} = \pm\sqrt{\frac{31}{4}}$$

$$x + \frac{5}{2} = \pm\frac{\sqrt{31}}{2}$$

$$\frac{-5}{2} \qquad \frac{-5}{2}$$

$$x = \frac{-5 \pm \sqrt{31}}{2}$$

$$\frac{1}{2}b = \frac{1}{2}(-5) = \frac{-5}{2}$$

$$(\frac{1}{2}b)^2 = (\frac{-5}{2})^2 = \frac{25}{4}$$

33. $3x^2 + 6x = 6$

$$\frac{1}{3}(3x^2 + 6x) = \frac{1}{3}(6)$$

$$x^2 + 2x = 2$$

$$x^2 + 2x + 1 = 2 + 1$$

$$(x + 1)^2 = 3$$

$$x + 1 = \pm\sqrt{3}$$

$$-1 \qquad -1$$

$$x = -1 \pm \sqrt{3}$$

$$\frac{1}{2}b = \frac{1}{2}(2) = 1$$

$$(\frac{1}{2}b)^2 = 1^2 = 1$$

35. $x^2 + 4x = 0$

$$x^2 + 4x + 4 = 0 + 4$$

$$(x + 2)^2 = 4$$

$$x + 2 = \pm\sqrt{4}$$

$$x + 2 = \pm 2$$

$x + 2 = 2$ $\quad$ $x + 2 = -2$

$-2 \quad -2$ $\quad$ $-2 \quad -2$

$x = 0$ $\quad$ $x = -4$

$$\frac{1}{2}b = \frac{1}{2}(4) = 2$$

$$(\frac{1}{2}b)^2 = 2^2 = 4$$

37. $2x^2 - 4x = 0$

$$\frac{1}{2}(2x^2 - 4x) = \frac{1}{2}(0)$$

$$x^2 - 2x = 0$$

$$x^2 - 2x + 1 = 0 + 1$$

$$(x - 1)^2 = 1$$

$$x - 1 = \pm\sqrt{1}$$

$$x - 1 = \pm 1$$

$x - 1 = 1$ $\quad$ $x - 1 = -1$

$+1 \quad +1$ $\quad$ $+1 \quad +1$

$x = 2$ $\quad$ $x = 0$

$$\frac{1}{2}b = \frac{1}{2}(-2) = -1$$

$$(\frac{1}{2}b)^2 = (-1)^2 = 1$$

39. x = a number.
$3x$ = three times the number.

$$x^2 + 3x = 4$$

$$x^2 + 3x - 4 = 0$$

$$(x + 4)(x - 1) = 0$$

$$x + 4 = 0 \qquad x - 1 = 0$$
$$-4 \quad -4 \qquad\qquad +1 \quad +1$$

$$x = -4 \qquad x = 1$$

41. x = a number.
$x + 3$ = three more than a number.

$$(x + 3)^2 = 9$$
$$x + 3 = \pm 3$$
$$x + 3 = 3 \qquad x + 3 = -3$$
$$-3 \quad -3 \qquad\qquad -3 \quad -3$$

$$x = 0 \qquad x = -6$$

43. x = a positive integer.
$x + 4$ = a larger positive integer.

$$x(x + 4) = 21$$

$$x^2 + 4x = 21$$
$$x^2 + 4x - 21 = 0$$
$$(x + 7)(x - 3) = 0$$
$$x + 7 = 0 \qquad x - 3 = 0$$
$$-7 \quad -7 \qquad\qquad +3 \quad +3$$

$$x = -7 \qquad x = 3$$
discard
$x = 3$, a positive integer
$x + 4 = 3 + 4 = 7$, a larger positive integer

45. a) $x^2 - 12x + 36$

b) $\left(\frac{-12}{2}\right)^2 = 36$

47. To determine if two equations represent parallel lines without graphing, write the equations in slope-intercept form and compare slopes. If the slopes are the same and the y-intercepts are different, the equations represent parallel lines.

49. $\sqrt{2x + 3} = 2x - 3$

$$(\sqrt{2x + 3})^2 = (2x - 3)^2$$

$$2x + 3 = 4x^2 - 12x + 9$$
$$0 = 4x^2 - 14x + 6$$
$$0 = 2(2x^2 - 7x + 3)$$
$$0 = 2(2x - 1)(x - 3)$$
$$2x - 1 = 0 \qquad x - 3 = 0$$
$$2x = 1 \qquad x = 3$$
$$x = \frac{1}{2}$$

Check:

$$\sqrt{2x + 3} = 2x - 3$$
$$\sqrt{2(3) + 3} = 2(3) - 3$$
$$\sqrt{6 + 3} = 6 - 3$$
$$\sqrt{9} = 3$$
$$3 = 3 \quad \text{True}$$

$$\sqrt{2\left(\frac{1}{2}\right) + 3} = 2\left(\frac{1}{2}\right) - 3$$
$$\sqrt{1 + 3} = 1 - 3$$
$$\sqrt{4} = -2$$
$$2 = -2 \quad \text{False}$$

3 is a solution, but 1/2 is not a solution.

Just for Fun

1. $x^2 + \frac{3}{5}x - \frac{1}{2} = 0$

$$x + \frac{3}{5}x = \frac{1}{2}$$

$$x^2 + \frac{3}{5}x + \frac{9}{100} = \frac{1}{2} + \frac{9}{100}$$

$$(x + \frac{3}{10})^2 = \frac{50 + 9}{100}$$

$$x + \frac{3}{10} = \pm\sqrt{\frac{59}{100}}$$

$$x + \frac{3}{10} = \pm \frac{\sqrt{59}}{10}$$

$$\frac{-3}{10} = \frac{-3}{10}$$

$$x = \frac{-3 \pm \sqrt{59}}{10}$$

$$\frac{1}{2} b = \frac{1}{2}(\frac{3}{5}) = \frac{3}{10}$$

$$(\frac{1}{2} b)^2 = (\frac{3}{10})^2 = \frac{9}{100}$$

2. $$x^2 - \frac{2}{3} x - \frac{1}{5} = 0$$

$$x^2 - \frac{2}{3} x = \frac{1}{5}$$

$$x^2 - \frac{2}{3} x + \frac{1}{9} = \frac{1}{5} + \frac{1}{9}$$

$$(x - \frac{1}{3})^2 = \frac{9 + 5}{45}$$

$$x - \frac{1}{3} = \pm \sqrt{\frac{14}{45} \cdot \frac{5}{5}}$$

$$x - \frac{1}{3} = \pm \sqrt{\frac{70}{225}}$$

$$x - \frac{1}{3} = \pm \frac{\sqrt{70}}{15}$$

$$x = \frac{1}{3} \pm \frac{\sqrt{70}}{15}$$

$$x = \frac{5 \pm \sqrt{70}}{15}$$

$$b(\frac{1}{2}) = \frac{-2}{3} \cdot \frac{1}{2} = \frac{-1}{3}$$

$$(\frac{1}{2} b)^2 = (\frac{-1}{3})^2 = \frac{1}{9}$$

3. $$3x^2 + \frac{1}{2} x = 4$$

$$\frac{1}{3}(3x^2 + \frac{1}{2} x) = \frac{1}{3}(4)$$

$$x^2 + \frac{1}{6} x = \frac{4}{3}$$

$$x^2 + \frac{1}{6} x + \frac{1}{144} = \frac{4}{3} + \frac{1}{144}$$

$$(x + \frac{1}{12})^2 = \frac{192 + 1}{144}$$

$$x + \frac{1}{12} = \pm\sqrt{\frac{193}{144}}$$

$$\frac{-1}{12} = \frac{-1}{12}$$

$$x = \frac{-1 \pm \sqrt{193}}{12}$$

$$\frac{1}{2} b = \frac{1}{2}(\frac{1}{6}) = \frac{1}{12}$$

$$(\frac{1}{2} b)^2 = (\frac{1}{12})^2 = \frac{1}{144}$$

4. $$ax^2 + bx + c = 0$$

$$\frac{1}{a}(ax^2 + bx + c) = \frac{1}{a}(0)$$

$$x^2 + \frac{b}{a} x + \frac{c}{a} = 0$$

$$x^2 + \frac{b}{a} x + (\frac{b}{2a})^2 = -\frac{c}{a} + (\frac{b}{2a})^2$$

$$(x + \frac{b}{2a})^2 = -\frac{c}{a} + \frac{b^2}{4a^2}$$

$$(x + \frac{b}{2a})^2 = \frac{b^2 - 4ac}{4a^2}$$

$$x + \frac{b}{2a} = \pm\sqrt{\frac{b^2 - 4ac}{4a^2}}$$

$$x = -\frac{b}{2a} \pm \frac{\sqrt{b^2 - 4ac}}{2a}$$

$$x = \frac{-b \pm \sqrt{b^2 - 4ac}}{2a}$$

Exercise Seet 10.3

1. $x^2 + 3x - 5 = 0$

$$\begin{aligned} d &= b^2 - 4ac \\ &= 3^2 - 4 \cdot 1 \cdot (-5) \\ &= 9 + 20 \\ &= 29 \end{aligned}$$

$d > 0$ Two real solutions

3. $3x^2 - 4x + 7 = 0$

$$\begin{aligned} d &= b^2 - 4ac \\ &= (-4)^2 - 4 \cdot 3 \cdot 7 \\ &= 16 - 84 \\ &= -68 \end{aligned}$$

$d < 0$ No solution

5. $5x^2 + 3x - 7 = 0$

$$\begin{aligned} d &= b^2 - 4ac \\ &= 3^2 - 4 \cdot 5 \cdot (-7) \\ &= 9 + 140 \\ &= 149 \end{aligned}$$

$d > 0$ Two real solutions

7. $4x^2 - 24x = -36$
$4x^2 - 24x + 36 = 0$

$$d = b^2 - 4ac$$

$$\begin{aligned} &= (-24)^2 - 4 \cdot 4 \cdot 36 \\ &= 576 - 576 \\ &= 0 \end{aligned}$$

$d = 0$ One solution

9. $x^2 - 8x + 5 = 0$

$$\begin{aligned} d &= b^2 - 4ac \\ &= (-8)^2 - 4 \cdot 1 \cdot 5 \\ &= 64 - 20 \\ &= 44 \end{aligned}$$

$d > 0$ Two real solutions

11. $-3x^2 + 5x - 8 = 0$

$$\begin{aligned} d &= b^2 - 4ac \\ &= (5)^2 - 4(-3)(-8) \\ &= 25 - 96 \\ &= -71 \end{aligned}$$

$d < 0$ No solution

13. $x^2 + 7x - 3 = 0$

$$\begin{aligned} d &= b^2 - 4ac \\ &= 7^2 - 4 \cdot 1 \cdot (-3) \\ &= 49 + 12 \\ &= 61 \end{aligned}$$

$d > 0$ Two real solutions

15. $4x^2 - 9 = 0$

$$\begin{aligned} d &= b^2 - 4ac \\ &= 0 - 4(4)(-9) \\ &= 144 \end{aligned}$$

$d > 0$ Two real solutions

17. $x^2 - 3x + 2 = 0$

$$x = \frac{-b \pm \sqrt{b^2 - 4ac}}{2a}$$

$$x = \frac{-(-3) \pm \sqrt{(-3)^2 - 4(1)(2)}}{2(1)}$$

$$x = \frac{3 \pm \sqrt{9 - 8}}{2}$$

$$x = \frac{3 \pm \sqrt{1}}{2}$$

$$x = \frac{3 \pm 1}{2}$$

$$x = \frac{3 + 1}{2} \qquad x = \frac{3 - 1}{2}$$

$$x = \frac{4}{2} \qquad x = \frac{2}{2}$$

$$x = 2 \qquad x = 1$$

19. $x^2 - 9x + 20 = 0$

$$x = \frac{-b \pm \sqrt{b^2 - 4ac}}{2a}$$

$$x = \frac{-(-9) \pm \sqrt{(-9)^2 - 4 \cdot 1 \cdot (20)}}{2(1)}$$

$$x = \frac{9 \pm \sqrt{81 - 80}}{2}$$

$$x = \frac{9 \pm \sqrt{1}}{2}$$

$$x = \frac{9 \pm 1}{2}$$

$$x = \frac{9 + 1}{2} \qquad x = \frac{9 - 1}{2}$$

$$x = \frac{10}{2} \qquad x = \frac{8}{2}$$

$$x = 5 \qquad x = 4$$

21. $x^2 + 5x - 24 = 0$

$$x = \frac{-b \pm \sqrt{b^2 - 4ac}}{2a}$$

$$x = \frac{-5 \pm \sqrt{5^2 - 4(1)(-24)}}{2(1)}$$

$$x = \frac{-5 \pm \sqrt{25 + 96}}{2}$$

$$x = \frac{-5 \pm \sqrt{121}}{2}$$

$$x = \frac{-5 \pm 11}{2}$$

$$x = \frac{-5 + 11}{2} \qquad x = \frac{-5 - 11}{2}$$

$$x = \frac{6}{2} \qquad x = \frac{-16}{2}$$

$$x = 3 \qquad x = -8$$

23. $x^2 = 13x - 36$
$x^2 - 13x + 36 = 0$

$$x = \frac{-b \pm \sqrt{b^2 - 4ac}}{2a}$$

$$x = \frac{-(-13) \pm \sqrt{(-13)^2 - 4(1)(36)}}{2(1)}$$

$$x = \frac{13 \pm \sqrt{169 - 144}}{2}$$

$$x = \frac{13 \pm \sqrt{25}}{2}$$

$$x = \frac{13 \pm 5}{2}$$

$$x = \frac{13 + 5}{2} \qquad x = \frac{13 - 5}{2}$$

$$x = \frac{18}{2} \qquad x = \frac{8}{2}$$

$$x = 9 \qquad x = 4$$

25. $x^2 - 25 = 0$

$$x = \frac{-b \pm \sqrt{b^2 - 4ac}}{2a}$$

$$x = \frac{-0 \pm \sqrt{0^2 - 4(1)(-25)}}{2(1)}$$

$$x = \frac{0 \pm \sqrt{100}}{2}$$

$$x = \pm \frac{10}{2}$$

$x = \frac{10}{2}$ $x = \frac{-10}{2}$

$x = 5$ $x = -5$

27. $x^2 - 3x = 0$

$x = \frac{-b \pm \sqrt{b^2 - 4ac}}{2a}$

$x = \frac{-(-3) \pm \sqrt{(-3)^2 - 4(1)(0)}}{2(1)}$

$x = \frac{3 \pm \sqrt{9 - 0}}{2}$

$x = \frac{3 \pm \sqrt{9}}{2}$

$x = \frac{3 \pm 3}{2}$

$x = \frac{3 + 3}{2}$ $x = \frac{3 - 3}{2}$

$x = \frac{6}{2}$ $x = \frac{0}{2}$

$x = 3$ $x = 0$

29. $p^2 - 7p + 12 = 0$

$p = \frac{-b \pm \sqrt{b^2 - 4ac}}{2a}$

$p = \frac{-(-7) \pm \sqrt{(-7)^2 - 4(1)(12)}}{2(1)}$

$p = \frac{7 \pm \sqrt{49 - 48}}{2}$

$p = \frac{7 \pm \sqrt{1}}{2}$

$p = \frac{7 + 1}{2}$ $p = \frac{7 - 1}{2}$

$p = \frac{8}{2}$ $p = \frac{6}{2}$

$p = 4$ $p = 3$

31. $2y^2 - 7y + 4 = 0$

$y = \frac{-b \pm \sqrt{b^2 - 4ac}}{2a}$

$y = \frac{-(-7) \pm \sqrt{(-7)^2 - 4(2)(4)}}{2(2)}$

$y = \frac{7 \pm \sqrt{49 - 32}}{4}$

$y = \frac{7 \pm \sqrt{17}}{4}$

33. $6x^2 = -x + 1$

$6x^2 + x - 1 = 0$

$x = \frac{-b \pm \sqrt{b^2 - 4ac}}{2a}$

$x = \frac{-1 \pm \sqrt{1^2 - 4(6)(-1)}}{2(6)}$

$x = \frac{-1 \pm \sqrt{1 + 24}}{12}$

$x = \frac{-1 \pm \sqrt{25}}{12}$

$x = \frac{-1 \pm 5}{12}$

$x = \frac{-1 + 5}{12}$ $x = \frac{-1 - 5}{12}$

$x = \frac{4}{12}$ $x = \frac{-6}{12}$

$x = \frac{1}{3}$ $x = \frac{-1}{2}$

35. $2x^2 - 4x - 1 = 0$

$x = \frac{-b \pm \sqrt{b^2 - 4ac}}{2a}$

$x = \frac{-(-4) \pm \sqrt{(-4)^2 - 4 \cdot 2(-1)}}{2(2)}$

$x = \frac{4 \pm \sqrt{16 + 8}}{4}$

$$x = \frac{4 \pm \sqrt{24}}{4}$$

$$x = \frac{4 \pm \sqrt{4 \cdot 6}}{4}$$

$$x = \frac{4 \pm 2\sqrt{6}}{4}$$

$$x = \frac{2(2 \pm \sqrt{6})}{4}$$

$$x = \frac{2 \pm \sqrt{6}}{2}$$

37. $2s^2 - 4s + 3 = 0$

$$s = \frac{-b \pm \sqrt{b^2 - 4ac}}{2a}$$

$$s = \frac{-(-4) \pm \sqrt{(-4)^2 - 4(2)(3)}}{2(2)}$$

$$s = \frac{4 \pm \sqrt{16 - 24}}{4}$$

$$s = \frac{4 \pm \sqrt{-8}}{4}$$

No real solution

39. $4x^2 = x + 5$

$4x^2 - x - 5 = 0$

$$x = \frac{-b \pm \sqrt{b^2 - 4ac}}{2a}$$

$$x = \frac{-(-1) \pm \sqrt{(-1)^2 - 4(4)(-5)}}{2(4)}$$

$$x = \frac{1 \pm \sqrt{1 + 80}}{8}$$

$$x = \frac{1 \pm \sqrt{81}}{8}$$

$$x = \frac{1 \pm 9}{8}$$

$$x = \frac{1 + 9}{8} \qquad x = \frac{1 - 9}{8}$$

$$x = \frac{10}{8} \qquad x = \frac{-8}{8}$$

$$x = \frac{5}{4} \qquad x = -1$$

41. $2x^2 - 7x = 9$

$2x^2 - 7x - 9 = 0$

$$x = \frac{-b \pm \sqrt{b^2 - 4ac}}{2a}$$

$$x = \frac{-(-7) \pm \sqrt{(-7)^2 - 4(2)(-9)}}{2(2)}$$

$$x = \frac{7 \pm \sqrt{49 + 72}}{4}$$

$$x = \frac{7 \pm \sqrt{121}}{4}$$

$$x = \frac{7 \pm 11}{4}$$

$$x = \frac{7 + 11}{4} \qquad x = \frac{7 - 11}{4}$$

$$x = \frac{18}{4} \qquad x = \frac{-4}{4}$$

$$x = \frac{9}{2} \qquad x = -1$$

43. $-2x^2 + 11x - 15 = 0$

$$x = \frac{-b \pm \sqrt{b^2 - 4ac}}{2a}$$

$$x = \frac{-11 \pm \sqrt{11^2 - 4(-2)(-15)}}{2(-2)}$$

$$x = \frac{-11 \pm \sqrt{121 - 120}}{-4}$$

$$x = \frac{-11 \pm \sqrt{1}}{-4}$$

$$x = \frac{-11 \pm 1}{-4}$$

$$x = \frac{-11 + 1}{-4} \qquad x = \frac{-11 - 1}{-4}$$

$$x = \frac{-10}{-4} \qquad x = \frac{-12}{-4}$$

$$x = +\frac{5}{2} \qquad x = +3$$

45. x = a positive integer.
x + 1 = next consecutive positive integer.

$$x(x + 1) = 20$$
$$x^2 + x - 20 = 0$$
$$(x + 5)(x - 4) = 0$$
$$x = -5 \quad x = 4$$
$$x = 4$$
$$x + 1 = 4 + 1$$
$$= 5$$

Two consecutive positive integers are 4 and 5.

47. x = width of rectangle.
2x - 3 = length of rectangle.

$$x(2x - 3) = 20$$
$$2x^2 - 3x - 20 = 0$$
$$(2x + 5)(x - 4) = 0$$
$$x = -\frac{5}{2} \quad x = 4$$

Width is 4 feet and length is 2x - 3 or 5 feet.

49. x = width of bark path.

$$40 \cdot 30 - (40 - 2x)(30 - 2x) = 296$$
$$120 - 120 + 140x - 4x^2 = 296$$
$$-\frac{1}{4}(-4x^2 + 140x - 296)$$
$$= -\frac{1}{4}(0)$$
$$x^2 - 35x + 74 = 0$$
$$(x + 37)(x - 2) = 0$$
$$x = -37 \quad x = 2$$

The width of the bark path is 2 feet.

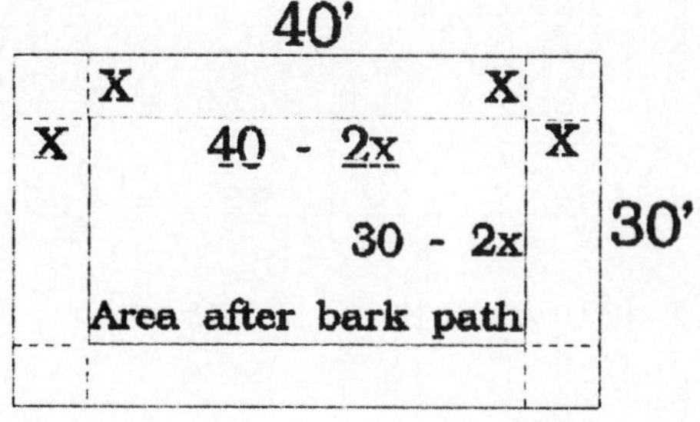

51. a) none
b) one
c) two

53. Since the discriminant follows the ± sign in the quadratic equation, you will have two solutions when its value is greater than 0. Its value is added to -b and subtracted from -b. There will be one solution when its value is 0, since to add or subtract 0 from -6 is the same number. When the discriminant is less than 0, there is no real solution. The square root of a negative number is not a real number.

Cumulative Review Exercises

55. By factoring:

a) $6x^2 + 11x - 35 = 0$
$(3x - 5)(2x + 7) = 0$

$$3x - 5 = 0 \qquad 2x + 7 = 0$$
$$3x = 5 \qquad 2x = -7$$
$$x = \frac{5}{3} \qquad x = -\frac{7}{2}$$

b) By completing the square:

$$6x^2 + 11x - 35 = 0$$
$$\frac{1}{6}(6x^2 + 11x - 35) = \frac{1}{6}(0)$$
$$x^2 + \frac{11}{6}x - \frac{35}{6} = 0$$
$$x^2 + \frac{11}{6}x + (\frac{11}{12})^2 = \frac{35}{6} + (\frac{11}{12})^2$$
$$(x + \frac{11}{12})^2 = \frac{35}{6} + \frac{121}{144}$$
$$(x + \frac{11}{12})^2 = \frac{840}{144} + \frac{121}{144}$$
$$(x + \frac{11}{12})^2 = \frac{961}{144}$$
$$x + \frac{11}{12} = \pm\sqrt{\frac{961}{144}}$$
$$x = -\frac{11}{12} \pm \frac{31}{12}$$
$$x = -\frac{11}{12} + \frac{31}{12} \qquad x = -\frac{11}{12} - \frac{31}{12}$$
$$x = \frac{20}{12} = \frac{5}{3} \qquad x = -\frac{42}{12} = -\frac{7}{2}$$

c) Quadratic formula:

$$6x^2 + 11x - 35 = 0$$
$$x = \frac{-b \pm \sqrt{b^2 - 4ac}}{2a}$$
$$x = \frac{-11 \pm \sqrt{(11)^2 - 4(6)(-35)}}{2(6)}$$
$$x = \frac{-11 \pm \sqrt{121 + 840}}{12}$$
$$x = \frac{-11 \pm \sqrt{961}}{12}$$
$$x = \frac{-11 \pm 31}{12}$$
$$x = \frac{-11 + 31}{12} \qquad x = \frac{-11 - 31}{12}$$
$$x = \frac{20}{12} = \frac{5}{3} \qquad x = \frac{-42}{12} = -\frac{7}{2}$$

57. a) Factoring:

$$6x^2 = 54$$
$$6x^2 - 54 = 0$$
$$6(x^2 - 9) = 0$$
$$6(x - 3)(x + 3) = 0$$
$$x - 3 = 0 \qquad x + 3 = 0$$
$$x = 3 \qquad x = -3$$

b) Cannot complete the square.

c) Quadratic formula:

$$6x^2 = 54$$
$$6x^2 - 54 = 0$$
$$x = \frac{-b \pm \sqrt{b^2 - 4ac}}{2a}$$
$$x = \frac{0 \pm \sqrt{0 - 4(6)(-54)}}{2(6)}$$
$$x = \frac{\pm\sqrt{1296}}{12} = \frac{\pm 36}{12}$$
$$x = \frac{36}{12} = 3 \qquad x = \frac{-36}{12} = -3$$

Just for fun

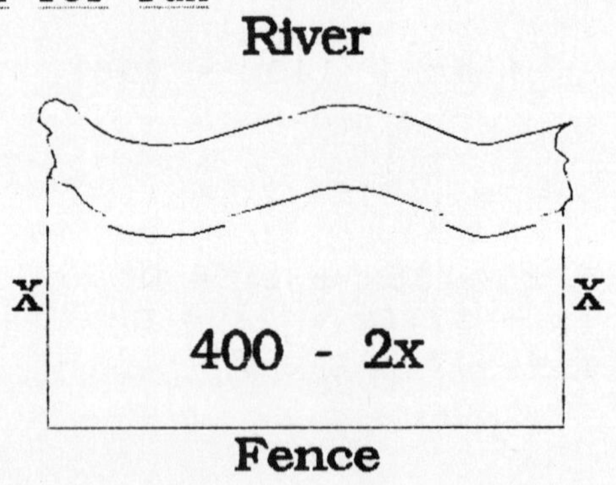

1. x = length of side.
$2x$ = length of two sides.
$400 - 2x$ = length of side.

$$x(400 - 2x) = 15{,}000$$
$$400x - 2x^2 = 15{,}000$$

$$-\frac{1}{2}(-2x^2 + 400x - 15{,}000) = 0$$

$$x^2 - 200x + 7{,}500 = 0$$
$$(x - 50)(x - 150) = 0$$

$x - 50 = 0$ $\quad x - 150 = 0$
$x = 50$ $\quad x = 150$

If $x = 50$.
$400 - 2x = 400 - 100 - 300$
300 ft. long by 50 ft. wide.

If $x = 150$.
$400 - 2x = 400 - 300 = 100$
150 ft. long by 100 ft. wide.

Exercise Set 10.4

1. $y = x^2 + 2x - 7$

$$x = \frac{-b}{2a}$$

$$x = \frac{-2}{2 \cdot 1}$$

$$x = -1$$

$$y = (-1)^2 + 2(-1) + 7$$
$$y = 1 - 2 - 7$$
$$y = -8$$

Axis of symmetry: $x = -1$

Vertex: $(-1, -8)$
Parabola opens up.

3. $y = -x^2 + 5x - 6$

$$x = \frac{-b}{2a}$$

$$x = \frac{-5}{2(-1)}$$

$$x = \frac{5}{2}$$

$$y = -\left(\frac{5}{2}\right)^2 + 5\left(\frac{5}{2}\right) - 6$$

$$y = \frac{-25}{4} + \frac{25}{2} - 6$$

$$y = \frac{-25 + 50 - 24}{4}$$

$$y = \frac{1}{4}$$

Axis of symmetry: $x = \frac{5}{2}$

Vertex: $\left(\frac{5}{2}, \frac{1}{4}\right)$

Parabola opens down.

5. $y = -3x^2 + 5x + 8$

$$x = \frac{-b}{2a}$$

$$x = \frac{-5}{2(-3)}$$

$$x = \frac{5}{6}$$

$$y = -3\left(\frac{5}{6}\right)^2 + 5\left(\frac{5}{6}\right) + 8$$

$$y = -3\left(\frac{25}{36}\right) + \frac{25}{6} + 8$$

$$y = \frac{-25 + 50 + 96}{12}$$

$y = \frac{121}{12}$

Axis of symmetry: $x = \frac{5}{6}$

Vertex: $(\frac{5}{6}, \frac{121}{12})$

Parabola opens down.

7. $y = -4x^2 - 8x - 12$

$x = \frac{-b}{2a}$

$x = \frac{-(-8)}{2(-4)}$

$x = \frac{8}{-8}$

$x = -1$

$y = -4(-1)^2 - 8(-1) - 12$
$y = -4 + 8 - 12$
$y = -8$

Axis of symmetry: $x = -1$
Vertex: $(-1, -8)$
Parabola opens down.

9. $y = 3x^2 - 2x + 2$

$x = \frac{-b}{2a}$

$x = \frac{-(-2)}{2 \cdot 3}$

$x = \frac{2}{6}$

$x = \frac{1}{3}$

$y = 3(\frac{1}{3})^2 - 2(\frac{1}{3}) + 2$

$y = 3(\frac{1}{9}) - \frac{2}{3} + 2$

$y = \frac{1 - 2 + 6}{3}$

$y = \frac{5}{3}$

Axis of symmetry: $x = \frac{1}{3}$

Vertex: $(\frac{1}{3}, \frac{5}{3})$

Parabola opens up.

11. $y = 4x^2 + 12x - 5$

$x = \frac{-b}{2a}$

$x = \frac{-12}{2 \cdot 4}$

$x = \frac{-12}{8}$

$x = \frac{-3}{2}$

$y = 4(\frac{-3}{2})^2 + 12(\frac{-3}{2}) - 5$

$y = 4(\frac{9}{4}) - \frac{36}{2} - 5$

$y = 9 - 18 - 5$
$y = -14$

Axis of symmetry: $x = -\frac{3}{2}$

Vertex: $(\frac{-3}{2}, -14)$

Parabola opens up.

13. $y = x^2 - 1$

$x = \frac{-b}{2a}$ $\quad y = 0^2 - 1$

$y = -1$

$x = \frac{0}{2(1)}$

$x = 0$ $\quad$ Roots: 1, -1

Axis of symmetry: $x = 0$
Vertex: $(0, -1)$
Since $a > 0$, the parabola opens up.

$y = x^2 - 1$

Let $x = -3$ $\quad y = (-3)^2 - 1$
$= 9 - 1 = 8$

Let $x = -2$ $\quad y = (-2)^2 - 1$
$= 4 - 1 = 3$

Let $x = -1$ $\quad y = (-1)^2 - 1$
$= 1 - 1 = 0$

Let $x = 0$ $\quad y = (0)^2 - 1$
$= 0 - 1 = -1$

Let $x = 1$ $\quad y = (1)^2 - 1$
$= 1 - 1 = 0$

Let $x = 2$ $\quad y = (2)^2 - 1$
$= 4 - 1 = 3$

Let $x = 3$ $\quad y = (3)^2 - 1$
$= 9 - 1 = 8$

x	y
-3	8
-2	3
-1	0
0	-1
1	0
2	3
3	8

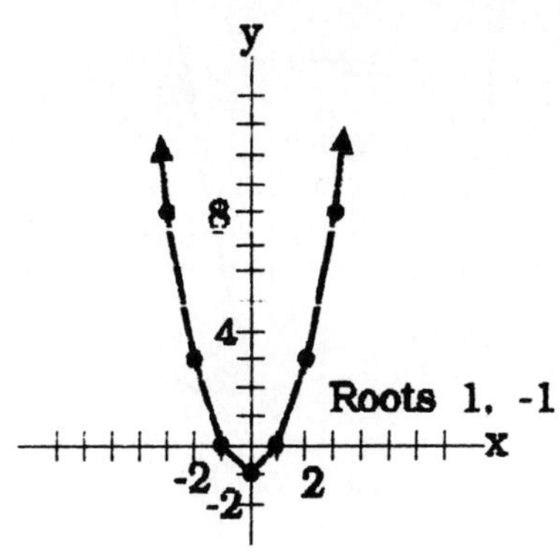

15. $y = -x^2 + 3$

$x = \frac{-b}{2a}$ $\quad y = -(0)^2 + 3$

$x = \frac{0}{2(-1)}$ $\quad y = 3$

$x = 0$

Roots $\sqrt{3}$, $-\sqrt{3}$

Axis of symmetry: $x = 0$
Vertex: (0,3)
Parabola opens down.

Let $x = 0$ $\quad y = (0)^2 + 3 = 3$
Let $x = 1$ $\quad y = -(1)^2 + 3$
$= 2$
Let $x = 2$ $\quad y = -(2)^2 + 3$
$= -1$

Make use of symmetry to complete the graph.

x	y
0	3
1	2
2	-1

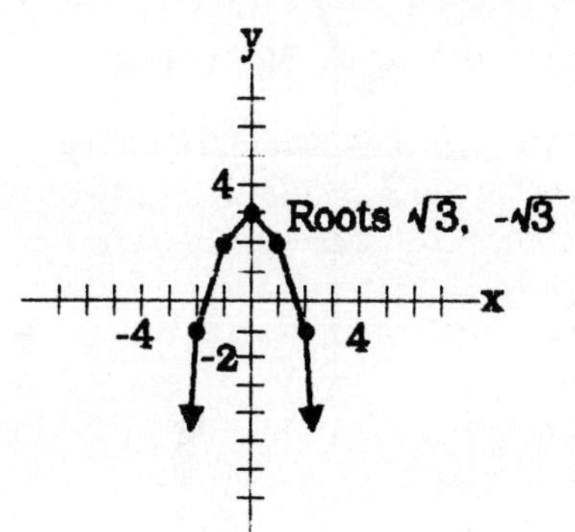

17. $y = x^2 + 2x + 3$

$x = \frac{-b}{2a}$ $\quad y = (-1)^2 + 2(-1) + 3$

$x = \frac{-2}{2 \cdot 1}$ $\quad y = 1 - 2 + 3$

$x = -1$ $\quad y = 2$

No roots

Axis of symmetry: $x = -1$
Vertex: (-1,2)
Parabola opens up.

Let $x = -1$ $\quad y = (-1)^2$

$+ 2(-1) + 3$
$= 1 - 2 + 3$
$= 2$

Let $x = 0$ $\quad y = (0)^2 + 2(0) + 3$
$= 0 + 0 + 3$
$= 3$

Let $x = 1$ $\quad y = (1)^2 + 2(1) + 3$
$= 1 + 2 + 3$
$= 6$

Make use of symmetry to complete the graph.

x	y
-1	2
0	3
1	6

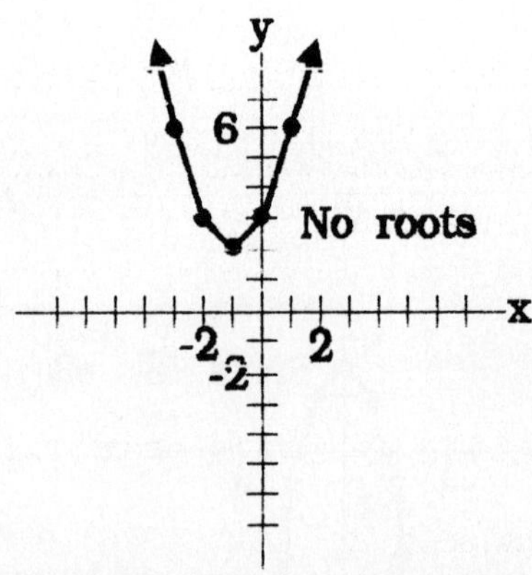

19. $y = x^2 + 2x - 15$

$x = \frac{-b}{2a}$ $\quad y = (-1)^2 + 2(-1) - 15$

$x = \frac{-2}{2(1)}$ $\quad y = 1 - 2 - 15$

$x = -1$ $\quad y = -16$

Roots 3, -5

Axis of symmetry: $x = -1$
Vertex: $(-1, -16)$
Since $a > 0$, the parabola opens upward.

$y = x^2 + 2x - 15$

Let $x = -1$ $\quad y = (-1)^2 + 2(-1) - 15$
$= 1 - 2 - 15$
$= -16$

Let $x = -2$ $\quad y = (-2)^2 + 2(-2) - 15$
$= 4 - 4 - 15$
$= -15$

Let $x = -3$ $\quad y = (-3)^2 + 2(-3) - 15$
$= 9 - 6 - 15$
$= -12$

Let $x = -4$ $\quad y = (-4)^2 + 2(-4) - 15$
$= 16 - 8 - 15$
$= -7$

Let $x = -5$ $\quad y = (-5)^2 + 2(-5) - 15$
$= 25 - 10 - 15$
$= 0$

Let $x = -6$ $\quad y = (-6)^2 + 2(-6) - 15$
$= 36 - 12 - 15$
$= 9$

Make use of symmetry to complete the graph.

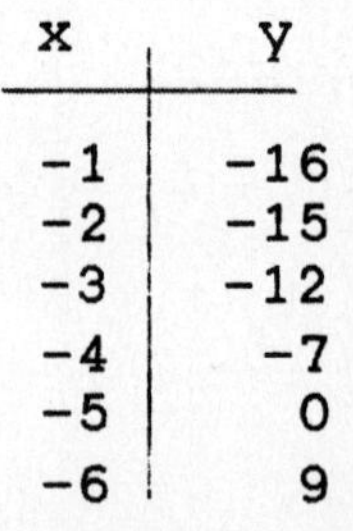

x	y
-1	-16
-2	-15
-3	-12
-4	-7
-5	0
-6	9

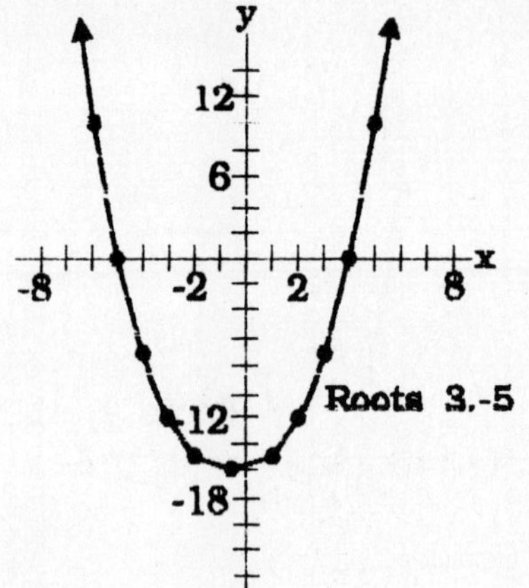

21. $y = -x^2 + 4x - 5$

$x = \frac{-b}{2a}$ $\quad y = -(2)^2 + 4(2) - 5$

$x = \frac{-4}{2(-1)}$ $\quad y = -4 + 8 - 5$

$x = \frac{-4}{-2}$ $\quad y = -1$

$x = 2$

No roots

Axis of symmetry: $x = 2$

Vertex: $(2,-1)$

Parabola opens down.

$y = -x^2 + 4x - 5$

Let $x = 2$ $\quad y = -(2)^2 + 4(2) - 5$
$= -4 + 8 - 5$
$= -1$

Let $x = 3$ $\quad y = -(3)^2 + 4(3) - 5$
$= -9 + 12 - 5$
$- 2$

Let $x = 4$ $\quad y = -(4)^2 + 4(4) - 5$
$= -16 + 16 - 5$
$= -5$

Make use of symmetry to complete this graph.

x	y
2	-1
3	-2
4	-5

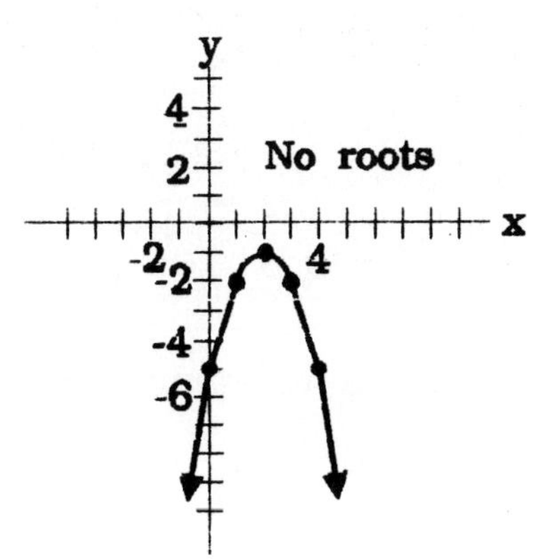

23. $y = x^2 - x - 12$

$x = \frac{-b}{2a}$ $\quad y = (\frac{1}{2})^2 - \frac{1}{2} - 12$

$x = \frac{-(-1)}{2(1)}$ $\quad y = \frac{1}{4} - \frac{1}{2} - 12$

$x = \frac{1}{2}$ $\quad y = \frac{1 - 2 - 48}{4}$

$y = \frac{-49}{4}$

Roots -3, 4

Axis of symmetry: $x = \frac{1}{2}$

Vertex: $(\frac{1}{2}, \frac{-49}{4})$

Parabola opens up.

$y = x^2 - x - 12$

Let $x = \frac{1}{2}$ $\quad y = (\frac{1}{2})^2 - \frac{1}{2} - 12$

$= \frac{1}{4} - \frac{2}{4} - \frac{48}{4}$

$= \frac{-49}{4}$

Let $x = 1$ $\quad y = (1)^2 - 1 - 12$
$y = 1 - 1 - 12$
$= -12$

Let $x = 2$ $\quad y = (2)^2 - 2 - 12$
$= 4 - 2 - 12$
$= -10$

Let $x = 3$ $\quad y = (3)^2 - 3 - 12$
$= 9 - 3 - 12$
$= -6$

Let $x = 4$ $\quad y = (-4)^2 - 4 - 12$
$= 16 - 4 - 12$
$= 0$

Make use of symmetry to complete the graph.

x	y
$\frac{1}{2}$	$\frac{-49}{2}$
1	-12
2	-10
3	-6
4	0

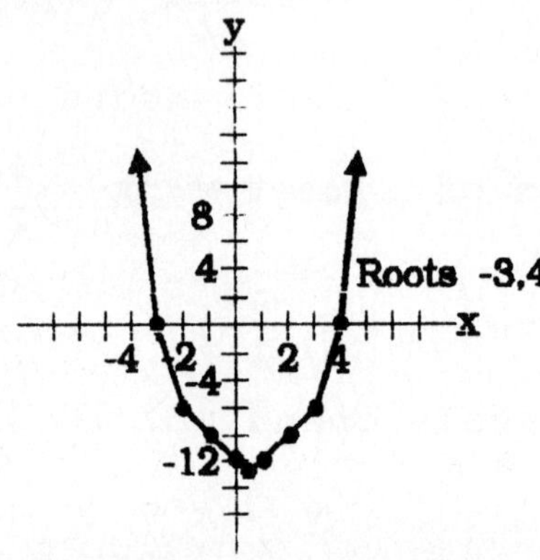

25. $y = x^2 - 6x + 9$

$x = \frac{-b}{2a}$ $\quad y = (3)^2 - 6(3) + 9$

$x = \frac{-(-6)}{2(1)}$ $\quad y = 9 - 18 + 9$

$x = \frac{6}{2}$ $\quad y = 0$

$x = 3$

Root 3

Axis of symmetry: $x = 3$

Vertex: (3,0)

Parabola opens up.

$y = x^2 - 6x + 9$

Let $x = 0$ $\quad y = (0)^2 - 6(0) + 9$
$= 0 - 0 + 9$
$= 9$

Let $x = 1$ $\quad y = (1)^2 - 6(1) + 9$
$= 1 - 6 + 9$
$= 4$

Let $x = 2$ $\quad y = (2)^2 - 6(2) + 9$
$= 4 - 12 + 9$
$= 1$

Let $x = 3$ $\quad y = (3)^2 - 6(3) + 9$
$= 9 - 18 + 9$
$= 0$

Make use of symmetry to complete the graph.

x	y
0	9
1	4
2	1
3	0

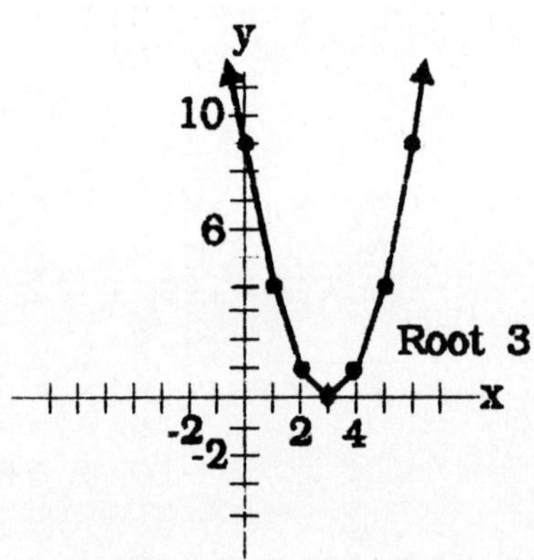

27. $y = -x^2 + 5x$

$x = \frac{-b}{2a}$ $\quad y = -(\frac{5}{2})^2 + 5(\frac{5}{2})$

$x = \frac{-5}{2(-1)}$ $\quad y = \frac{-25}{4} + \frac{25}{2}$

$x = \frac{5}{2}$ $\quad y = \frac{-25 + 50}{4}$

$y = \frac{25}{4}$

Roots 0,5

Axis of symmetry: $x = \frac{5}{2}$

Vertex: $(\frac{5}{2}, \frac{25}{4})$

Parabola opens down.

$$y = -x^2 + 5x$$

Let $x = \frac{5}{2}$ $\quad y = -(\frac{5}{2})^2 + (\frac{5}{2})$

$$y = \frac{-25}{4} + \frac{50}{4} = \frac{25}{4}$$

Let $x = 3$ $\quad y = -(3)^2 + 5(3) = -9 + 15 = 6$

Let $x = 4$ $\quad y = -(4)^2 + 5(4) = -16 + 20 = 4$

Let $x = 5$ $\quad y = -(5)^2 + 5(5) = -25 + 25 = 0$

Make use of symmetry to complete this graph.

x	y
$\frac{5}{2}$	$\frac{25}{4}$
3	6
4	4
5	0

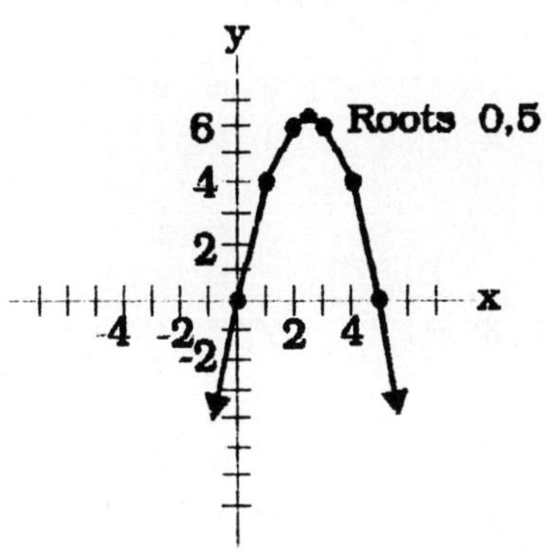

29. $y = 2x^2 - 6x + 4$

$x = \frac{-b}{2a}$ $\quad y = 2(\frac{3}{2})^2 - 6(\frac{3}{2}) + 4$

$x = \frac{-(-6)}{2(2)}$ $\quad y = \frac{-9}{2} - \frac{18}{2} + \frac{8}{2}$

$x = \frac{6}{4}$ $\quad y = \frac{-9 - 18 + 8}{2}$

$x = \frac{3}{2}$ $\quad y = -\frac{1}{2}$

Roots 1,2

Parabola opens up.

$$y = 2x^2 - 6x + 4$$

Let $x = 0$ $\quad y = 2(0)^2 - 6(0) + 4 = 0 - 0 + 4 = 4$

Let $x = 2$ $\quad y = 2(2)^2 - 6(2) + 4 = 8 - 12 + 4 = 0$

Let $x = 4$ $\quad y = 2(4)^2 - 6(4) + 4 = 32 - 24 + 4 = 12$

Make use of symmetry to complete this graph.

x	y
0	4
2	0
4	12

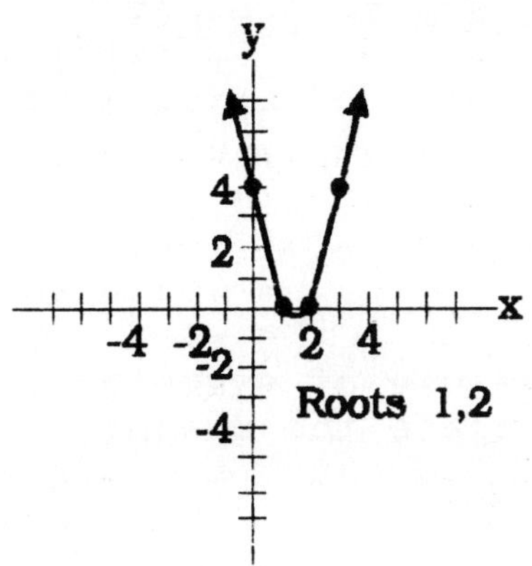

31. $y = -x^2 + 11x - 28$

$x = \frac{-b}{2a}$ $y = -(\frac{11}{2})^2 + 11(\frac{11}{2}) - 28$

$x = \frac{-11}{2(-1)}$ $y = \frac{-121}{4} + \frac{121}{2} - \frac{56}{2}$

$x = \frac{11}{2}$ $y = \frac{-121+242-112}{4}$

$y = \frac{9}{4}$

Roots 4,7

Axis of symmetry: $x = \frac{11}{2}$

Vertex: $(\frac{11}{2}, \frac{9}{4})$

Parabola opens down.

$y = -x^2 + 11x - 28$

Let x = 6 $y = -(6)^2 + 11(6) - 28$
$= -36 + 66 - 28$
$= 2$

Let x = 7 $y = -(7)^2 + 11(7) - 28$
$= -49 + 77 - 28$
$= 0$

Let x = 8 $y = -(8)^2 + 11(8) - 28$
$= -64 + 88 - 28$
$= -4$

Make use of symmetry to complete the graph.

x	y
$\frac{11}{2}$	$\frac{9}{4}$
6	2
7	0
8	-4

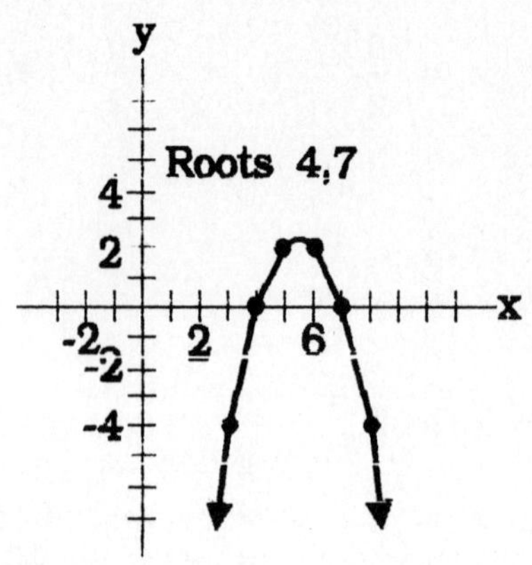

33. $y = x^2 - 2x - 15$

$x = \frac{-b}{2a}$ $y = (1)^2 - 2(1) - 15$

$x = \frac{-(-2)}{2(1)}$ $y = 1 - 2 - 15$

$x = \frac{2}{2}$ $y = -16$

$x = 1$

Roots 5,-3

Axis of symmetry: $x = 1$

Vertex: (1,-16)

Parabola opens up.

$y = x^2 - 2x - 15$

Let x = 1 $y = (1)^2 - 2(1) - 15$
$= 1 - 2 - 15$
$= -16$

Let x = 2 $y = (2)^2 - 2(2) - 15$
$= 4 - 4 - 15$
$= -15$

Let x = 3 $y = (3)^2 - 2(3) - 15$
$= 9 - 6 - 15$
$= -12$

Let x = 4 $y = (4)^2 - 2(4) - 15$
$= 16 - 8 - 15$
$= -7$

Let x = 5 $y = (5)^2 - 2(5) - 15$
$= 25 - 10 - 15 = 0$

Make use of symmetry in completeing the graph.

x	y
1	-16
2	-15
3	-12
4	-7
5	0

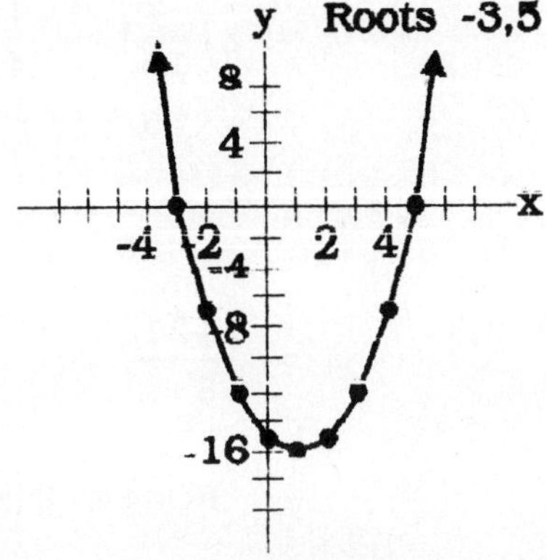

35. $y = -2x^2 + 7x - 3$

$x = \frac{-b}{2a}$ $\quad y = -2(\frac{7}{4})^2 + 7(\frac{7}{4}) - 3$

$x = \frac{-7}{2(-2)}$ $\quad y = -2(\frac{49}{16}) + \frac{49}{4} - 3$

$x = \frac{7}{4}$ $\quad y = \frac{-49}{8} + \frac{98}{8} - \frac{24}{8}$

roots $3, \frac{1}{2}$

Axis of symmetry: $x = \frac{7}{4}$

Vertex: $(\frac{7}{4}, \frac{25}{8})$

Parabola opens down.

$y = -2x^2 + 7x - 3$

Let $x = \frac{7}{4}$ $\quad y = -2(\frac{7}{4})^2 + 7(\frac{7}{4}) - 3$

$= -2(\frac{49}{16}) + \frac{49}{4} - 3 = \frac{25}{8}$

Let $x = 2$ $\quad y = -2(2)^2 + 7(2) - 3$

$= -2(4) + 14 - 3 = 3$

Let $x = 3$ $\quad y = -2(3)^2 + 7(3) - 3$

$= -2(9) + 21 - 3 = 0$

Let $x = 4$ $\quad y = -2(4)^2 + 7(4) - 3$

$= -2(16) + 28 - 3 = -7$

Make use of symmetry to complete the graph.

x	y
$\frac{7}{4}$	$\frac{25}{8}$
2	3
3	0
4	-7

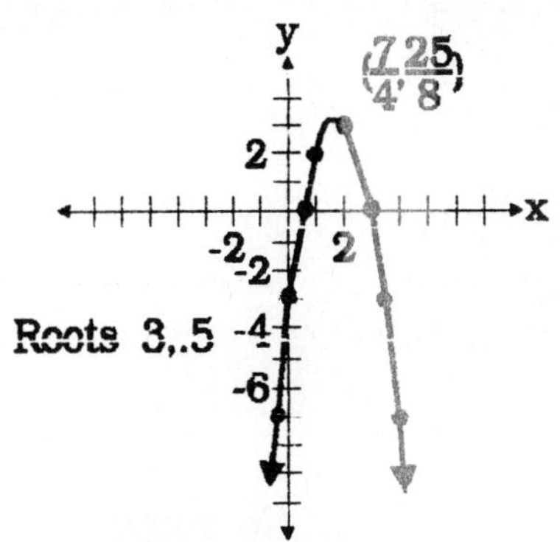

37. $y = -2x^2 + 3x - 2$

$x = \frac{-b}{2a}$ $y = -2(\frac{3}{4})^2 + 3(\frac{3}{4}) - 2$

$x = \frac{-3}{2(-2)}$ $y = -2(\frac{9}{16}) + \frac{9}{4} - 2$

$x = \frac{3}{4}$ $y = \frac{-9 + 18 - 16}{8}$

$y = \frac{-7}{8}$

No roots

Axis of symmetry: $x = \frac{3}{4}$

Vertex: $(\frac{3}{4}, \frac{-7}{8})$

Parabola opens down.

$y = -2x^2 + 3x - 2$

Let $x = 1$ $y = -2(1)^2 + 3(1) - 2$
$= -2 + 3 - 2$
$= -1$

Let $x = 2$ $y = -2(2)^2 + 3(2) - 2$
$= -8 + 6 - 2$
$= -4$

Make use of symmetry to complete the graph.

x	y
$\frac{3}{4}$	$\frac{-7}{8}$
1	-1
2	-4

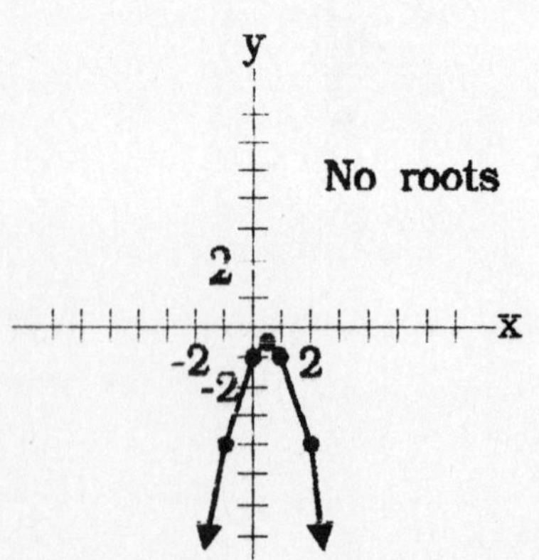

39. $y = 2x^2 - x - 15$

$x = \frac{-b}{2a}$ $y = 2(\frac{1}{4})^2 - \frac{1}{4} - 15$

$x = \frac{-(-1)}{2(2)}$ $y = 2(\frac{1}{16}) - \frac{1}{4} - 15$

$x = \frac{1}{4}$ $y = \frac{1 - 2 - 120}{8}$

$y = \frac{-121}{8}$

Roots $3, \frac{-5}{2}$

Axis of symmetry: $x = \frac{1}{4}$

Vertex: $(\frac{1}{4}, \frac{-121}{8})$

Parabola opens up.

$y = 2x^2 - x - 15$

Let $x = \frac{1}{4}$ $y = 2(\frac{1}{4})^2 - \frac{1}{4} - 15$
$= 2(\frac{1}{16}) - \frac{1}{4} - 15 = \frac{-121}{8}$

Let $x = 1$ $y = 2(1)^2 - 1 - 15$
$= 2(1) - 1 - 15$
$= -14$

Let $x = 2$ $y = 2(2)^2 - 2 - 15$
$= 2(4) - 2 - 15 = -9$

Let $x = 3$ $y = 2(3)^2 - 3$

$$- 15$$
$$= 2(9) - 3$$
$$- 15$$
$$= 0$$

Make use of symmetry to complete the graph.

x	y
$\frac{1}{4}$	$\frac{-121}{8}$
1	-14
2	-9
3	0

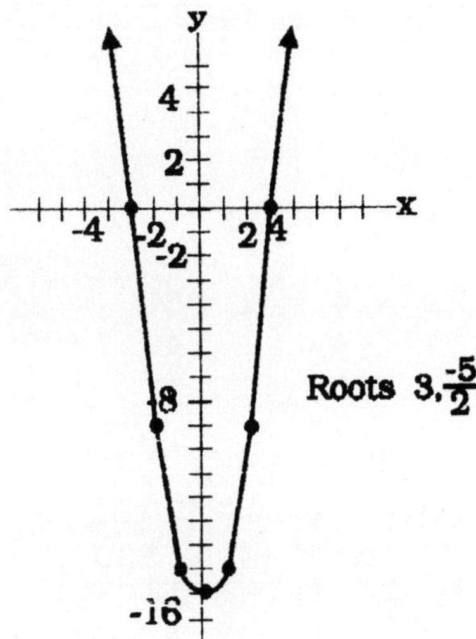

41. $\left(\frac{-b}{2a}, \frac{4ac - b^2}{4a}\right)$

43. a) The values of x where the graph crosses the x axis.

b) Set y = 0 and solve for x.

45. $x = \frac{-b}{2a}$

47. The parabola will have 2 x-intercepts. The vertex is below the x-axis, and the parabola opens upward.

49. The parabola has no x-intercepts. The vertex is below the x-axis, and the parabola opens downward.

51. a) The graphs are reflections of each other.

$$x^2 - 2x - 8$$
$$= -(-x^2 + 2x - 8)$$

b)

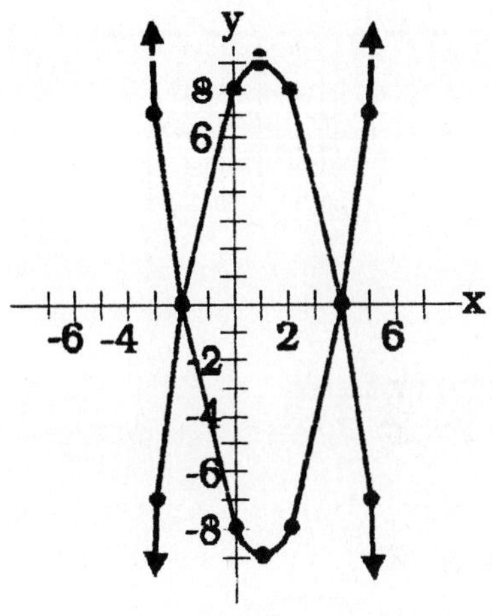

Cumulative Review Exercises

53. $\frac{1}{3}(x + 6) = 3 - \frac{1}{4}(x - 5)$

$$\frac{1}{3}x + 2 = 3 - \frac{1}{4}x + \frac{5}{4}$$

$$12\left(\frac{1}{3}x + 2\right)$$
$$= 12\left(3 - \frac{1}{4}x + \frac{5}{4}\right)$$

$$4x + 24 = 36 - 3x + 15$$
$$4x + 24 = -3x + 51$$
$$7x + 24 = 51$$
$$7x = 27$$

$$x = \frac{27}{7}$$

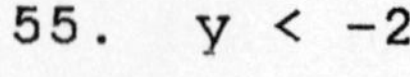

55. $y < -2$

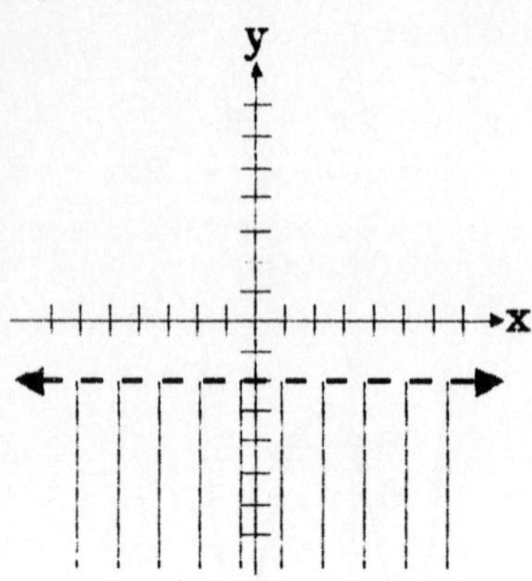

Review Exercises

1. $x = 25$

$$\sqrt{x^2} = \pm\sqrt{25}$$

$$x = \pm 5$$

3. $2x^2 = 12$

$$\frac{2x^2}{2} = \frac{12}{2}$$

$$x^2 = 6$$

$$\sqrt{x^2} = \pm\sqrt{6}$$

$$x = \pm\sqrt{6}$$

5. $x^2 - 4 = 16$

$$\underline{\quad +4 \quad +4}$$

$$x^2 = 20$$

$$\sqrt{x^2} = \pm\sqrt{20}$$

$$\sqrt{x^2} = \pm\sqrt{4 \cdot 5}$$

$$x = \pm 2\sqrt{5}$$

7. $3x^2 + 8 = 32$

$$\underline{\quad -8 \quad -8}$$

$$\frac{3x^2}{3} = \frac{24}{3}$$

$$x^2 = 8$$

$$\sqrt{x^2} = \pm\sqrt{8}$$

$$\sqrt{x^2} = \pm\sqrt{4 \cdot 2}$$

$$x = \pm 2\sqrt{2}$$

9. $(2x + 4)^2 = 30$

$$2x + 4 = \pm\sqrt{30}$$

$$\underline{\quad -4 \quad -4}$$

$$2x = -4 \pm \sqrt{30}$$

$$\frac{2x}{2} = \frac{-4 \pm \sqrt{30}}{2}$$

$$x = \frac{-4 \pm \sqrt{30}}{2}$$

11. $x^2 - 10x + 16 = 0$

$$x^2 - 10x + 25 = -16 + 25$$

$$(x - 5)^2 = \pm\sqrt{9}$$

$$x - 5 = \pm 3$$

$$x - 5 = 3 \qquad x - 5 = -3$$

$$\underline{\quad +5 \quad +5} \qquad \underline{\quad +5 \quad +5}$$

$$x = 8 \qquad x = 2$$

$$\frac{1}{2}b = \frac{1}{2}(-10) = -5$$

$(\frac{1}{2}\ b)^2 = (-5)^2 = 25$

13. $x^2 - 14x + 13 = 0$

$$x^2 - 14x = -13$$
$$x^2 - 14x + 49 = -13 + 49$$
$$(x - 7)^2 = 36$$
$$x - 7 = \pm\sqrt{36}$$
$$x - 7 = \pm 6$$

$x - 7 = 6$ $\quad$ $x - 7 = -6$

$+7 \quad +7$ $\quad$ $+7 \quad +7$

$x = 13$ $\quad$ $x = 1$

$\frac{1}{2}\ b = \frac{1}{2}(-14) = -7$

$(\frac{1}{2}\ b)^2 = (-7)^2 = 49$

15. $x^2 - 3x - 54 = 0$

$$x^2 - 3x = 54$$
$$x^2 - 3x + \frac{9}{4} = 54 + \frac{9}{4}$$
$$(x - \frac{3}{2})^2 = \frac{216 + 9}{4}$$
$$x - \frac{3}{2} = \pm\sqrt{\frac{225}{4}}$$
$$x - \frac{3}{2} = \pm\frac{15}{2}$$

$x - \frac{3}{2} = \frac{15}{2}$ $\quad$ $x - \frac{3}{2} = \frac{-15}{2}$

$+\frac{3}{2} \quad +\frac{3}{2}$ $\quad$ $+\frac{3}{2} \quad +\frac{3}{2}$

$x = \frac{18}{2}$ $\quad$ $x = \frac{-12}{2}$

$x = 9$ $\quad$ $x = -6$

$\frac{1}{2}\ b = \frac{1}{2}(-3) = \frac{-3}{2}$

$(\frac{1}{2}\ b)^2 = (\frac{-3}{2})^2 = \frac{9}{4}$

17. $x^2 + 2x - 5 = 0$

$$x^2 + 2x = 5$$
$$x^2 + 2x + 1 = 5 + 1$$
$$(x + 1)^2 = 6$$
$$x + 1 = \pm\sqrt{6}$$

$-1 \quad -1$

$$x = -1 \pm \sqrt{6}$$

$\frac{1}{2}\ b = \frac{1}{2}(2) = 1$

$(\frac{1}{2}\ b)^2 = 1^2 = 1$

19. $2x^2 - 8x = 64$

$$\frac{1}{2}(2x^2 - 8x) = \frac{1}{2}(64)$$
$$x^2 - 4x + 4 = 32 + 4$$
$$(x - 2)^2 = 36$$
$$x - 2 = \pm\sqrt{36}$$
$$x - 2 = \pm 6$$

$x - 2 = 6$ $\quad$ $x - 2 = -6$

$+2 \quad +2$ $\quad$ $+2 \quad +2$

$x = 8$ $\quad$ $x = -4$

$\frac{1}{2}\ b = \frac{1}{2}(-4) = -2$

$(\frac{1}{2}\ b) = (-2)^2 = 4$

21. $4x^2 + 2x - 12 = 0$

$$\frac{1}{4}(4x^2 + 2x - 12) = \frac{1}{4}(0)$$

$$x^2 + \frac{1}{2}x - 3 = 0$$

$$x^2 + \frac{1}{2}x + \frac{1}{16} = 3 + \frac{1}{16}$$

$$(x + \frac{1}{4})^2 = \frac{48 + 1}{16}$$

$$x + \frac{1}{4} = \pm\sqrt{\frac{49}{16}}$$

$$x + \frac{1}{4} = \pm \frac{7}{4}$$

$$x + \frac{1}{4} = \frac{7}{4} \qquad x + \frac{1}{4} = \frac{-7}{4}$$

$$-\frac{1}{4} - \frac{1}{4} \qquad -\frac{1}{4} - \frac{1}{4}$$

$$x = \frac{6}{4} \qquad x = \frac{-8}{4}$$

$$x = \frac{3}{2} \qquad x = -2$$

$$\frac{1}{2}b = \frac{1}{2}(\frac{1}{2}) = \frac{1}{4}$$

$$(\frac{1}{2}b)^2 = (\frac{1}{4})^2 = \frac{1}{16}$$

23. $3x^2 - 4x - 20 = 0$

$d = b^2 - 4ac$
$= (-4)^2 - 4(3)(-20)$
$= 16 + 240$
$= 256$

$d > 0$ two solutions

25. $2x^2 + 6x + 7 = 0$

$d = b^2 - 4ac$
$= 6^2 - 4(2)(7)$
$= 36 - 56$
$= -20$

$d < 0$ no real solution

27. $x^2 - 12x = 36$
$x^2 - 12x + 36 = 0$

$d = b^2 - 4ac$

$= (-12)^2 - 4(1)(36)$
$= 144 - 144$
$= 0$

$d = 0$ one solution

29. $-3x^2 - 4x + 8 = 0$

$d = b^2 - 4ac$
$= (-4)^2 - 4(-3)(8)$
$= 16 + 96$
$= 112$

$d > 0$ two solutions

31. $x^2 - 9x + 14 = 0$

$$x = \frac{-b \pm \sqrt{b^2 - 4ac}}{2a}$$

$$x = \frac{-(-9) \pm \sqrt{(-9)^2 - 4(1)(14)}}{2(1)}$$

$$x = \frac{9 \pm \sqrt{81 - 56}}{2}$$

$$x = \frac{9 \pm \sqrt{25}}{2}$$

$$x = \frac{9 \pm 5}{2}$$

$$x = \frac{9 + 5}{2} \qquad x = \frac{9 - 5}{2}$$

$$x = \frac{14}{2} \qquad x = \frac{4}{2}$$

$$x = 7 \qquad x = 2$$

33. $x^2 = 7x - 10$
$x^2 - 7x + 10 = 0$

$$x = \frac{-b \pm \sqrt{b^2 - 4ac}}{2a}$$

$$x = \frac{-(-7) \pm \sqrt{(-7)^2 - 4(1)(10)}}{2(1)}$$

$$x = \frac{7 \pm \sqrt{49 - 40}}{2}$$

$$x = \frac{7 \pm \sqrt{9}}{2}$$

$$x = \frac{7 \pm 3}{2}$$

$$x = \frac{7 + 3}{2} \qquad x = \frac{7 - 3}{2}$$

$$x = \frac{10}{2} \qquad x = \frac{4}{2}$$

$$x = 5 \qquad x = 2$$

35. $x^2 - 18 = 7x$

$$x^2 - 7x - 18 = 0$$

$$x = \frac{-b \pm \sqrt{b^2 - 4ac}}{2a}$$

$$x = \frac{-(-7) \pm \sqrt{(-7)^2 - 4(1)(-18)}}{2(1)}$$

$$x = \frac{7 \pm \sqrt{49 + 72}}{2}$$

$$x = \frac{7 \pm \sqrt{121}}{2}$$

$$x = \frac{7 \pm 11}{2}$$

$$x = \frac{7 + 11}{2} \qquad x = \frac{7 - 11}{2}$$

$$x = \frac{18}{2} \qquad x = \frac{-4}{2}$$

$$x = 9 \qquad x = -2$$

37. $6x^2 + x - 15 = 0$

$$x = \frac{-b \pm \sqrt{b^2 - 4ac}}{2a}$$

$$x = \frac{-1 \pm \sqrt{1^2 - 4(6)(-15)}}{2(6)}$$

$$x = \frac{-1 \pm \sqrt{1 + 360}}{12}$$

$$x = \frac{-1 \pm \sqrt{361}}{12}$$

$$x = \frac{-1 \pm \sqrt{19}}{12}$$

$$x = \frac{-1 + 19}{12} \qquad x = \frac{-1 - 19}{12}$$

$$x = \frac{18}{12} \qquad x = \frac{-20}{12}$$

$$x = \frac{3}{2} \qquad x = \frac{-5}{3}$$

39. $-2x^2 + 3x + 6 = 0$

$$x = \frac{-b \pm \sqrt{b^2 - 4ac}}{2a}$$

$$x = \frac{-3 \pm \sqrt{3^2 - 4(-2)(6)}}{2(-2)}$$

$$x = \frac{-3 \pm \sqrt{9 + 48}}{-4}$$

$$x = \frac{-3 \pm \sqrt{57}}{-4}$$

$$x = \frac{-1(-3 \pm \sqrt{57})}{-1(-4)}$$

$$x = \frac{3 \pm \sqrt{57}}{4}$$

41. $3x^2 - 4x + 6 = 0$

$$x = \frac{-b \pm \sqrt{b^2 - 4ac}}{2a}$$

$$x = \frac{-(-4) \pm \sqrt{(-4)^2 - 4(3)(6)}}{2(3)}$$

$$x = \frac{4 \pm \sqrt{16 - 72}}{6}$$

$$x = \frac{4 \pm \sqrt{-56}}{6}$$

No real solution

43. $2x^2 + 3x = 0$

$$x = \frac{-b \pm \sqrt{b^2 - 4ac}}{2a}$$

$$x = \frac{-3 \pm \sqrt{3^2 - 4(2)(0)}}{2(2)}$$

$$x = \frac{-3 \pm \sqrt{9 - 0}}{4}$$

$$x = \frac{-3 \pm \sqrt{9}}{4}$$

$$x = \frac{-3 \pm 3}{4}$$

$$x = \frac{-3 - 3}{4} \qquad x = \frac{-3 + 3}{4}$$

$$x = \frac{-6}{4} \qquad x = 0$$

$$x = \frac{-3}{2}$$

45. $x^2 - 11x + 24 = 0$
$(x - 8)(x - 3) = 0$

$$\begin{array}{rr} x - 8 = 0 & x - 3 = 0 \\ +8 \quad +8 & +3 \quad +3 \\ \hline x = 8 & x = 3 \end{array}$$

47. $x^2 = -3x + 40$
$x^2 + 3x - 40 = 0$
$(x + 8)(x - 5) = 0$

$$\begin{array}{rr} x + 8 = 0 & x - 5 = 0 \\ -8 \quad -8 & +5 \quad +5 \\ \hline x = -8 & x = 5 \end{array}$$

49. $x^2 - 4x - 60 = 0$
$(x - 10)(x + 6) = 0$

$$\begin{array}{rr} x - 10 = 0 & x + 6 = 0 \\ +10 \quad +10 & -6 \quad -6 \\ \hline x = 10 & x = -6 \end{array}$$

51. $x^2 + 11x - 12 = 0$
$(x + 12)(x - 1) = 0$

$$\begin{array}{rr} x + 12 = 0 & x - 1 = 0 \\ -12 \quad -12 & +1 \quad +1 \\ \hline x = -12 & x = 1 \end{array}$$

53. $x^2 + 6x = 0$
$x(x + 6) = 0$

$$\begin{array}{rr} x = 0 & x + 6 = 0 \\ & -6 \quad -6 \\ \hline & x = -6 \end{array}$$

55. $2x^2 = 9x - 10$
$2x^2 - 9x + 10 = 0$
$(2x - 5)(x - 2) = 0$

$$\begin{array}{rr} 2x - 5 = 0 & x - 2 = 0 \\ +5 \quad +5 & +2 \quad +2 \\ \hline \frac{2x}{2} = \frac{5}{2} & x = 2 \end{array}$$

$$x = \frac{5}{2}$$

57. $x^2 + 3x - 6 = 0$

$$x^2 + 3x = 6$$

$$x^2 + 3x + \frac{9}{4} = 6 + \frac{9}{4}$$

$$(x + \frac{3}{2})^2 = \frac{24 + 9}{4}$$

$$x + \frac{3}{2} = \pm\sqrt{\frac{33}{4}}$$

$$x + \frac{3}{2} = \pm \frac{\sqrt{33}}{2}$$

$$\frac{-3}{2} \quad \frac{-3}{2}$$

$$x = \frac{-3 \pm \sqrt{33}}{2}$$

$$\frac{1}{2} b = \frac{1}{2}(3) = \frac{3}{2}$$

$$(\frac{1}{2} b)^2 = (\frac{3}{2})^2 = \frac{9}{4}$$

59. $-3x^2 - 5x + 8 = 0$

$(3x + 8)(-x + 1) = 0$

$3x + 8 = 0 \qquad -x + 1 - 0$

$\quad -8 \quad -8 \qquad +x \quad +x$

$\frac{3x}{3} = \frac{-8}{3} \qquad 1 = x$

$x = \frac{-8}{3}$

61. $2x^2 - 5x = 0$

$x(2x - 5) = 0$

$x = 0 \qquad 2x - 5 = 0$

$\qquad +5 \quad +5$

$\frac{2x}{2} = \frac{5}{2}$

$x = \frac{5}{2}$

63. $y = x^2 - 2x - 3$

$x = \frac{-b}{2a} \qquad y = 1^2 - 2(1) - 3$

$x = \frac{-(-2)}{2 \cdot 1} \qquad y = 1 - 2 - 3$

$x = \frac{2}{2} \qquad y = -4$

$x = 1$

Axis of symmetry: $x = 1$

Vertex: $(1,-4)$

Parabola opens up

65. $y = x^2 + 7x + 12$

$x = \frac{-b}{2a} \qquad y = (\frac{-7}{2})^2 + 7(\frac{-7}{2}) + 12$

$x = \frac{-7}{2(1)} \qquad y = \frac{49}{4} - \frac{49}{2} + 12$

$x = \frac{-7}{2} \qquad y = \frac{49 - 98 + 48}{4}$

$y = \frac{-98 + 97}{4}$

$y = \frac{-1}{4}$

Axis of symmetry: $x = -\frac{7}{2}$

Vertex: $(\frac{-7}{2}.\frac{-1}{4})$

Parabola opens up

67. $y = x^2 - 3x$

$x = \frac{-b}{2a} \qquad y = (\frac{3}{2})^2 - 3(\frac{3}{2})$

$x = \frac{-(-3)}{2' 1} \qquad y = \frac{9}{4} - \frac{9}{2}$

$x = \frac{3}{2} \qquad y = \frac{9 - 18}{4}$

$y = \frac{-9}{4}$

Axis of symmetry: $x = \frac{3}{2}$

Vertex: $(\frac{3}{2},\frac{-9}{4})$

Parabola opens up

69. $y = -x^2 - 8$

$x = \frac{-b}{2a}$ $y = -(0)^2 - 8$

$x = \frac{0}{2(-1)}$ $y = -8$

$x = 0$

Axis of symmetry: $x = 0$

Vertex: $(0,-8)$

Parabola opens down

71. $y = -x^2 - x + 20$

$x = \frac{-b}{2a}$ $y = -(\frac{-1}{2})^2 - (\frac{-1}{2})$
$+ 20$

$x = \frac{-(-1)}{2(-1)}$ $y = \frac{-1}{4} + \frac{1}{2} + 20$

$x = \frac{-1}{2}$ $y = \frac{-1 + 2 + 80}{4}$

$y = \frac{81}{4}$

Axis of symmetry: $x = - \frac{1}{2}$

Vertex: $(\frac{-1}{2}, \frac{81}{4})$

Parabola opens down

73. $y = x^2 + 6x$ $y = (-3)^2 + 6(-3)$

$x = \frac{-b}{2a}$ $y = 9 - 18$

$x = \frac{-6}{2 \cdot 1}$ $y = -9$

$x = -3$

Roots 0,-6

Axis of symmetry: $x = -3$

Vertex: $(-3,-9)$

Parabola opens up

Let $x = -3$ $y = (-3)^2 + 6(-3)$
$= 9 - 18 = -9$

Let $x = -2$ $y = (-2)^2 + 6(-2)$
$= 4 - 12 = -8$

Let $x = -1$ $y = (-1)^2 + 6(-1)$
$= 1 - 6 = -5$

Let $x = 0$ $y = (0)^2 + 6(0)$
$= 0 + 0 = 0$

Let $x = 1$ $y = (1)^2 + 6(1)$
$= 1 + 6 = 7$

Make use of symmetry to complete the graph.

x	y
-3	-9
-2	-8
-1	-5
0	0
1	7

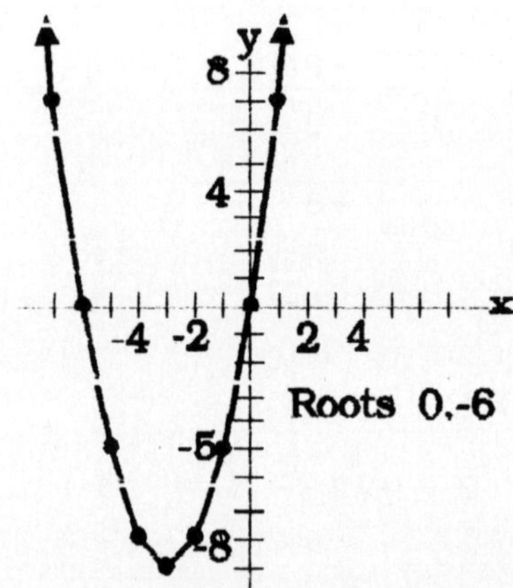

75. $y = x^2 + 2x - 8$ $y = (-1)^2 + 2(-1) - 8$

$x = \frac{-b}{2a}$ $y = 1 - 2 - 8$

$x = \frac{-2}{2 \cdot 1}$ $y = -9$

$x = -1$

Roots -4,2

Axis of symmetry: $x = -1$

Vertex: $(-1,-9)$

Parabola opens up

Let $x = -1$ $y = (-1)^2$

$+ 2(-1) - 8$
$= 1 - 2 - 8$
$= -9$

Let $x = 0$ $y = (0)^2 + 2(0) - 8$
$= 0 + 0 - 8$
$= -8$

Let $x = 1$ $y = (1)^2 + 2(1) - 8$
$= 1 + 2 - 8$
$= -5$

Let $x = 2$ $y = (2)^2 + 2(2) - 8$
$= 4 + 4 - 8$
$= 0$

Make use of symmetry to complete the graph.

x	y
-1	-9
0	-8
1	-5
2	0

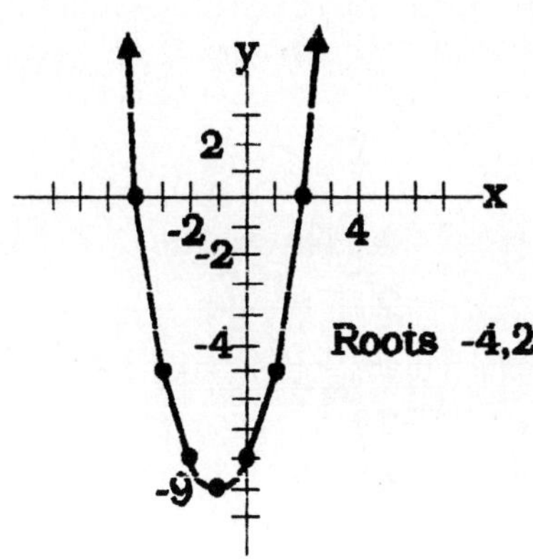

77. $y = x^2 + 5x + 4$ $y = (\frac{-5}{2})^2 + 5(\frac{-5}{2}) + 4$

$x = \frac{-b}{2a}$ $y = \frac{25}{4} - \frac{25}{2} + 4$

$x = \frac{-5}{2 \cdot 1}$ $y = \frac{25 - 50 + 16}{4}$

$x = \frac{-5}{2}$ $y = \frac{-9}{4}$

Roots $-1, -4$

Axis of symmetry: $x = -\frac{5}{2}$

Vertex: $(\frac{-5}{2}, \frac{-9}{4})$

Parabola opens up

Let $x = \frac{-5}{2}$ $y = (\frac{-5}{2})^2 + 5(\frac{-5}{2}) + 4$

$= \frac{25}{4} - \frac{25}{2} + 4$

$= \frac{-9}{4}$

Let $x = -2$ $y = (-2)^2 + 5(-2) + 4$
$= 4 - 10 + 4$
$= -2$

Let $x = -1$ $y = (-1)^2 + 5(-1) + 4$
$= 1 - 5 + 4$
$= 0$

Let $x = 0$ $y = (0)^2 + 5(0) + 4$
$= 0 + 0 + 4$
$= 4$

Make use of symmetry to complete the graph.

x	y
$\frac{-5}{2}$	$\frac{-9}{4}$
-2	-2
-1	0
0	4

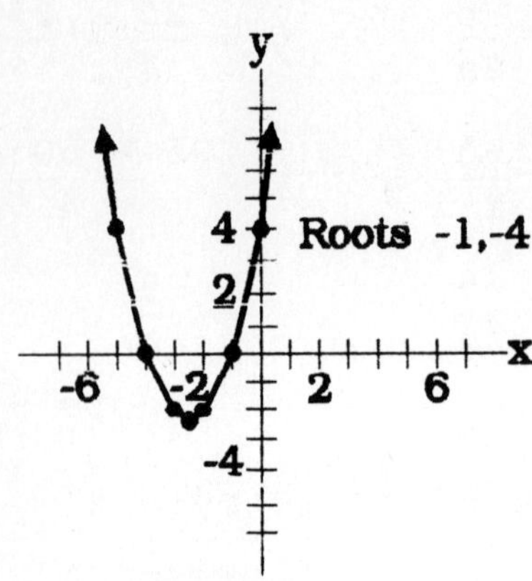

79. $y = -2x^2 + 3x - 2$

$x = \frac{-b}{2a}$

$x = \frac{-3}{2(-2)}$

$x = \frac{3}{4}$

$y = -2(\frac{3}{4})^2 + 3(\frac{3}{4}) - 2$

$y = -2(\frac{9}{16}) + \frac{9}{4} - 2$

$y = \frac{-9 + 18 - 16}{8}$

$y = \frac{-7}{8}$

No roots

Axis of symmetry: $x = \frac{3}{4}$

Vertex: $(\frac{3}{4}, \frac{-7}{8})$

Parabola opens down

Let $x = -1$ $y = -2(-1)^2 + 3(-1) - 2 = -2(1) - 3 - 2 = -7$

Let $x = 0$ $y = -2(0)^2 + 3(0) - 2 = 0 + 0 - 2 = -2$

Let $x = \frac{3}{4}$ $y = -2(\frac{3}{4})^2 + 3(\frac{3}{4}) - 2 = -2(\frac{9}{16}) + \frac{9}{4} - 2 = \frac{-7}{8}$

Let $x = 1$ $y = -2(1)^2 + 3(1) - 2 = -2(1) + 3 - 2 = -1$

Let $x = 2$ $y = -2(2)^2 + 3(2) - 2 = -2(4) + 6 - 2 = -4$

Make use of symmetry to complete the graph.

x	y
-1	-7
0	-2
$\frac{3}{4}$	$\frac{-7}{8}$
1	-1
2	-4

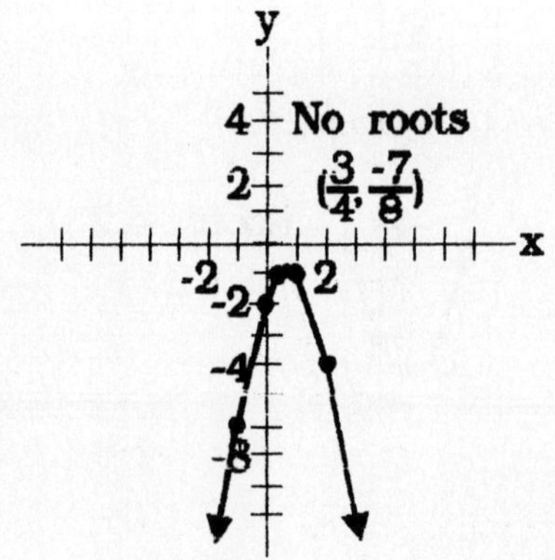

81. $y = 4x^2 - 8x + 6$

$x = \frac{-b}{2a}$

$y = 4(1)^2 - 8(1) + 6$

$y = 4 - 8 + 6$

$x = \frac{-(-8)}{2(4)}$ $y = 2$

$x = \frac{8}{8}$

$x = 1$ No roots

Axis of symmetry: $x = 1$

Vertex: (1,2)

Parabola opens up

Let $x = 0$ $y = 4(0)^2 - 8(0) + 6$
$= 0 + 0 + 6$
$= 6$

Let $x = 1$ $y = 4(1)^2 - 8(1) + 6$
$= 4 - 8 + 6$
$= 2$

Let $x = 2$ $y = 4(2)^2 - 8(2) + 6$
$= 4(4) - 16 + 6 = 6$

Make use of symmetry to complete the graph.

x	y
0	6
1	2
2	6

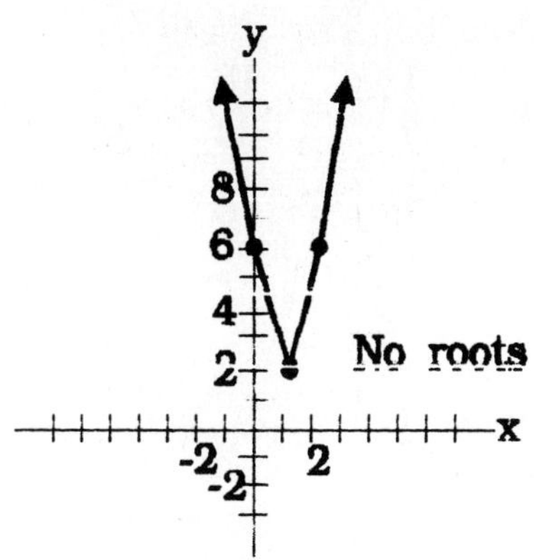

83. $y = -x^2 - 6x - 4$ $y = -(-3)^2 - 6(-3) - 4$

$x = \frac{-b}{2a}$ $y = -9 + 18 - 4$

$x = \frac{-(-6)}{2(-1)}$ $y = 5$

$x = \frac{6}{-2}$

$x = -3$

Roots $-3 - \sqrt{5}$, $-3 + \sqrt{5}$

Axis of symmetry: $x = -3$

Vertex: (-3,5)

Parabola opens down

Let $x = -3$ $y = -(-3)^2 - 6(-3) - 4$
$= -9 + 18 - 4$
$= 5$

Let $x = -2$ $y = -(-2)^2 - 6(-2) - 4$
$= -4 + 12 - 4$
$= 4$

Let $x = -1$ $y = -(-1)^2 - 6(1) - 4$
$= -1 + 6 - 4$
$= 1$

Let $x = 0$ $y = -(0)^2 - 6(0) - 4$
$= 0 + 0 - 4$
$= -4$

Make use of symmetry to complete the graph.

x	y
-3	5
-2	4
-1	1
0	-4

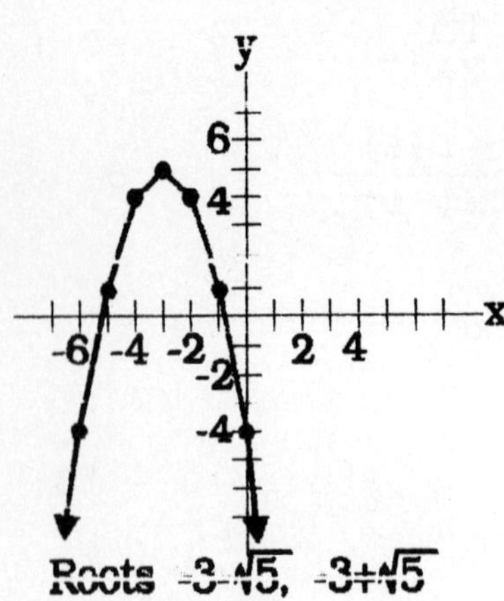

85. x = a positive integer.
x + 3 = a second positive integer.

$$x(x + 3) = 88$$
$$x^2 + 3x - 88 = 0$$
$$(x + 11)(x - 8) = 0$$
$$x + 11 = 0 \quad x - 8 = 0$$
$$-11 \quad -11 \qquad +7 \quad +7$$
$$x = -11 \qquad x = 8$$

Discard, not positive

x = 8, first positive integer
x + 3 = 8 + 3 = 11, second positive integer

Practice Test

1. $x^2 + 1 = 21$
$$-1 = -1$$
$$x^2 = 20$$
$$\sqrt{x^2} = \pm\sqrt{4 \cdot 5}$$
$$x = \pm 2\sqrt{5}$$

2. $(2x - 3)^2 = 35$
$$2x - 3 = \pm\sqrt{35}$$
$$+3 \quad +3$$
$$2x = 3 \pm \sqrt{35}$$
$$\frac{2x}{2} = \frac{3 \pm \sqrt{35}}{2}$$
$$x = \frac{3 \pm \sqrt{35}}{2}$$

3. $x^2 - 4x = 60$
$$x^2 - 4x + 4 = 60 + 4$$
$$(x - 2)^2 = 64$$
$$(x - 2)^2 = \pm\sqrt{64}$$
$$x - 2 = \pm 8$$
$$x - 2 = 8 \quad x - 2 = -8$$
$$x = 10 \qquad x = -6$$

$$\frac{1}{2} b = \frac{1}{2}(-4) = -2$$
$$(\frac{1}{2} b)^2 = (-2)^2 = 4$$

4. $x^2 = -x + 12$
$$x^2 + x = 12$$
$$x^2 + x + \frac{1}{4} = 12 + \frac{1}{4}$$
$$(x + \frac{1}{2})^2 = \frac{48}{4} + \frac{1}{4}$$
$$(x + \frac{1}{2})^2 = \frac{49}{4}$$
$$\sqrt{(x + \frac{1}{2})^2} = \pm\sqrt{\frac{49}{4}}$$
$$x + \frac{1}{2} = \pm\frac{7}{2}$$
$$x + \frac{1}{2} = \frac{7}{2} \qquad x + \frac{1}{2} = \frac{-7}{2}$$
$$x = \frac{6}{2} \qquad x = \frac{-8}{2}$$

$x = 3 \qquad x = -4$

$\frac{1}{2} b = \frac{1}{2}(1) = \frac{1}{2}$

$(\frac{1}{2} b)^2 = (\frac{1}{2})^2 = \frac{1}{4}$

5. $x^2 - 5x - 6 = 0$

$a = 1, b = -5, c = -6$

$$x = \frac{-b \pm \sqrt{b^2 - 4ac}}{2a}$$

$$= \frac{-(-5) \pm \sqrt{(-5)^2 - 4(1)(-6)}}{2(1)}$$

$$= \frac{5 \pm \sqrt{25 + 24}}{2}$$

$$= \frac{5 \pm \sqrt{49}}{2}$$

$$= \frac{5 \pm 7}{2}$$

$$x = \frac{5 + 7}{2} \qquad x = \frac{5 - 7}{2}$$

$$= \frac{12}{2} \qquad = \frac{-2}{2}$$

$$= 6 \qquad = -1$$

6. $2x^2 + 5 = -8x$

$2x^2 + 8x + 5 = 0$

$a = 2, b = 8, c = 5$

$$x = \frac{-b \pm \sqrt{b^2 - 4ac}}{2a}$$

$$= \frac{-(8) \pm \sqrt{(8)^2 - 4(2)(5)}}{2(2)}$$

$$= \frac{-8 \pm \sqrt{64 - 40}}{4}$$

$$= \frac{-8 \pm \sqrt{24}}{4}$$

$$= \frac{-8 \pm \sqrt{4}\sqrt{6}}{4}$$

$$= \frac{-8 \pm 2\sqrt{6}}{4}$$

$$= \frac{2(-4 \pm \sqrt{6})}{4}$$

$$x = \frac{-4 + \sqrt{6}}{2} \qquad x = \frac{-4 - \sqrt{6}}{2}$$

7. $3x^2 - 5x = 0$

$x(3x - 5) = 0$

$x = 0 \qquad 3x - 5 = 0$

$3x = 5$

$x = \frac{5}{3}$

8. $2x^2 + 9x = 5$

$2x^2 + 9x - 5 = 0$

$(2x - 1)(x + 5) = 0$

$2x - 1 = 0 \qquad x + 5 = 0$

$x = \frac{1}{2} \qquad x = -5$

9. $3x^2 - 4x + 2 - 0$

$a = 3, b = -4, c = 2$

Discriminant

$b^2 - 4ac = (-4)^2 - 4(3)(2)$

$= 16 - 24$

$= -8$

Since $b^2 - 4ac < 0$, there is no real solution.

10. $y = -x^2 + 3x + 8$

$x = \frac{-b}{2a} \qquad y = -(\frac{3}{2})^2 + 3(\frac{3}{2}) + 8$

$x = \frac{-3}{2(-1)} \qquad y = -\frac{9}{4} + \frac{9}{2} + 8$

$x = \frac{3}{2} \qquad y = \frac{-9 + 18 + 32}{4}$

$y = \frac{41}{4}$

Axis of symmetry: $x = \frac{3}{2}$

vertex: $(\frac{3}{2}, \frac{41}{4})$

Parabola opens down

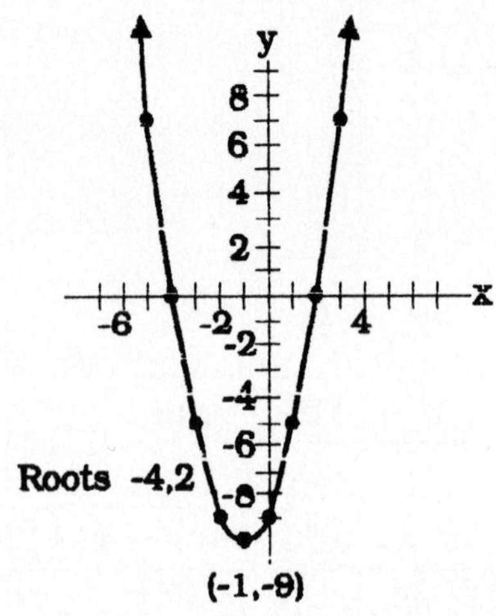

11. $y = x^2 + 2x - 8$ $\quad y = (-1)^2 + 2(-1) - 8$

$x = \frac{-b}{2a}$ $\quad y = 1 - 2 - 8$

$x = \frac{-2}{2(1)}$ $\quad y = -9$

$x = \frac{-2}{2}$

$x = -1$

Roots -4,2

Axis of symmetry: $x = -1$
Vertex: $(-1,-9)$
Parabola opens up

$y = x^2 + 2x - 8$

Let $x = 0$ $\quad y = (0)^2 - 8 = 0 + 0 - 8 = -8$

Let $x = 1$ $\quad y = (1)^2 + 2(1) - 8 = 1 + 2 - 8 = -5$

Let $x = 2$ $\quad y = (2)^2 + 2(2) - 8 = 4 + 4 - 8 = 0$

Let $x = 3$ $\quad y = (3)^2 + 2(3) - 8 = 9 + 6 - 8 = 7$

Make use of symmetry to complete the graph.

x	y
-1	-9
0	-8
1	-5
2	0
3	7

12. $y = -x^2 + 6x - 9$ $\quad y = -(3)^2 + 6(3) - 9$

$x = \frac{-b}{2a}$ $\quad y = -9 + 18 - 9$

$x = \frac{-6}{2(-1)}$ $\quad y = 0$

$x = \frac{6}{2}$

$x = 3$

Root 3

Axis of symmetry: $x = 3$
Vertex: $(3,0)$
Parabola opens down

$y = x^2 + 6x - 9$

Let $x = 4$ $\quad y = -(4)^2 + 6(4) - 9 = -16 + 24 - 9 = -1$

Let $x = 5$ $\quad y = -(5)^2 + 6(5) - 9 = -25 + 30 - 9 = -4$

Let $x = 6$ $\quad y = -(6)^2 + 6(6) - 9 = -36 + 36 - 9 = -9$

Make use of symmetry to complete the graph.

x	y
3	0
4	-1
5	-4
6	-9

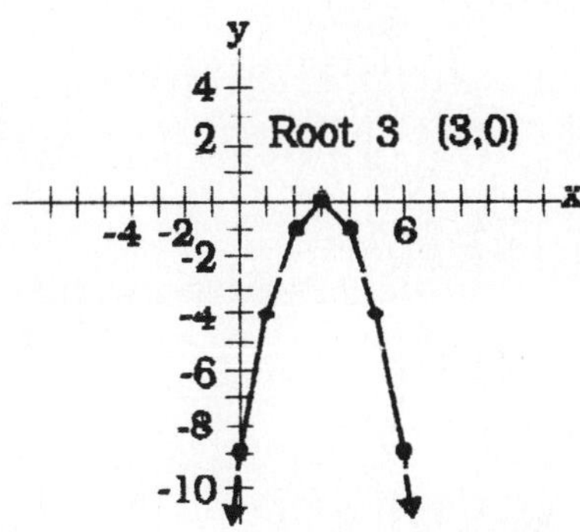

13. Let x = width.
Let $3x + 1$ = length.

(length) × (width)
= Area of a rectangle

$$x(3x + 1) = 30$$
$$3x^2 + x = 30$$
$$3x^2 + x - 30 = 0$$
$$(3x + 10)(x - 3) = 0$$

$3x + 10 = 0$ $\quad$ $x - 3 = 0$
$3x = -10$ $\quad$ $x = 3$

$$x = \frac{-10}{3}$$

Since dimensions are positive values, -10/3 is not used. The width is 3 feet and the length is 3(3) + 1 = 10 feet.

Cumulative Review Test

1. $-x^2y + y^2 - 3xy$
$= -(-3)^2(4) + (4)^2 - 3(-3)(4)$
$= -(9)(4) + 16 + 36$
$= -36 + 16 + 36$
$= 16$

2. $\frac{1}{4}x + \frac{3}{5}x = \frac{1}{3}(x + 2)$

$\frac{1}{4}x + \frac{3}{5}x = \frac{1}{3}x + \frac{2}{3}$

$60(\frac{1}{4}x + \frac{3}{5}x) = 60(\frac{1}{3}x + \frac{2}{3})$

$15x + 36x = 20x + 40$
$51x = 20x + 40$
$31x = 40$
$x = \frac{40}{31}$

3. $\frac{3}{8} = \frac{2}{x}$

$3x = 16$

$x = \frac{16}{3} = 5\frac{1}{3}$ in.

4. $2(x - 3) \leq 6x - 5$
$2x - 6 \leq 6x - 5$
$-4x - 6 \leq -5$
$-4x \leq 1$
$x \geq -\frac{1}{4}$

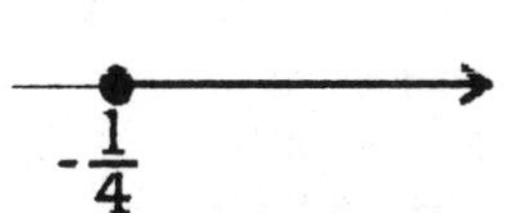

5. $A = \frac{m + n + P}{3}$

$3A = m + n + P$
$3A - m - n = P$

6. $(6x^2y^4)^3(2x^4y^5)^2$

$= (6^3x^6y^{12})(2^2x^8y^{10})$

$= 216 \cdot 4\ x^{14}y^{22}$

$= 864x^{14}y^{22}$

7.

$$\begin{array}{r} x+4 \;-\; 3/x+2 \\ \hline x+2\,/\,x^2 + 6x + 5 \\ -x^2 \mp 2x \\ \hline 4x + 5 \\ -4x \mp 8 \\ \hline -3 \end{array}$$

8. $2x^2 - 3xy - 4xy + 6y^2$
$= x(2x - 3y) - 2y(2x - 3y)$
$= (x - 2y)(2x - 3y)$

9. $4x^2 - 14x - 8$

$= 2(2x^2 - 7x - 4)$
$= 2(2x + 1)(x - 4)$

10. $\dfrac{4}{a^2 - 16} + \dfrac{2}{(a - 4)^2}$

$= \dfrac{4}{(a - 4)(a + 4)}$

$+ \dfrac{2}{(a - 4)^2}$

$= \dfrac{4(a - 4)}{(a - 4)^2(a + 4)}$

$+ \dfrac{2(a + 4)}{(a - 4)^2(a + 4)}$

$= \dfrac{4a - 16 + 2a + 8}{(a - 4)^2(a + 4)}$

$= \dfrac{6a - 8}{(a - 4)^2(a + 4)}$

11. $x + \dfrac{48}{x} = 14$

$x(x + \dfrac{48}{x}) = 14x$

$x^2 + 48 = 14x$
$x^2 - 14x + 48 = 0$
$(x - 6)(x - 8) = 0$
$x - 6 = 0 \quad x - 8 = 0$
$x = 6 \quad\quad x = 8$

12. $3x + 5y = 10$

x	y
0	2
$3\frac{1}{3}$	0
5	-1

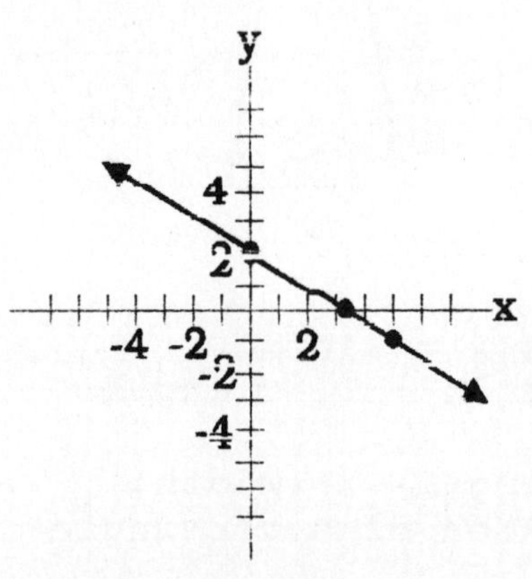

14. $\sqrt{\dfrac{3x^2y^2}{54x}} = \sqrt{\dfrac{xy^3}{18}} = \dfrac{\sqrt{xy^3}}{\sqrt{18}} \quad \dfrac{\sqrt{2}}{\sqrt{2}}$

$= \dfrac{\sqrt{2xy^3}}{\sqrt{36}} = \dfrac{\sqrt{y^2}\ \sqrt{2xy}}{6}$

$= \dfrac{y\sqrt{2xy}}{6}$

15. $2\sqrt{28} - 3\sqrt{7} + \sqrt{63}$

$= 2\sqrt{4}\ \sqrt{7} - 3\sqrt{7} + \sqrt{9}\ \sqrt{7}$

$= 4\sqrt{7} - 3\sqrt{7} + 3\sqrt{7}$

$= 4\sqrt{7}$

16. $\sqrt{x^2 + 5} = x + 1$

$(\sqrt{x^2 + 5})^2 = (x + 1)^2$

$$x^2 + 5 = x^2 + 2x + 1$$
$$5 = 2x + 1$$
$$4 = 2x$$
$$2 = x$$

Check:

$$\sqrt{2^2 + 5} = 2 + 1$$
$$\sqrt{4 + 5} = 3$$
$$\sqrt{9} = 3$$

True $3 = 3$

Solution: $x = 2$

17. $2x^2 - 4x - 5 = 0$

$$x = \frac{-b \pm \sqrt{b^2 - 4ac}}{2a}$$
$$x = \frac{4 \pm \sqrt{4^2 - 4(2)(-5)}}{2(2)}$$
$$x = \frac{4 \pm \sqrt{16 + 40}}{4}$$
$$x = \frac{4 \pm \sqrt{56}}{4} = \frac{4 \pm 2\sqrt{14}}{4}$$
$$= \frac{2(2 \pm \sqrt{14})}{4} = \frac{2 \pm \sqrt{14}}{2}$$

18.

$$\frac{4 \text{ pounds}}{500 \text{ square feet}} = \frac{x \text{ pounds}}{3200 \text{ square feet}}$$
$$\frac{4}{500} = \frac{x}{3200}$$
$$500x = 4(3200)$$
$$500x = 12{,}800$$
$$x = 25.6 \text{ pounds}$$

19. Let x = the width.
$3x - 3$ = the length.

$$2l + 2w = P$$
$$2(3x - 3) + 2x = 74$$
$$6x - 6 + 2x = 74$$
$$8x - 6 = 74$$
$$8x = 80$$
$$x = 10$$

Width: 10 ft.

length: $3x - 3 = 3(10) - 3$
$= 30 - 3 = 27$ feet

20.

	distance	rate	time
Jogs	2	x+3	$\frac{2}{x+3}$
walks	2	x	$\frac{2}{x}$

Let x = rate Willie walks.
$x + 3$ = rate Willie jogs.

$$\frac{2}{x+3} + \frac{2}{x} = 1$$

$$x(x + 3)\left[\frac{2}{x + 3} + \frac{2}{x}\right] = x(x + 3)$$
$$2x + 2(x + 3) = x^2 + 3x$$
$$2x + 2x + 6 = x^2 + 3x$$
$$4x + 6 = x^2 + 3x$$
$$0 = x^2 - x - 6$$
$$0 = (x-3)(x+2)$$

$x + 2 = 0$ $x - 3 = 0$
$x = -2$ $x = 3$

Rate cannot be negative, $x \neq -2$.
Walks: 3 mph
Jogs: $x + 3 = 3 + 3 = 6$mph